AutoDesk Inventor®10
Essentials Plus

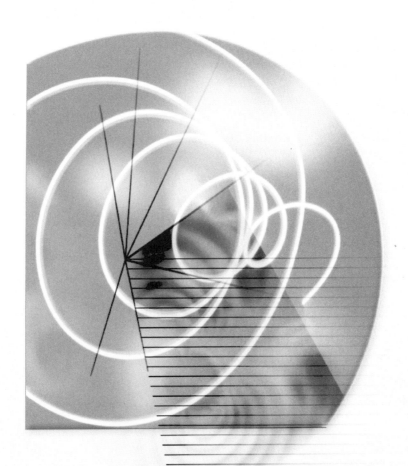

AutoDesk Inventor®10 Essentials Plus

DANIEL T. BANACH
TRAVIS J. JONES
ALAN J. KALAMEJA

autodesk® Press

THOMSON™
DELMAR LEARNING

Australia • Canada • Mexico • Singapore • Spain • United Kingdom • United States

autodesk®Press

Autodesk Inventor® Essentials Plus
Daniel T. Banach, Travis Jones, and Alan J. Kalameja

Autodesk Press Staff:

Vice President, Technology and Trades SBU:
Alar Elken

Editorial Director:
Sandy Clark

Senior Acquisitions Editor:
James DeVoe

Senior Development Editor:
John Fisher

Marketing Director:
Dave Garza

Channel Manager:
Dennis Williams

Marketing Coordinator:
Stacey Wiktorek

Production Manager:
Andrew Crouth

Production Editor:
Stacy Masucci

Editorial Assistant:
Tom Best

Cover image courtest of iStockPhoto

COPYRIGHT © 2006 by Delmar Learning, a part of the Thomson Corporation. Delmar Learning, Thomson and the Thomson logo are trademarks used herein under license. Autodesk, AutoCAD and the AutoCAD logo are registered trademarks of Autodesk. Delmar Learning uses "Autodesk Press" with permission from Autodesk for certain purposes.

Printed in Canada
1 2 3 4 5 XX 07 06 05

For more information contact
Delmar Learning
Executive Woods
5 Maxwell Drive, PO Box 8007,
Clifton Park, NY 12065-8007
Or find us on the World Wide Web at
http://www.delmar.com

All rights reserved. No part of this work covered by the copyright hereon may be reproduced in any form or by any means—graphic, electronic, or mechanical, including photocopying, recording, taping, Web distribution, or information storage and retrieval systems—without the written permission of the publisher.

For permission to use material from the text or product, contact us by
Tel. 1 (800) 730-2214
Fax 1 (800) 730-2215
www.thomsonrights.com

Library of Congress Cataloging-in-Publication Data:

ISBN 1-4180-1698-5

NOTICE TO THE READER

Publisher does not warrant or guarantee any of the products described herein or perform any independent analysis in connection with any of the product information contained herein. Publisher does not assume, and expressly disclaims, any obligation to obtain and include information other than that provided to it by the manufacturer.

The reader is expressly warned to consider and adopt all safety precautions that might be indicated by the activities herein and to avoid all potential hazards. By following the instructions contained herein, the reader willingly assumes all risks in connection with such instructions.

The publisher makes no representation or warranties of any kind, including but not limited to, the warranties of fitness for particular purpose or merchantability, nor are any such representations implied with respect to the material set forth herein, and the publisher takes no responsibility with respect to such material. The publisher shall not be liable for any special, consequential, or exemplary damages resulting, in whole or part, from the readers' use of, or reliance upon, this material.

CONTENTS

ACKNOWLEDGMENTS XIX
INTRODUCTION.. XXI

CHAPTER 1 GETTING STARTED
CHAPTER OBJECTIVES.. 1
GETTING STARTED WITH AUTODESK INVENTOR 1
    Getting Started.. 2
    New.. 3
    Open.. 3
    Projects... 4
PROJECTS IN AUTODESK INVENTOR 4
    Project Setup 4
    Project File Search Options 5
    Included File and Workgroup 6
    Workspace.. 6
    Workgroup Search Paths 7
    Library Search Paths... 7
    Frequently Used Subfolders.. 7
    Folder Options 8
    Options... 8
    Project File Search Order 9
    Resolving File Links.. 10
    Creating Projects 10
    What type of project are you creating? 11
    Editing Projects... 14
    Autodesk Vault.. 14
EXERCISE 1-1 PROJECTS 15
FILE INFORMATION... 19
    File Types 19
    Multiple Document Environment.. 20
    Save Options 20
APPLICATION OPTIONS... 21
DESIGN SUPPORT SYSTEM... 22
USER INTERFACE... 24

TOOLBARS, PANEL BAR, AND COMMAND BAR..25
    Command Entry...25
    Toolbars and the Panel Bar...26
    Shortcut Menus..26
    Windows Shortcuts..26
    Autodesk Inventor Shortcut Keys..26
CUSTOMIZING SHORTCUTS..27
    Commands Tab..28
    Rollup Command Dialogs...29
    Undo and Redo..30
VIEWPOINT OPTIONS...30
    Isometric View..30
    Camera Views...31
    Shadow..31
    View Tools..32
EXERCISE 1-2 VIEWING A MODEL...33
CHAPTER SUMMARY...38
CHECKING YOUR SKILLS..39

CHAPTER 2 SKETCHING AND PART APPLICATION OPTIONS
CHAPTER OBJECTIVES...40
SKETCHING AND PART APPLICATION OPTIONS.......................................40
    Sketch Options...40
    Part Options...42
UNITS...44
TEMPLATES..45
CREATING A PART..46
    Sketches and Default Planes...47
    New Sketch...48
STEP 1–SKETCH THE OUTLINE OF THE PART..49
    Sketching Overview..49
    Sketching Tools..50
    Precise Input..55
    Selecting Objects..55
    Deleting Objects...56
    Measure Tools..56
EXERCISE 2-1 CREATING A SKETCH WITH LINES......................................59
EXERCISE 2-2 CREATING A SKETCH WITH TANGENCIES........................61
STEP 2–CONSTRAINING THE SKETCH..64
    Constraint Types..65
    Adding Constraints..66
    Snaps...67
    Dragging a Sketch...67
    Showing and Deleting Constraints..68
EXERCISE 2-3 ADDING AND DISPLAYING CONSTRAINTS.......................69

STEP 3–ADDING DIMENSIONS	72
General Dimensioning	72
Entering and Editing a Dimension Value	75
AUTO DIMENSION	78
EXERCISE 2-4 DIMENSIONING A SKETCH	79
OPENING/IMPORTING AUTOCAD FILES	82
Opening an AutoCAD File or AutoCAD Mechanical File	83
IMPORTING/INSERTING AUTOCAD DATA INTO A SKETCH	88
IMPORTING OTHER FILE TYPES	89
SAT	89
STEP	89
PRO/E	89
DXF	90
IGES	90
EXERCISE 2-5 INSERTING AN AUTOCAD FILE	90
CHAPTER SUMMARY	92
APPLYING YOUR SKILLS	93
Skill Exercise 2-1	93
Skill Exercise 2-2	93
CHECKING YOUR SKILLS	94

CHAPTER 3 CREATING AND EDITING SKETCHED FEATURES

CHAPTER OBJECTIVES	96
UNDERSTANDING FEATURES	96
Consumed and Unconsumed Sketches	97
USING THE BROWSER FOR CREATING AND EDITING	97
SWITCHING ENVIRONMENTS	100
FEATURE TOOLS	102
EXTRUDING A SKETCH	104
Shape	105
Match Shape	107
More	108
EXERCISE 3-1 EXTRUDING A SKETCH	108
REVOLVING A SKETCH	110
Shape	111
Operation	111
Extents	112
Output	112
Match Shape	113
CENTERLINES AND DIAMETRIC DIMENSIONS	113
EXERCISE 3-2 REVOLVING A SKETCH	114
EDITING A FEATURE	116
Editing Feature Size	118
3D GRIPS	119
Move Feature	121

Renaming Features and Sketches	122
Feature Color	123
Deleting a Feature	124
Failed Features	124
EDITING A FEATURE SKETCH	124
EXERCISE 3-3 EDITING FEATURES AND SKETCHES	126
SKETCHED FEATURES	127
DEFINING THE ACTIVE SKETCH PLANE	128
Face Cycling	129
EXERCISE 3-4 SKETCH PLANES	130
PROJECTING PART EDGES	133
Direct Model Edge Referencing	133
PROJECT EDGES	134
CHAPTER SUMMARY	136
APPLYING YOUR SKILLS	136
Skill Exercise 3-1	136
Skill Exercise 3-2	137
Skill Exercise 3-3	138
CHECKING YOUR SKILLS	140

CHAPTER 4 Creating Placed Features

CHAPTER OBJECTIVES	142
FILLETS	142
Constant Tab	144
Variable Tab	147
Setbacks Tab	148
CHAMFERS	149
Method	150
Edge and Face	151
Distance and Angle	152
Edge Chain and Setback	152
EXERCISE 4-1 CREATING FILLETS AND CHAMFERS	152
HOLES	155
Editing Hole Features	156
Holes Dialog Box	157
Placement	157
Hole Option	158
Hole Type	159
Hole Centers	161
EXERCISE 4-2 CREATING HOLES	162
THREADS	166
Location Tab	167
Specification Tab	168
EXERCISE 4-3 CREATING THREADS	168

SHELLING 170
    Remove Faces 172
    Thickness 172
    Direction 172
    Unique Face Thickness 172
EXERCISE 4-4 SHELLING A PART 173
FACE DRAFT 175
    Draft Type 176
    Pull Direction and Fixed Plane 176
    Faces 176
    Draft Angle 176
EXERCISE 4-5 CREATING FACE DRAFTS 177
WORK FEATURES 180
    Creating a Work Axis 181
    Creating Work Points 181
EXERCISE 4-6 CREATING WORK AXES 183
CREATING WORK PLANES 185
    Feature Visibility 187
EXERCISE 4-7 CREATING WORK PLANES 188
PATTERNS 190
    Rectangular Patterns 192
    Circular Patterns 195
    Linear Patterns–Pattern Along a Path 197
EXERCISE 4-8 CREATING RECTANGULAR PATTERNS 197
EXERCISE 4-9 CREATING CIRCULAR PATTERNS 199
EXERCISE 4-10 CREATING PATH PATTERNS 200
CHAPTER SUMMARY 202
APPLYING YOUR SKILLS 203
    Skill Exercise 4-2 203
CHECKING YOUR SKILLS 204

CHAPTER 5 Creating and Editing Drawing Views
CHAPTER OBJECTIVES 206
DRAWING OPTIONS, CREATING A DRAWING, AND DRAWING TOOLS .. 207
    Drawing Options 207
    Creating a Drawing 208
    Drawing Tools 209
    Drawing Views Panel Bar 210
    Drawing Annotation Panel Bar 210
DRAWING SHEET PREPARATION 212
BORDER CREATION 214
TITLE BLOCKS 214
    Inserting a Default Title Block 214
    Edit Property Fields Dialog Box 215
    Creating a New Title Block 216

Reorder and Sort by Name for Drawing Resources	217
STYLES	217
Style Name/Value	217
Sub-Styles	218
Controlling Style Library Changes	218
Documents, Style Libraries, and Templates	219
Managing Styles	221
Creating a New Style	221
Edit Styles Manually in Documents	222
Setting Object Defaults	223
Overriding an Object's Style	224
EXERCISE 5-1 SHEETS, BORDERS, AND TITLE BLOCKS	224
TEMPLATES	229
CREATING DRAWING VIEWS	229
Selecting Drawing View Commands	230
USING THE DRAWING VIEW DIALOG BOX	230
Base Views	234
Projected Views	235
EXERCISE 5-2 CREATING A MULTIVIEW DRAWING	236
Auxiliary Views	241
Section Views	242
Detail Views	248
Broken Views	250
Break-Out Views	253
Section Depth	263
Editing the Section View Depth	265
Creating Draft Views	266
Creating Perspective Views	268
EXERCISE 5-3 COMPLEX DRAWING VIEW TECHNIQUES	270
EDITING DRAWING VIEWS	276
Moving Drawing Views	276
Editing Drawing View Properties	276
Displaying Tangent Edges	278
Deleting Drawing Views	278
Drawing View Alignments	279
EXERCISE 5-4 EDITING DRAWING VIEWS	280
DIMENSIONS	283
Retrieving Model Dimensions	284
Dimension Visibility	287
Changing Model Dimension Values	287
General Dimensions	288
Adding General Dimensions to a Drawing View	289
General Dimensions–Linear	289
General Dimensions–Radius	290
General Dimensions–Diameter	291

    General Dimensions–Angular 291
    Moving and Centering Dimension Text 292
    Editing General Dimension Extension Lines 292
    Editing Model Extension Lines 293
    Editing the Arrow Terminator 293
    Selecting Drawing Objects 294
ANNOTATIONS 296
    Center Marks and Centerlines..296
    Automated Centerlines from Models...298
    Text and Leaders 301
    Text Positioning...301
    Additional Annotation Tools...304
HOLE AND THREAD NOTES..304
    Hole Note Styles..305
    Editing Hole Notes 306
EXERCISE 5-5 CREATING TEXT AND DIMENSION STYLES.....................307
EXERCISE 5-6 ADDING DIMENSIONS AND ANNOTATIONS..................310
REVIEW SEQUENCE FOR CREATING DRAWING VIEWS.........................318
CHAPTER SUMMARY...318
APPLYING YOUR SKILLS..319
    Skill Exercise 5-1 319
    Skill Exercise 5-2 320
CHECKING YOUR SKILLS 321

CHAPTER 6 CREATING AND DOCUMENTING ASSEMBLIES
CHAPTER OBJECTIVES...322
CREATING ASSEMBLIES 323
ASSEMBLY OPTIONS 324
    Assembly Tools 324
THE ASSEMBLY BROWSER 327
BOTTOM-UP APPROACH 328
TOP-DOWN APPROACH...329
    Occurrences..330
    Multiple Documents and Drag & Drop Components 331
    Active Component 331
    Open and Edit 332
GROUNDED COMPONENTS 333
INSERTING MULTIPLE PARTS IN A SINGLE OPERATION........................333
SUBASSEMBLIES 334
RESTRUCTURING COMPONENTS 335
REORDERING COMPONENTS..336
ASSEMBLY CONSTRAINTS 336
    Types of Constraints 338
    Assembly Constraint Types 339
    Motion Constraint Types...344

- Transitional Constraint 346
- Applying Assembly Constraints 346
- Moving and Rotating Components 348

EDITING ASSEMBLY CONSTRAINTS 350
OTHER CONSTRAINT TOOLS 352
- Find Other Half 352
- Constraint Tool Tip 353
- Constraint Offset Value Modification 353

EXERCISE 6-1 ASSEMBLING PARTS 354
DESIGNING PARTS IN-PLACE 364
EXERCISE 6-2 DESIGNING PARTS IN THE ASSEMBLY CONTEXT 365
ASSEMBLY BROWSER TOOLS 372
- In-Place Activation 372
- Visibility Control 374

ADAPTIVITY 374
- Assembly Tab Options 375
- Underconstrained Adaptive Features 376
- Adaptive Feature Properties 378
- Adaptive Subassemblies 382
- Adapting the Sketch or Feature 383

EXERCISE 6-3 CREATING ADAPTIVE PARTS 384
ENABLED COMPONENTS 391
PATTERNING COMPONENTS 391
- Component Patterns 391
- Assembly Patterns 394
- Additional Component Pattern Options 395

EXERCISE 6-4 PATTERNING COMPONENTS 398
ANALYSIS TOOLS 403
- Center of Gravity 403
- Interference Checking 404

EXERCISE 6-5 ANALYZING AN ASSEMBLY 405
DRIVING CONSTRAINTS 409
EXERCISE 6-6 DRIVING CONSTRAINTS 412
DESIGN VIEW REPRESENTATIONS 415
- Creating a New Design View Representation 418
- Increasing Performance Through Design View Representations ... 420
- Visibility Overrides 421
- Importing Design View Representations from Other Assemblies ... 421
- Creating a Private Design View Representation 424
- Associative Design Views 426
- Creating Drawing Views from Design View Representations 427

FLEXIBLE ASSEMBLIES 430
POSITIONAL REPRESENTATIONS 432
- Creating Drawing Views from Positional Representations 438

EXERCISE 6-7 POSITIONAL REPRESENTATIONS 439

CREATING OVERLAY VIEWS..446
DOCUMENTING ASSEMBLIES USING PRESENTATION FILES......................448
    Creating Presentation Views..449
    Tweaking Components..450
    Animation...453
    Changing the Animation Sequence..455
    Presentation Highlighting..456
    Setting Units in a Presentation File..456
EXERCISE 6-8 CREATING PRESENTATION VIEWS..458
CREATING DRAWING VIEWS FROM
   ASSEMBLIES AND PRESENTATION FILES..463
CREATING BALLOONS..464
    Item Numbering Dialog Box Options..465
    Auto Ballooning...466
    Split Balloons...469
    Creating Balloons with User-Defined Symbols..................................471
    Editing Balloons...471
PARTS LISTS..478
    Editing BOM Data..479
PARTS LIST TOOLS..480
    Parts List Operations (Top Icons)...480
    Creating Custom Parts...482
    Spreadsheet View..482
    Nested Parts Lists..483
EXERCISE 6-9 CREATING ASSEMBLY DRAWINGS..484
CHAPTER SUMMARY...493
APPLYING YOUR SKILLS..494
    Skill Exercise 6-1...494
    Assembly Tips:..495
    Skill Exercise 6-2...496
CHECKING YOUR SKILLS...498

CHAPTER 7 ADVANCED SKETCHING AND CONSTRAINING TECHNIQES
CHAPTER OBJECTIVES...500
CONSTRUCTION GEOMETRY...500
ELLIPSES..502
2D SPLINES...503
    Bowtie..503
    Fit Method..505
    Insert Point...506
    Close Spline..507
    Display Curvature...507
    Spline Tension..508
    Constrain and Dimension a Spline...508
EXERCISE 7-1 COMPLEX SKETCHING..508

PATTERN SKETCHES..514
SHARED SKETCHES...516
MIRROR SKETCHES AND SYMMETRY CONSTRAINT.....................517
SLICE GRAPHICS..519
SKETCH ON ANOTHER PART'S FACE 520
EXERCISE 7-2 PROJECTING EDGES AND
   SKETCHING ON ANOTHER PART'S FACE 521
DIMENSION DISPLAY, RELATIONSHIPS, AND EQUATIONS.............524
    Dimension Display..525
    Dimension Relationships...526
    Equations..526
PARAMETERS 527
    User Parameters 529
    Linked Parameters...530
EXERCISE 7-3 AUTO DIMENSION, RELATIONSHIPS, AND PARAMETERS.....533
CHAPTER SUMMARY..539
APPLYING YOUR SKILLS 539
    Skill Exercise 7-1 539
CHECKING YOUR SKILLS 544

CHAPTER 8 Advanced Part Modeling Techniques
CHAPTER OBJECTIVES...546
USING OPEN PROFILES 547
RIB AND WEB FEATURES..549
    Shape..550
    Thickness..550
    Extents..550
    Creating Ribs and Webs..551
    Rib Networks...552
EXERCISE 8-1 CREATING RIBS AND WEBS...................................552
EMBOSS TEXT AND PROFILES...556
    Step 1 - Creating Text 556
    Step 2 - Embossing Text..557
EXERCISE 8-2 CREATING TEXT AND EMBOSS FEATURES 559
SWEEP FEATURES 563
    Shape..564
    More...564
    Creating a Sweep Feature...565
    3D Sketching...566
EXERCISE 8-3 CREATING SWEEP FEATURES.................................575
COIL FEATURES...578
    Coil Shape Tab 578
    Coil Size Tab..580
    Coil Ends Tab...581

LOFT FEATURES	582
Curves Tab	584
Conditions Tab	585
Transition Tab	586
EXERCISE 8-4 CREATING LOFT FEATURES	588
PART SPLIT AND FACE SPLIT	591
Method	592
Faces	593
EXERCISE 8-5 SPLITTING A PART	593
COPYING FEATURES	596
MIRRORING FEATURES	600
SUPPRESSING FEATURES	601
CONDITIONALLY SUPPRESS FEATURES	603
REORDERING FEATURES	604
FEATURE ROLLBACK	605
FILE PROPERTIES	606
General	607
Summary	607
Project	607
Status	608
Custom	608
Save	608
Physical	608
CENTER OF GRAVITY	608
Part Materials	609
OVERRIDE MASS AND VOLUME PROPERTIES	610
IPARTS	612
Creating iParts	613
Editing iParts	619
iPart Placement	621
Standard iPart Libraries	622
Custom iParts	623
EXERCISE 8-6 CREATING AND PLACING IPARTS	624
IFEATURES	630
Create iFeatures	631
Insert iFeatures	634
Editing iFeatures	637
Inserting Table-Driven iFeatures	639
EXERCISE 8-7 CREATING AND PLACING IFEATURES	640
PUBLISHING PARTS AND FEATURES	644
Publishing Parts	644
Publishing Features	645
PLACING PUBLISHED PARTS AND FEATURES	646
Placing a Published Part	646
Placing a Published Feature	648

DESIGN ACCELERATOR	649
CHAPTER SUMMARY	650
APPLYING YOUR SKILLS	652
Skill Exercise 8-1	652
CHECKING YOUR SKILLS	654

CHAPTER 9 Sheet Metal Design

CHAPTER OBJECTIVES	657
INTRODUCTION TO SHEET METAL DESIGN	657
Sheet Metal Fabrication	658
SHEET METAL PARTS	661
Sheet Metal Design Methods	661
Creating a Sheet Metal Part	662
SHEET METAL TOOLS	663
Sheet Metal Styles	664
Face	669
Contour Flange	673
Flange	676
EXERCISE 9-1 CREATING SHEET METAL PARTS	678
HEM	685
EXERCISE 9-2 HEMS	687
Fold	690
Bend	692
EXERCISE 9-3 MODIFYING SHEET METAL PARTS	694
CUT	701
EXERCISE 9-4 CUT ACROSS BEND	703
CORNER SEAM	707
Corner Round	710
Corner Chamfer	711
PunchTool	712
EXERCISE 9-5 PUNCH TOOL	714
FLAT PATTERN	719
Common Tools	721
DETAILING SHEET METAL DESIGNS	722
EXERCISE 9-6 DOCUMENTING SHEET METAL DESIGNS	723
CHAPTER SUMMARY	729
APPLYING YOUR SKILLS	730
Skill Exercise 09-1	730
CHECKING YOUR SKILLS	731
INDEX	733

INTRODUCTION

INTRODUCTION

Welcome to the *Autodesk Inventor 10 Essentials Plus* manual. This manual provides a thorough coverage of the features and functionalities offered in Autodesk Inventor 10. It is designed to help you become productive quickly, and to encourage you to use the self-paced learning activities in the Autodesk Inventor Design Support System (DSS).

Each chapter in this manual is organized with the following elements:

Objectives   Describes the content and learning objectives.

Topic Coverage   Presents a concise, thorough review of the topic.

Exercises   Presents the workflow for a specific tool or process through illustrated, step-by-step instructions.

Chapter Summary   Summarizes, in table format, the tools and processes discussed in the chapter.

Applying Your Skills   Checks your skills and understanding of the material covered in the chapter using challenge exercises. These exercises describe a design challenge, but do not provide step-by-step instructions.

Checking Your Skills   Tests your understanding of the material using True/False or multiple-choice questions.

NOTE TO THE LEARNER

Autodesk Inventor is designed for easy learning. The DSS provides you with ongoing support, as well as access to online documentation.

As described above, each chapter in this manual has the same instructional design, making it is easy to follow and understand. Each exercise is task-oriented and is based on real-world mechanical engineering examples.

WHO SHOULD USE THIS MANUAL?

The manual is designed to be used in instructor-led courses, although you may also find it helpful as a self-paced learning tool.

RECOMMENDED COURSE DURATION

Four days (32 hours) to seven days (56 hours) are recommended, although you may use the manual for specific Autodesk Inventor topics that may have a duration of only a few hours.

USER PREREQUISITES

It is recommended that you have a working knowledge of Microsoft®Windows NT®4.0, Windows 2000®, or Windows XP®, and a working knowledge of parametric solid modeling concepts.

MANUAL OBJECTIVES

The primary objective of this manual is to provide instruction on how to create part and assembly models, document those designs with drawing views, and automate the design process.

Upon completion of all chapters in this manual, you will be proficient in the following:

- Basic and advanced part modeling techniques
- Drawing view creation techniques
- Assembly modeling techniques
- Sheet metal design

While working through these materials, we encourage you to make use of the Autodesk Inventor DSS, where you may find solutions to additional design problems that are not addressed specifically in this manual.

MANUAL DESCRIPTION

This manual provides the foundation for a hands-on course that covers basic and advanced Autodesk Inventor features used to create, edit, document, and print parts and assemblies. You learn about the part and assembly modeling tools through online and print documentation and through the real-world exercises in this manual.

ESSENTIALS EXERCISE FILES

The files for the exercises can be installed from the CD-ROM attached to the back cover of your book.

INSTALLING THE EXERCISE FILES

To install the exercise data files:

1. Browse to the *Autodesk Inventor 10 Essentials Plus Exercises* CD-ROM and double-click *Setup.exe*.

2. Follow the onscreen instructions.

3. The installation program tests your system to ensure that Autodesk Inventor 10 is installed. By default, an Essentials Exercises folder will be created in C:\IV10 Essentials Plus folder, unless you use the Browse button to specify a different folder.

4. All required files will be copied to the Essentials Plus Exercise folder.

PROJECTS

Most engineers work on several projects at a time, with each project consisting of a number of files. To accommodate this, Autodesk Inventor uses projects to help organize related files and maintain links between files.

Each project has a *project file* that stores the paths to all files related to the project. When you attempt to open a file, Autodesk Inventor uses the paths in the current project file to locate other necessary files.

For convenience, a project file is provided with the *exercises*.

USING THE PROJECT FILE

Before starting the exercise, you must complete the following steps:

1. Start Autodesk Inventor.

2. In the Getting Started dialog box, click Projects in the What to Do section.

3. In the Projects window, right-click the background and select Browse. Navigate to the folder where you installed the Essentials Exercises and then double-click the *C:\IV10 Essentials Plus\Exercises\IV10 Essentials Plus Exercises.ipj* file.

4. Double-click the *IV10 Essentials Plus Exercises* project in the Projects window to make it the active project.

5. You can now start using the exercises.

 NOTE Projects are reviewed in more detail in the Getting Started section in Chapter 1.

TERMS AND PHRASES

To help you to understand Autodesk Inventor better, the following sections explain a few of the terms and phrases that will be used in this book.

Parametric modeling: Parametric modeling is the ability to drive the size of the geometry by dimensions, and it is sometimes referred to as "dimension-driven design." For example, if you increase the length of a plate from 5 to 6, change the 5=

dimension to 6=, and the geometry will be updated. Think of it as the geometry being along for a ride, driven by the dimensions. This is the opposite of associative dimensions. As lines, arcs, and circles are drawn, they are created to the exact length or size; when they are dimensioned, the dimension reflects the exact value of the geometry. If you need to change the size of the geometry, you stretch the geometry, and the dimension automatically gets updated. Think of this as the dimension being along for a ride, driven by the geometry.

Feature-based parametric modeling: Feature-based means that as you create your model, each hole, fillet, chamfer, extrusion, and so on is an independent feature whose dimensional values you can change without having to re-create the feature.

Adaptivity: Adaptivity allows parts to have a physical relationship between each other, such as when one plate is created with a dimensioned cutout, and another plate is created with the same cutout, but no dimensions are created. You can constrain the second cutout to match the size of the dimensioned cutout. When the dimensions change on the first cutout, the second cutout will be updated or adapt to show the change.

Bi-directional associativity: This means the model and the drawing views are linked. If the model changes, the drawing views will be updated automatically. Additionally, if the dimensions in a drawing view change, the model is updated and the drawing views are updated based on the updated part.

Bottom-up assembly: This refers to an assembly whose parts were created in individual part files outside of the assembly and then referenced into the assembly.

Top-down assembly: This refers to an assembly whose parts were created while working in the assembly file.

CHAPTER 1

Getting Started

This chapter provides a look at the user interface, projects, application options, starting tools (commands), and instructions on how to view parts in Autodesk Inventor.

CHAPTER OBJECTIVES

In this chapter, you will gain an understanding of

- The Getting Started screen
- Reasons for which a project file is used
- How to create a project file for a single user
- Autodesk Vault
- Different file types used in Autodesk Inventor
- Application options
- The Help system
- The user interface
- How to issue commands
- Different viewing tools

GETTING STARTED WITH AUTODESK INVENTOR

The Autodesk Inventor startup screen looks similar to the following image. If the Autodesk Inventor Launchpad screen is not present when you start up the program, click Getting Started on the File menu. There are four startup options on the left side of the screen:

- Getting Started
- New
- Open
- Projects

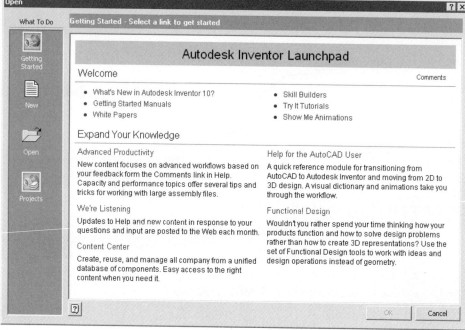

Figure 1-1

GETTING STARTED

The information available on the Getting Started page helps reinforce what you learn in this book. Use Getting Started to find information about

- Learning how to use Autodesk Inventor

Welcome

In this section you will find information about

- What's new in the current release
- Autodesk Inventor Basics
- Information about styles and standards
- Online exercises
- Tutorials for using Autodesk Inventor
- Step-by-step animations for using Autodesk Inventor

Expand Your Knowledge

In this section you will find information about

- Advanced functionality in Autodesk Inventor
- Solutions written in response to user questions

- Using the Content Center
- Using Online tutorials
- Presentations to help you transition from AutoCAD to Autodesk InventorUsing the knowledge content tools

NEW

Click New when you want to start a new file. The Open dialog box will appear. Begin by selecting one of the drafting standards named on the tabs, and then select a template for a new part, assembly, presentation file, sheet metal part, or drawing, as shown in the following image. Create a subdirectory in the *Autodesk\Inventor(version number)\templates* directory and add a file to it. A new templates tab with the same name as the subdirectory is created automatically.

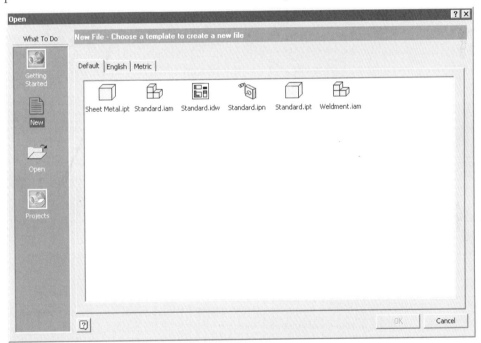

Figure 1-2

OPEN

Click Open to open an existing Autodesk Inventor file. The directory that opens by default is set in the current project file. You can open files from other directories that are not defined in the current project file, but it is not recommended. Part, drawing, and assembly relationships may not be resolved when you reopen an assembly containing components outside the locations defined in the current project file.

PROJECTS

Click Projects to create a new project, make an existing project current, or modify the content of an existing project. Projects are files that contain search paths to find files that are needed for a given project. With these search paths, all the needed files can be located when you open an assembly, drawing, or presentation file. There is no limit to the number of projects you can create, but only one project can be active at any given time. Project concepts are introduced in the next section.

PROJECTS IN AUTODESK INVENTOR

Almost every design you create in Autodesk Inventor involves more than a single file. Each part, assembly, presentation, and drawing created is stored in a separate file. Each of these files has its own unique file extension. There are many times when a design will reference other files. An assembly file, for example, will reference a number of individual part files and/or additional subassemblies. When you open the parent or top-level assembly, it must contain information that allows Autodesk Inventor to locate each of the referenced files. Autodesk Inventor uses a project file to organize and manage these file-location relationships.

You can structure the file locations for a design project in many ways. A single-person design shop has different needs than a large manufacturing company or a design team with multiple designers working on the same project. In addition to project files, Autodesk Inventor includes a basic check-out and check-in file-reservation mechanism that controls file access for multi-user design teams. For more information on multi-user environments consult the online Help system.

Autodesk Inventor always has a project named Default. Specifically, if all the files defining a design are located in a single folder, or in a folder tree where each referenced part is located with its parent or in a subfolder underneath the parent, the Default project may be all that is required.

 NOTE It is recommended that files in different folders never have the same name to avoid the possibility of Autodesk Inventor unintentionally resolving a reference to an incorrect file of the same name.

PROJECT SETUP

Always plan your project folder structure before you start a design to reduce the possibility of file resolution problems later in the design process. A typical project might consist of parts and assemblies unique to the project, standard components that are unique to your company, and off-the-shelf components such as fasteners, fittings, or electrical components.

PROJECT FILE SEARCH OPTIONS

Before you create a project, you need to understand how Autodesk Inventor stores cross-file reference information and how it resolves that information to find the referenced file. Autodesk Inventor stores the file name, a subfolder path to the file, and a library name (optionally) as the three fundamental pieces of information about the referenced file.

When you use the Default project file, the subfolder path is located relative to the folder containing the referencing file. It may be empty, or go deeper in the subfolder hierarchy, but it can never be located at a level above the parent folder.

If you create your own project file and specify project search locations, the subfolder path is applied initially to your specified project locations. If Autodesk Inventor does not find the file there, it will continue to look relative to the folder containing the source file.

A project file can have zero or more file search path locations defined. These paths are searched in a specific order when a document, such as an assembly, needs to find referenced components, such as the part files in the assembly.

You create projects in the Projects dialog box that is shown in the following image. The Projects dialog box is divided into two panes. The top pane lists shortcuts to the project files that have been active previously.

Double-click on a project's name to make it the active project. Only one project file can be active in Autodesk Inventor at a time. The bottom portion reflects information about the active project, an explanation of which follows.

The bottom pane of the dialog box lists information about the highlighted, or active, project. To see more information about the project click the more button (>>) on the bottom-right side of the dialog box.

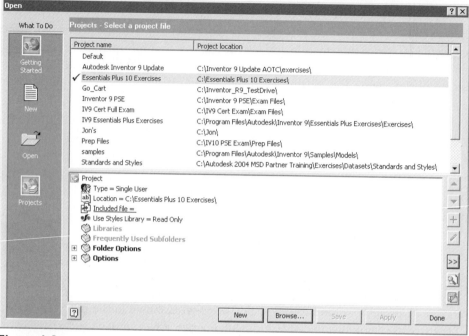

Figure 1-3

The following image shows the additional entries that appear after clicking the more button (>>).

Figure 1-4

INCLUDED FILE AND WORKGROUP

An included file and workgroup are used when implementing a multi-user design environment. These values typically are not defined outside of this type of environment. For more information on multi-user environments, consult the online Help system.

WORKSPACE

A workspace is a location where you edit your personal copy of design files in single-user, semi-isolated, and vault modes. For single-user and vault modes, the workspace should be the only defined editable location. Only one designer should use a project with a defined workspace in a single session of Autodesk Inventor at a time.

WORKGROUP SEARCH PATHS

Some design environments require more file management sophistication than the Default project provides. The files that compose the design, for example, may not be in a single folder tree and may even be on different disks or servers. You may also want to keep your drawings in one folder, presentations in another, main assemblies in another, etc. A project addresses this need by allowing you to define a search path for each location where files are to be stored.

When defining a path to a folder on another machine, it is better to define a full UNC path starting with the server name (*server*\...) and not to use shared network drives.

NOTE You do not need a separate workgroup location defined for each folder; rather, it is recommended that you only define one workgroup location for the root folder of a folder tree containing the Autodesk Inventor files for one project or design. In most cases in which the Default project file is insufficient, a project file with only one workgroup location will suffice. It gives you more flexibility in organizing the subfolders of your project than the Default project does.

LIBRARY SEARCH PATHS

Library search paths are treated differently than other search paths. You use them to store standard company and third-party models that are not edited normally. In a networked environment, you store these standard components on the network and define the location as a library search path. Files loaded from a library search path cannot be modified; they are recognized by Autodesk Inventor as read-only files. If you create a reference to a file that exists under a library search path, Autodesk Inventor will store the library name with the reference information. It uses that name to scope the reference specifically to the associated library path.

NOTE The folders need not exist prior to adding them to the project file search paths. Named paths added in the Project File Editor are created automatically if they do not exist. Consult the online Help system for more information on projects for collaborative environments and search paths.

FREQUENTLY USED SUBFOLDERS

In this area add the subfolders that you need to access regularly; these subfolders need to be subfolders of a project search path. These paths will be listed in the Locations section of the Open dialog box as shown in the following image. Click on the desired frequently used subfolder in the Locations portion of the Open dialog box to access files in the folder.

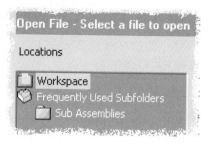

Figure 1-5

FOLDER OPTIONS

In this folder (see the following image, top) you see where Autodesk Inventor will look for the following:

Style Library The location where Autodesk Inventor will look for style definitions for the project.

Templates The location where Autodesk Inventor will look for template files for the project.

Content Center Files The location where Autodesk Inventor will look for Content Library files for the project.

OPTIONS

In this folder (see the following image, bottom) you define the following:

Old Versions To Keep On Save Sets the number of edited versions of a file you want to keep. The files will be saved in the subdirectory OldVersions.

Using Unique File Names Allows you to choose whether all the files in the project will have unique file names. Library files are not affected by this option.

Name Displays the name of the current project. To edit the name, right-click Name, select Edit from the menu, and then type in a new name. The name updates in the Select Project pane, but does not update the folder name.

Shortcut Displays the name of the shortcut to the active project. The shortcut is saved in the Projects Folder.

Owner Identifies the project owner, typically the lead engineer or CAD administrator. To add or edit the owner's name, click Owner, click Edit Selected Item on the right side of the dialog box, and type in the owner's name.

Release ID Identifies the version of the released project data. In the event that the project is used as a library by another project, the release ID can identify which project to use.

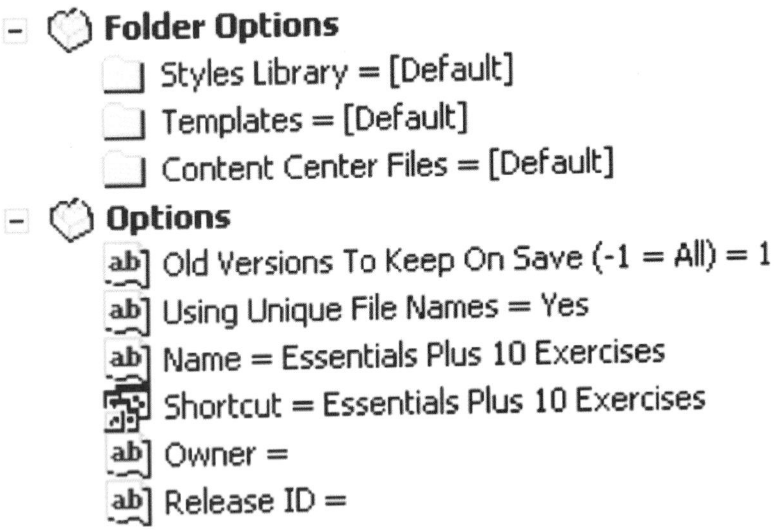

- 📁 **Folder Options**
    - 📄 Styles Library = [Default]
    - 📄 Templates = [Default]
    - 📄 Content Center Files = [Default]
- 📁 **Options**
    - 🔤 Old Versions To Keep On Save (-1 = All) = 1
    - 🔤 Using Unique File Names = Yes
    - 🔤 Name = Essentials Plus 10 Exercises
    - 🔤 Shortcut = Essentials Plus 10 Exercises
    - 🔤 Owner =
    - 🔤 Release ID =

Figure 1-6

PROJECT FILE SEARCH ORDER

Project file search paths are named paths to local or network folders that contain the documents (or components) associated with the project. When opening a document that references other files, Autodesk Inventor searches the paths in the following order and loads the first file that matches the reference information stored in the parent document.

Library Components

- Library Path

Library components are treated differently than other referenced files. The library name is stored along with the file name in the referencing file. When the referencing file is opened, only the path associated with this library name is searched for the referenced component. If the file is not found there, no other project locations are searched.

All Other Referenced Components

- Workspace
- Local Search Paths in the order listed in the project file
- Workgroup Search Paths in the order listed in the project file

In all cases, if the file is not found by the search outlined above, Autodesk Inventor additionally searches the folder containing the parent document. Also, with each location searched, Autodesk Inventor appends the subfolder information stored with the reference to the search folder location and looks for the file there. If not found there, Autodesk Inventor then looks for the file in the search folder location.

RESOLVING FILE LINKS

When a part or assembly (subassembly) is placed in a higher-level assembly, the information required to locate the referenced file is saved in the assembly document. This information helps locate the referenced components the next time you open the assembly. The same is true when placing a view of a component or presentation in a drawing, creating a presentation of an assembly, or deriving a part from another component. If the file cannot be located, the Resolve Link dialog box appears as shown in the following image. From here, you can locate or skip the file manually.

NOTE The search order for referenced files follows a defined set of rules. If there are multiple referenced files of the same name in the searched paths, Autodesk Inventor will load the first one found.

NOTE When creating a new project, close all Autodesk Inventor files and use descriptive names for search paths and directory shortcut names.

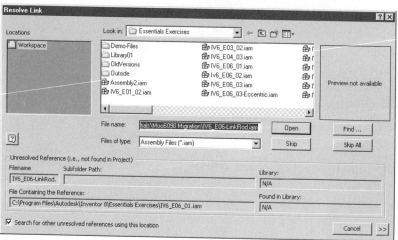

Figure 1-7

CREATING PROJECTS

To create a new project or edit an existing project, use the Autodesk Inventor Project File Editor. The Project File Editor displays a list of shortcuts to previously active projects. A project file has an *.ipj* file extension and typically is stored in the home folder for the design-specific documents, while a shortcut to the project file is stored in the Projects Folder. The Projects Folder is specified on the Files tab of the Options dialog box, as shown in the following image. All projects with a shortcut in the Projects Folder are listed in the top pane of the Project File Editor.

Figure 1-8

To create a project, follow these steps:

1. Click the Project icon in the What To Do area of the Getting Started menu, or click Projects on the File menu.

 TIP You can also create a new project when the Autodesk Inventor application is not running. Click on the Microsoft Windows Start menu, and then click Programs > Inventor (version number) > Tools > Project Editor.

2. In the Projects dialog box, click the New button (at the bottom) to initiate the Inventor project wizard.

3. In the Inventor project wizard, follow the prompts to the following questions:

WHAT TYPE OF PROJECT ARE YOU CREATING?

If Autodesk Vault is installed you will prompted to create a New Vault project or a New Single-User project. If Autodesk Vault is not installed a New Single-User project type will automatically be selected without a prompt.

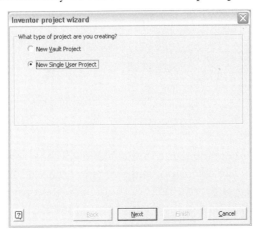

Figure 1-9

New Vault Project

This project type is used with Autodesk Vault and is grayed out until you install Autodesk Vault. It creates a project with one workspace and any needed library location(s), and it sets the multi-user mode to Vault. More information about Autodesk Vault appears later in this section.

New Single-User Project

This is the default project type, which is used when only one user will reference Autodesk Inventor files. It creates one workspace (where Autodesk Inventor files are stored) and any needed library location(s), and it sets the Project Type to Single User. No workgroup is defined but can be defined later.

The next section covers the steps for creating a new single-user project. For more information on projects, consult the online Help system or the book on projects that shipped in the box with Autodesk Inventor.

Creating a New Single-User Project

After clicking New Single-User Project (if Autodesk Vault is not installed it will be selected for you) click the Next button and specify the project file name and location on the second page of the Inventor project wizard as shown in the following image.

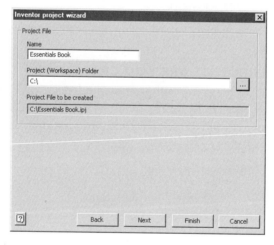

Figure 1-10

Name

Enter a descriptive project file name in the Name edit box. The project file will use this name with an *.ipj* file extension, and will appear in the folder indicated in the Project (Workspace) Folder edit box. A shortcut to the project will appear in the Projects folder specified on the Files tab of the Application Options dialog box.

Project (Workspace) Folder

This specifies the path to the home or top-level folder for the project. You can accept the suggested path, enter a path, or click the Browse button (...) to manually locate the path. The default home folder is a subfolder under My Documents (or wherever you last browsed) named to match the project file name.

Project File to be created

The full path name of the project file is displayed below the Location edit box.

 NOTE You can specify the home or top-level folder as the location of your workspace, but it is preferable that the project file (*.ipj*) be the only Autodesk Inventor file stored in the home folder. This makes it easier to create other project locations (such as library and workspace folders) as subfolders without creating nesting situations that can lead to confusion.

Click the next button at the bottom of the Inventor project wizard and specify the project library search paths.

You can add library search paths from existing project files to this new project file. The library search paths from every project with a shortcut in your Projects Folder are listed on the left in the Inventor project wizard dialog box, as shown in the following image. You can add and remove libraries from the New Project area by clicking on their names in either the All Projects or New Project area and then clicking the arrows in the middle of the dialog box. The libraries listed by default in the All Projects list will match those in the project file you selected prior to starting the New Project process.

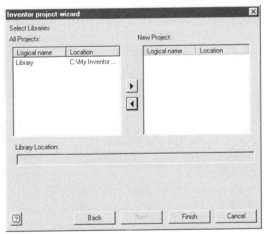

Figure 1-11

Click the Finish button to create the project. The new project will appear in the Open dialog box. Double-click on a project's name in the Open dialog box to make it the active project. A check mark will appear to the left of the active project, as shown in the following image.

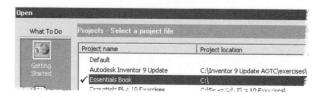

Figure 1-12

EDITING PROJECTS

To edit a project, follow these steps:

1. Click the Project icon in the What To Do area of the Getting Started screen. If you are already working in Autodesk Inventor, click Projects on the File menu (any open Autodesk Inventor files need to be closed). If Autodesk Inventor is not running, you can edit a project from the standalone Inventor Project Editor. To start the editor, click on the Microsoft Windows Start menu and then click Programs > Inventor (version number) > Tools > Project Editor, or in Microsoft Windows Explorer, right-click an *.ipj* file and select Edit.

2. In the top portion of the Project Editor dialog box, double-click the name of the project to make it the active project.

3. In the lower portion of the Project Editor dialog box, right-click the section to edit and select an option from the menu. You can also select the area to edit, and then click the Add or Edit icon on the right side of the dialog box.

4. To edit the order in which directories are searched, select the directory path to move, and click the up or down arrow on the right side of the dialog box.

AUTODESK VAULT

Autodesk Vault is available on the Autodesk Inventor CD. The Autodesk Vault enhances the data-management process by managing more than just Autodesk Inventor files and by tracking file versions as well as team member access. Controlling access to data, tracking modifications, and communicating the design history are important aspects of managing collaborative data. When working with a vault project, your data files are stored in a central repository that records the entire development history of the design. The vault manages Inventor and non-Inventor files alike. In order to modify a file, it first must be checked out of the vault. When the file is checked back into the vault, the modifications are stored as the most recent version for the project and the previous version is sequentially indexed as part of the living history of the design.

 NOTE For more information about Autodesk Inventor projects and Autodesk Vault, review the book Managing Your Data, shipped in the box with Autodesk Inventor or *Managing Your Data.pdf*, which is located on the Autodesk Inventor 10 CD.

Chapter 1   Getting Started   15

EXERCISE 1-1 Projects

In this exercise, you create a project file for a single-user project. You then examine how the project file provides access to referenced files in an Autodesk Inventor assembly. Prior to creating a new project file, close all files currently open in Autodesk Inventor.

1. From the main menu, select File > Projects. You can also access the Project File Editor from the Open dialog box.

2. In the upper pane of the Project File Editor, click the Samples project.

3. Examine the search paths and frequently used subfolders associated with this project. Click the more button (>>) on the right side of the dialog box to display the Workspace.

4. Double-click the Default project. The default project contains no search paths. You base the new project on the default project.

5. You now create a new project file based on the default project file. Click the New button at the bottom of the Project File Editor.

6. If you have Autodesk Vault installed click New Single User Project and click Next, if Autodesk vault is not installed skip this step.

7. On the second page of the Inventor project wizard: Enter **Single User Demo** in the Name edit box.

8. Click the browse button (...) to the right of the Project (Workspace) Folder: edit box.

9. Browse to and open the C:*IV10 Essentials Plus**Exercises* folder. The project file (.ipj) will be placed in this folder. The selected folder is the top-level folder for the project. It is good practice to place project component files (parts, assemblies, drawings) in folders below the top-level folder, but not in the top-level folder.

Figure 1-13

10. Click Next to add a library search path. If the New Project library list is not blank, select all items in the list and click the Delete selected libraries tool between the two lists.

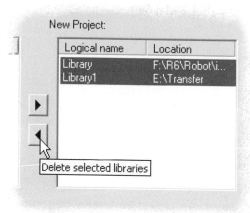

Figure 1-14

11. Click Finish. The new project file is highlighted in the upper pane of the Project File Editor. The search paths for the new project are listed in the lower pane. If desired click the more button (>>) on the right side of the dialog box to display the workspace path.

12. Activate the new project by double clicking on **Single User Demo** in upper pane of the Project File Editor.

13. Click the more button (>>) on the bottom-right side of the dialog box.

14. Expand Workspace and notice that the Workspace search path is listed as a period ".." The "." denotes that the workspace location is relative to the location where the project file is saved.

Figure 1-15

15. Click Frequently Used Subfolders.

16. Click the plus button (+) on the right side of the dialog box.

17. For the name of the folder enter **Demo Files**.

18. For the path navigate to the Demo Files folder: C:*IV10 Essentials Plus**Exercises*\Demo Files and then click OK.

Figure 1-16

19. Click Save in the Project File Editor.

20. Click the Open tool on the left side of the Open dialog box.

21. In the Locations area of the Open dialog box double click on *Demo Files* under Frequently Used Subfolders as shown in the following image.

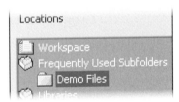

Figure 1-17

22. Double click on the file *ESS_E01_01.iam* to open an existing assembly from a frequently used subfolder.

23. The Resolve Link dialog box will appear. The parts that cannot be located are Library parts. Click the Skip button twice.

24. Click the OK button to remove the alert box.

25. Click the Yes button to update the assembly. The file contains a model of a pivot assembly as shown in the following image.

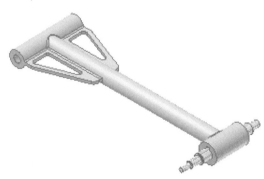

Figure 1-18

26. Close the file without saving your work.

27. Close all other Autodesk Inventor files.
28. From the main menu, select File > Projects or access the Project File Editor from the Open dialog box.

The missing files are located in a Library that you will add to the project. Right-click Libraries then select Add Path.

Figure 1-19

29. Navigate to and select C:*IV10 Essentials Plus\Essentials Library Demo* folder.
30. Click Save in the Project File Editor.
31. Click the Open tool to open the assembly that contains components in a library.
32. If needed, double click on the Frequently Used Subfolder *Demo Files*.
33. Double click on the file *ESS_E01_01.iam*.
34. Click Yes button to update the assembly. The pivot assembly will open without the Resolve Link dialog box appearing.

Figure 1-20

35. You must change the active project file to successfully complete the remaining exercises. Close all files currently open in Autodesk Inventor.
36. Select File > Projects from the main menu.
37. Double click *IV10 Essentials PlusExercises* in the upper pane. If the project file does not show up in the list, click the Browse button (...) at the bottom of the dialog box and navigate to select C:*IV10 Essentials Plus/Exercises\IV10 Essentials Plus Exercises.ipj*.

38. Right-click on the project Single User Demo in the upper pane and select delete.

Figure 1-21

39. End of exercise.

FILE INFORMATION

While creating parts, assemblies, presentation files, and drawing views, the data is stored in separate files with different file extensions. This section describes the different file types and the options for creating them.

FILE TYPES

The following are the main file types that you can create in Autodesk Inventor, their file extensions, and a description of their uses.

Part (.ipt)

Part files contain only one part, which can be either 2D or 3D.

Assembly (.iam)

Assembly files can consist of a single part, multiple parts, or subassemblies. The parts themselves are saved to their own part file and are referenced (linked) in the assembly file. See Chapters 6 and 10 for more information about assemblies.

Presentation (.ipn)

Presentation files show parts of an assembly exploded in different states. A presentation file is associated with an assembly, and any changes made to the assembly will be updated in the presentation file. A presentation file can be animated, showing how parts are assembled or disassembled. The presentation file extension is *ipn*, but you save animations as AVI files. See Chapter 6 for more information about presentation files.

Sheet Metal (.ipt)

Sheet metal files are part files that have the sheet metal environment loaded. In the sheet metal environment, you can create sheet metal parts and flat patterns. You can create a sheet metal part while in a regular part. This requires that you load the sheet metal environment manually. See Chapter 11 for more information about creating sheet metal parts.

Drawing (.idw)

Drawing files can contain 2D projected drawing views of parts, assemblies, and/or presentation files. You can add dimensions and annotations to drawing views. The parts and assemblies in drawing files are linked, like the parts and assemblies in assembly and presentation files. See Chapters 5 and 9 for more information about drawing views.

Project (.ipj)

Project files are ASCII-based text files that contain search paths to locations of all the files in the project. The search paths are used to find the files in a project.

iFeature (.ide)

iFeature files can contain complete parts, 3D features, or 2D sketches that can be inserted into a part file. You can place size limits and ranges on iFeatures to enhance their functionality. See Chapter 12 for more information about creating iFeatures.

Design Views (.idv)

Private Design Views are configuration files in which the following information about an assembly file is saved: component visibility, component selection status, color settings, zoom magnification, and viewing angle. See Chapter 10 for more information about creating design views.

MULTIPLE DOCUMENT ENVIRONMENT

You can open multiple Autodesk Inventor files at the same time in a single Autodesk Inventor session. To switch between the open documents, click the file on the Windows menu. The files can also be arranged to fit the screen or to appear cascaded. If the files are arranged or cascaded, click a file to activate it. Only one file can be active at a time.

SAVE OPTIONS

There are three options on the File menu for saving your files: Save, Save Copy As, and Save All, as shown in the following image.

Figure 1-22

Save

The Save command saves the current document with the same name and to the location where you created it. If this is the first time that a new file is saved, you are prompted for a file name and file location.

NOTE To run the Save command, click the Save icon on the standard toolbar, use the shortcut keys CTRL-S, or click Save on the File menu.

Save Copy As

Use the Save Copy As command to save the active document with a new name and location, if required. A new file is created, but is not made active.

Save All

Use the Save All command to save all open documents and all of their dependents. The files are saved with the same name to the location where you created them. The first time that a new file is saved, you will be prompted for a file name and file location.

APPLICATION OPTIONS

Autodesk Inventor can be customized to your preferences. On the Tools menu, click Application Options to open the Options dialog box, as shown in the following image. You set options on each of the tabs to control specific actions in the Autodesk Inventor software, as outlined below. Each tab is covered in more detail in the pertinent sections throughout this book. For more information about application options, see the online Help system.

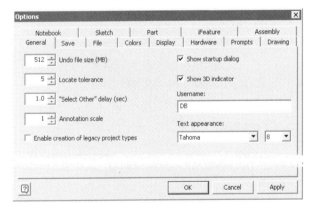

Figure 1-23

General
Set general options for how Autodesk Inventor operates.

Save
Set how files are saved.

File
Set where files are located.

Colors
Change the color scheme, and color of the background on your screen. Also determine if reflections and textures will be displayed

Display
Adjust how the parts look. Your video card and your requirements affect the appearance of parts on your screen. Experiment with different settings to achieve maximum video performance.

Hardware
Adjust the interaction between your video card and the Autodesk Inventor software. The software is dependent upon your video card. Take time to make sure you are running a supported video card and the recommended video drivers. If you experience video-related issues, experiment with the options on the Hardware tab. For more information about video drivers, click Graphics Drivers on the Help menu.

Prompts
Modify the response given to messages that are displayed.

Drawing
Specify the way drawings are created and displayed.

Notebook
Specify how the Engineer's Notebook is displayed.

Sketch
Modify how sketch data is created and displayed.

Part
Change how parts are created.

iFeature
Adjust where iFeatures data is stored.

Assembly
Specify how assemblies are controlled and behave.

DESIGN SUPPORT SYSTEM

The online Help system in Autodesk Inventor goes beyond basic command definition by offering assistance while you design. The following help mechanisms make up the Design Support System:

- Help Topics
- Visual Syllabus

- What's New
- Tutorials
- Design Doctor
- Sketch Doctor
- Autodesk Online (Skill Builders)

To access the online Help system, use one of these methods:

- Press the F1 key and the Help system gives you help with the operation that is active.
- Click an option on the Help menu.
- Click a Help option on the right side of the standard toolbar.
- In any dialog box, click the ? icon.
- Click on How To on a shortcut menu initiated within an active command.

The visual syllabus and tutorials guide you through the process to complete chosen tasks. To start the visual syllabus, click Visual Syllabus on the standard toolbar. To start a tutorial, click Tutorials on the Help menu. The following image shows the first dialog box that appears after you click Visual Syllabus.

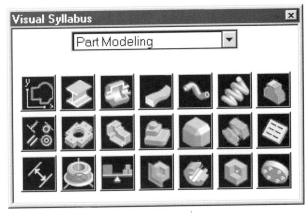

Figure 1-24

Design Doctor and Sketch Doctor appear on your screen when a problem exists with the current operation. The following image shows the Sketch Doctor examining a problem sketch.

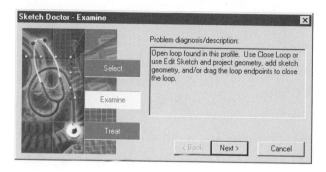

Figure 1-25

Another way to access Design Doctor and Sketch Doctor is to right-click on a sketch in the Browser and click Sketch Doctor from the menu or click on the cross (+) on the right side of the standard toolbar when it appears red, or when the red cross appears in a dialog box. This alerts you to a problem. Design Doctor or Sketch Doctor opens and provides options for resolving the problem.

USER INTERFACE

The default sketch environment of a part (*.ipt*) in the Autodesk Inventor application window looks similar to the following image.

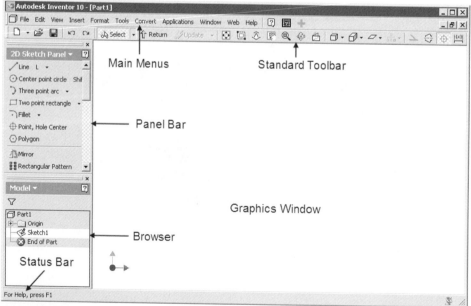

Figure 1-26

The screen is divided into the following areas:

Main Menus   Access tools via text menus.

Standard Toolbar   Access basic Windows and Autodesk Inventor tools.

Command Bar   View the current settings. Perform quick edit of objects.

Panel Bar   Activate most of the Autodesk Inventor tools. The set of tools in the Panel Bar changes to reflect the environment in which you are working.

Browser Toolbar   Change how the Browser looks. The default settings are used for this book.

Browser   Shows the history of how the contents in the file were created. The Browser can also be used to edit objects.

Graphics Window   Displays the graphics of the current file.

Status Bar   View text messages about where you are.

TOOLBARS, PANEL BAR, AND COMMAND BAR

Toolbars, the Panel Bar, and the Browser can be moved to another docked or undocked location. To move a toolbar, click the two horizontal lines on the top of the toolbar, keep the mouse button depressed, and drag the toolbar to a new location. You get a preview of what the toolbar looks like in the new location. To turn toolbar visibility on and off, click Toolbar on the View menu.

The tools in the Panel Bar adjust to the environment in which you are working, such as a part, an assembly, or sheet metal. You can change the status of the Panel Bar manually. Click the title line of the Panel Bar that has a down arrow, or right-click in the Panel Bar and select another environment from the menu.

The tool icons in the Panel Bar are displayed in non-Expert mode by default, with a line of text describing the icon. In Expert mode, the text is removed. To turn Expert mode on (text is removed), either click the title area of the panel, or right-click in the Panel Bar, and select Expert, as shown in the following image.

Figure 1-27

COMMAND ENTRY

There are several methods to issue commands in Autodesk Inventor. In the following sections you will learn how to start a command. There are no right or wrong methods for starting a command; with experience, you will develop your own preference.

TOOLBARS AND THE PANEL BAR

In the last section, you learned how to control toolbars and the Panel Bar. To use a tool from a toolbar, move the cursor over the desired tool and a tool tip appears with the name of the tool. The following image shows the Line tool from the Sketch toolbar and the tool tip. Some of the icons in the Panel Bar have a small down arrow in the lower right corner. Select the arrow to see additional tools. To activate a tool, move the cursor over a tool icon and click.

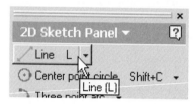

Figure 1-28

SHORTCUT MENUS

Autodesk Inventor also uses shortcut menus. Shortcut menus are text menus that pop up when you press the right mouse button. The shortcut menus are context sensitive; you open a menu with tools or options that are relevant to the current task.

WINDOWS SHORTCUTS

Another method for activating tools is to use Windows shortcut keys. Windows shortcut keys use a two-key combination. Press the two keys at the same time to activate the following operations:

CTRL-C Copy

CTRL-N Create a new document

CTRL-O Open a document

CTRL-P Print the active document

CTRL-S Save the current document

CTRL-V Paste

CTRL-Y Redo

CTRL-Z Undo

AUTODESK INVENTOR SHORTCUT KEYS

Autodesk Inventor has keystrokes called shortcut keys that are preprogrammed. To start a command via a shortcut key, press the desired preprogrammed key(s). The following table shows a few of the predefined shortcut keys. In the next section you will learn how to modify shortcut keys.

Shortcut Keys

Key	Description
F1	Access help.
F2	Pan the screen.
F3	Zoom into or out of the objects on the screen.
F4	Rotate the objects on the screen.
F5	Return to the previous view.
SHIFT+F5	Move to the next view.
B	Add a balloon to a drawing.
C	In the assembly environment, add an assembly constraint using the Assembly Constraint dialog box.
D	Add a general dimension to a sketch or drawing.
E	Extrude a sketch.
F	Add a feature control frame to a drawing.
H	Create holes using the Hole dialog box.
L	Draw a line.
O	Add an ordinate dimension.
P	Place a component into the current assembly.
R	Revolve a sketch.
S	Make a plane the active sketch.
T	Add a tweak to a part in the current presentation file.
ESC	Abort a command.
DELETE	Delete the selected objects.
BACKSPACE	Clears the last sketch selection as long as the command is active.
SHIFT+right-click	Access the Selection tool menu.
SHIFT+Rotate tools	Auto rotate or put the contents of the screen into hands-free rotation. Left-click to stop the rotation.
SPACE BAR	While in the rotate command, toggle the common view (glass box) on and off.
TAB	Alternate between input fields.
CTRL+SHIFT	Add and remove objects from a selection set.
CTRL+ENTER	Disable inferencing when committing precise input sketch points.
CTRL	Override the creation of constraints while sketching.

CUSTOMIZING SHORTCUTS

You can customize many of the predefined command shortcuts. Customization of Autodesk Inventor is performed by clicking Customize on the Tools menu.

COMMANDS TAB

The Commands tab gives you access to all of the tools that are available in Autodesk Inventor, and it allows you to modify the shortcut that is used to start the tool. The following image shows the Commands tab of the Customize dialog box.

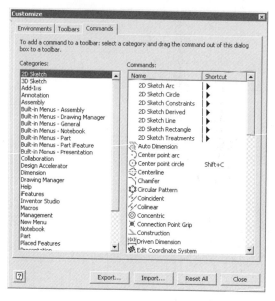

Figure 1-29

When you select an item from the Categories list, the tools that are available in that category are displayed on the right side of the dialog box. A black triangle next to the tool indicates that it has additional tools available in a fly-out menu. The Constraints tool, for example, contains all of the constraint types.

To add a new shortcut or change an existing shortcut, follow these steps:

1. Click in the Shortcut area of the tool that you want to change. The following image shows what the dialog box looks like when you click in the Shortcut area of the Auto Dimension tool.

2. On the keyboard, press the combination of key(s) that will become the shortcut. A single letter or number can be used. You can also use the SHIFT, CTRL, and ALT keys and a letter or number. The ALT key can be combined with the SHIFT and/or CTRL key(s) and a letter or number. The ALT key cannot be used with a single letter, however.

3. Press ENTER on the keyboard to create the shortcut.

To delete a shortcut, click on the shortcut and then press DELETE or BACKSPACE on your keyboard to remove the shortcut.

NOTE To switch an existing shortcut to a different, new tool, you must first delete the shortcut from the existing tool.

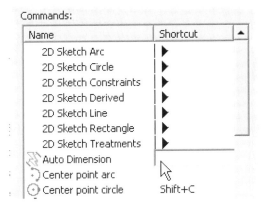

Figure 1-30

General Options The Export, Import, Reset All, and Close buttons reside at the bottom of the dialog box and are available regardless of which tab is active. These buttons function as follows:

Export Exports an XML file containing customized settings to a selected name and location.

Import Imports an XML file containing customized settings.

Reset All Resets all customized settings to their original installation setting.

Close Closes the dialog box.

ROLLUP COMMAND DIALOGS

You can control whether a dialog box appears normal or rolled up to show only the name of the box when the cursor is not located in the dialog box itself. To roll up a dialog box, click the left icon (push pin) in the upper-right corner of the dialog box (see the following image). The dialog box then shows only the horizontal title bar, as shown in the image on the right. To maximize the dialog box, move the cursor over the title bar.

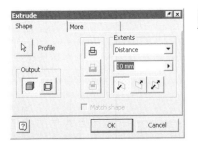 

Figure 1-31

UNDO AND REDO

You may want to undo an action that you just performed, or undo an undo. The Undo tool backs up Autodesk Inventor one function at a time. If you undo too far, you can use the Redo tool to move forward one step at a time. The Zoom, Rotate, and Pan tools do not affect the Undo and Redo tool. To start the tools, either use the shortcut keys— CTRL-Y for Redo or CTRL-Z for Undo—or select the tool from the standard toolbar, as shown in the following image. The Undo tool is to the left, and the Redo tool is to the right.

 NOTE To set the Undo file size allocation, click Tools Application Options. On the General tab of the Options dialog box, change the Maximum size of Undo file (MB).

Figure 1-32

VIEWPOINT OPTIONS

When you work on a 2D sketch, the default view is looking straight down at the XY plane (plane view). When you work in 3D, it is helpful to view objects from a different viewpoint, and to zoom in and out or pan the objects on the screen. The next section guides you through the most common methods for viewing objects from different perspectives and viewpoints. As you use these tools, the physical objects remain unmoved. Your perspective or viewpoint of the objects is what creates the perceived movement of the part. If you are in an operation while a viewing command is issued, the operation resumes after the transition to the new view is completed.

ISOMETRIC VIEW

Change to an isometric viewpoint by pressing the F6 key or by right-clicking in the graphics window, and then selecting Isometric View from the menu, as shown in the following image. The view on the screen transitions to a predetermined isometric view. You can redefine the isometric view with the Common View (Glass Box) option. Common View is covered later in this section.

Figure 1-33

CAMERA VIEWS

You can set the camera viewpoint to be either orthographic (lines are projected perpendicular to the plane of projection) or perspective (geometry on the screen converges to a vanishing point similar to the way the human eye sees). By default, the orthographic camera is set. To change the camera, click either the Orthographic Camera icon from the standard toolbar or the Perspective Camera icon which is the option just below Orthographic Camera as shown in the following image. For clarity, this book shows all the images in the Orthographic Camera view. For more information about options for setting the Perspective Camera, access the Help system.

Figure 1-34

SHADOW

To give your model a realistic look, you can choose to have shadows displayed on the ground. From the standard toolbar, click either Ground Shadow (as shown in the following image), No Ground Shadow, or X-Ray Ground Shadow. The default is No Ground Shadow. With shadows on, only model features cast shadows. Work geometry, sketches, origin indicators, engineer's notes, and trails in presentation documents do not cast shadows. X-Ray Ground Shadows are the same as Ground Shadows, but they also show detail information about hidden features.

Figure 1-35

VIEW TOOLS

To help zoom, pan, and rotate the geometry on the screen, use the View tools on the standard toolbar. Descriptions of the View tools follow.

Viewing Tools

Button	Tool	Function
	Zoom All	Maximizes the screen with all parts that are in the current file. The screen transitions to the new view. Issue the Zoom All tool or press the Home key.
	Zoom Window	Zooms in on an area that is designated by two points. Issue the Zoom Window tool or press **SHIFT** + F3, and select the first point. With the mouse button depressed, move the cursor to the second point. A rectangle appears representing the window. When the correct window is displayed on the screen, release the mouse button and the view transitions to it.
	Zoom In-Out	Zooms in or out from the parts. Issue the Zoom In-Out tool, or press the F3 key. Then, in the graphics window, press and hold the left mouse key. Move the mouse toward you to make the parts appear large, and away from you to make the parts appear smaller. If you have a mouse with a wheel, roll the wheel toward you and the parts appear larger; roll the wheel away from you and the parts appear smaller.
	Pan View	Moves the view to a new location. Issue the Pan View tool or select the F2 key. Press and hold the left mouse button and the screen moves in the same direction that the cursor moves. If you have a mouse with a wheel, hold down the wheel, and the screen moves in the same direction that the cursor moves.
	Zoom Selected	Fills the screen with the maximum size of a selected face or faces. Either select the face or faces and then issue the Zoom Selected tool or select the END key, or launch the Zoom Selected tool or select the END key and then select the face or faces to which to zoom.
	Dynamic Rotate	Rotates objects dynamically. Issue the Dynamic Rotate tool; a circular image with lines at the quadrants and center appears. To rotate the parts freely, click a point inside the circle and keep the mouse button pressed as you move the cursor. The model rotates in the direction of the cursor movement. When you release the mouse button, the model stops rotating. To accept the view orientation, either press the ESC key or right-click and select Done from the menu. Click the outside of the circle to rotate the model about the center of the circle. To rotate the parts about the vertical axis, click one of the horizontal lines on the circle and, with the mouse button pressed, move the cursor sideways. To rotate the parts about the horizontal axis, click one of the vertical lines on the circle and, with the mouse button pressed, move the cursor upward or downward.
	Look At	Changes your viewpoint so you are looking parallel to a plane, or rotate the screen viewpoint to be horizontal to an edge. Issue the Look At tool or press the PAGE UP key, then select a plane or edge. The Look At tool can also be issued by selecting a plane or edge, and then right-clicking while the cursor is in the graphics window and selecting Look At from the standard toolbar.
	Common View (Glass Box)	Changes the viewpoint to a predetermined viewpoint. Issue the Dynamic Rotate tool and then either press the SPACE bar or right-click and select Common View from the menu. A cube appears with arrows pointing at each corner and face. Click one of the arrows on the cube and the viewpoint rotates to that perspective. If you are looking at the cube from a plan view, you can click an edge of the cube to rotate the cube 90\r¢ .

Viewing Tools (continued)

Button	Tool	Function
		To accept the view orientation, either press **ESC** or right-click and select Done from the menu. To change the default viewpoint that is used when Isometric View is selected from the menu, click the arrow on a corner that you want to be the default isometric view. Then right-click and select Redefine Isometric from the menu. The redefined isometric view is only applied to the current file.
F4	F4 Shortcut to Dynamic Rotation	Rotates objects dynamically. While working, press and hold down the F4 key. The circular image appears with lines at the quadrants and center. With the F4 key depressed, rotate the model. When you finish rotating the model, release the F4 key. If you are performing an operation while the F4 key is pressed, that operation will resume after you release the F4 key.
F5	Previous View	Returns to the view that was previously on the screen. Press the F5 key or right-click in the graphics window and select Previous View from the menu. Continue this process until the view to which you want to return is on the screen.
	Display Options	Accesses the three options from the standard toolbar to display 3D parts: Shaded Display, Hidden Edge Display, and Wireframe Display. You can choose which mode works best for you and switch between the modes as you see fit. Each display mode is described next.
	Shaded Display	Shades objects in the color or material that was assigned to them. Parts or faces that are behind other parts or faces are not displayed.
	Hidden Edge Display	Shades objects and displays the edges that are behind other parts or faces. In complex parts and assemblies, this display can be confusing.
	Wireframe Display	Displays only the outline of objects. In wireframe display, objects are not shaded.

EXERCISE 1-2 Viewing a Model

In this exercise, you use View Manipulation tools that make it easier to work on your designs.

1. Open C:*IV10 Essentials Plus**Exercises****ESS_E01_02.iam**, the file contains a model of a clamp.

2. Click the Zoom Window tool.

3. Click a point below one of the clamp grips, move the cursor to expand the preview window, and then click again to define the zoom window as shown in the following image.

Figure 1-36

4. Press F5 to return to the previous view.

5. Use the rotate tool to rotate a part or assembly. Click the Rotate tool. The 3D Rotate symbol appears in the window.

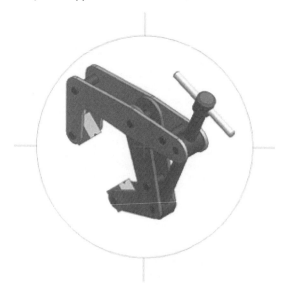

Figure 1-37

6. Move the cursor inside the rim of the 3D Rotate symbol, noting the cursor display.

7. Click and drag the cursor to rotate the model.

8. To return to the previous view, press F5.

9. Place the cursor on one of the horizontal handles of the 3D Rotate symbol, noting the cursor display.

10. Drag the cursor right or left to rotate the model about the Y axis.

11. To rotate the model about the X axis, drag one of the vertical handles.

12. Press Esc then right-click in the graphics window and click Isometric View or press F6.

13. If you have a wheel mouse, spin the wheel toward you to have the model appear larger and then spin the wheel away from you to have the model appear smaller.

14. Click the Zoom tool, click and hold the mouse button down, and then drag the cursor toward the bottom edge of the screen. Drag the cursor toward the top of the screen.

15. Right-click in the graphics window and click Isometric View or press F6.

 Tip: While you are performing another process, you can click on the viewing tools or press F2 to pan, F3 to zoom, or F4 to rotate the view.

16. In drafting, standard views of a model are defined using a glass box technique. The Rotate tool gives you similar results. Click the Rotate tool. The 3D Rotate symbol appears in the window.

17. Press the SPACE BAR or right-click in the graphics window and click Common View [Space] to display the glass box as shown in the following image.

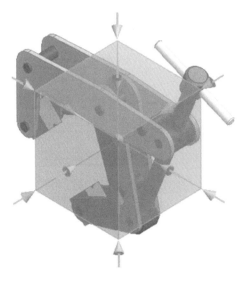

Figure 1-38

18. Move the cursor over an arrow, it turns red. Click the arrow to transition the model to the predefined view.

19. Click other arrows to review the standard view options.
    Click the arrow in the middle of the box as shown in the following image.

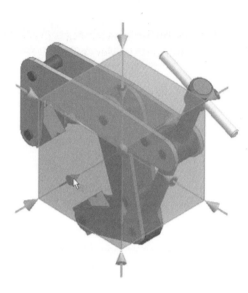

Figure 1-39

20. Click the top-left edge of the glass box as shown in the following image to rotate the view 90 degrees to the left.

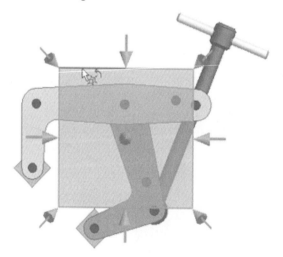

Figure 1-40

21. Click other edges to review the rotation.
22. Right-click in the graphics window, and then click Free Rotate [SPACE] to return to the Rotate tool.
23. Right-click in the graphics window, and then click Done to cancel Rotate.
24. View the model as an isometric.

25. View the model shaded, or with hidden edges visible or wire frame. Click the Hidden Edge Display tool and the view will change as shown in the following image.

Figure 1-41

26. Click the arrow beside Shaded Display, then click the Wireframe Display tool and the view will change as shown in the following image.

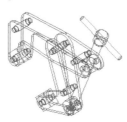

Figure 1-42

27. Click the Shaded Display tool.

28. Set the view mode to Perspective Camera mode and display shadows. Click the arrow beside Orthographic Camera, then click Perspective Camera. To change the perspective press the Ctrl, Shift and F3 keys and move the mouse up and down. If you have a wheel mouse press Ctrl, Shift and spin the wheel.

29. Click the arrow beside No Ground Shadow, then click Ground Shadow. Shadows are projected onto a floor plane below the model as shown in the following image.

Figure 1-43

30. Click the arrow beside No Ground Shadow, then click X-Ray Ground Shadow. Individual components are distinguishable as shown in the following image.

Figure 1-44

31. Rotate the model and review the shadow effects.

CHAPTER SUMMARY

To	Do This	Tool
Create a project	Click Projects on either the Autodesk Inventor File menu or in the What To Do area of the Getting Started page.	
Save the active document	Click Save on the File menu or on the standard toolbar.	
Save a copy of the active document with a new name	Click Save Copy As on the File menu.	
Save the active document and all of its dependents	Click Save All on the File menu.	
Set Autodesk Inventor preferences	Click Application Options in the Tools menu.	
Get help	Press the F1 key, click one of the options on the Help menu, or click the Help Topics or Visual Syllabus icon on the main menu bar.	
Undo or redo a command	Click the Undo or Redo icon on the standard toolbar.	
Manage Styles	Click Style tools on the Format menu.	
Change the current viewpoint to an isometric viewpoint	Right-click in the graphics window, and select Isometric View from the menu.	
Zoom or pan	On the standard toolbar, click one of the Zoom or Pan tools, or use the wheel on a wheel mouse.	
Dynamically rotate a part	Click the Dynamic Rotate tool on the standard toolbar, or use the F4 key.	

To	Do This	Tool
Change the display	Click the Shaded Display, Hidden Edge Display, or Wireframe Display icon on the standard toolbar.	

CHECKING YOUR SKILLS

Use these questions to test your knowledge of the material covered in this chapter.

1. Explain the reasons for which a project file is used.

2. Only one project can be active at any time.

3. List the sequence of the locations searched when a file is opened in Autodesk Inventor.

4. True__ False__ Autodesk Inventor stores the part, assembly information, and related drawing views in the same file.

5. True__ False__ Press and hold down the F4 key to dynamically rotate a part.

6. True__ False__ The Save Copy As command saves the active document with a new name and then makes it current.

7. List four ways to access the Help system.

8. Explain how to create a shortcut.

9. True__ False__ The Look At tool changes the viewpoint to an isometric view.

10. True__ False__ You can only edit a part while it is in shaded display.

CHAPTER 2

Sketching, Constraining, and Dimensioning

Most 3D parts in Autodesk Inventor start from a 2D sketch. This chapter first provides a look at the application options for sketching and part creation. It then covers the three steps in creating a 2D parametric sketch: sketching a rough 2D outline of a part, applying geometric constraints, and then adding parametric dimensions.

CHAPTER OBJECTIVES

After completing this chapter, you will be able to

- Change the sketch and part options as needed
- Sketch an outline of a part
- Create geometric constraints
- Dimension a sketch
- Create dimensions using the automatic dimensioning tool
- Change a dimension's value in a sketch
- Import AutoCAD DWG data

SKETCHING AND PART APPLICATION OPTIONS

Before you create a sketch, examine the sketch and part options in Autodesk Inventor that will affect sketching and part modeling. While learning Autodesk Inventor, refer back to these option settings to determine which ones work best for you—there are no right or wrong settings.

SKETCH OPTIONS

Autodesk Inventor sketching options can be customized to your preferences. Click Tools > Application Options and then click on the Sketch tab, as displayed in the following image. Descriptions of the Sketch options follow. These settings are global and affect all currently active Autodesk documents and Autodesk Inventor documents you open in the future.

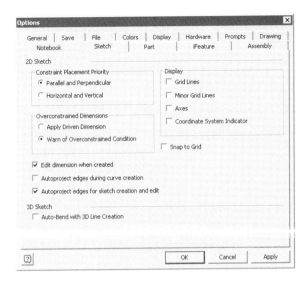

Figure 2-1

Constraint Placement Priority

- Parallel and Perpendicular

When checked, and a parallel or perpendicular condition exists while sketching, applies a parallel or perpendicular constraint before any other possible constraints that affect the geometry being created.

- Horizontal and Vertical

When checked, and a horizontal or vertical condition exists while sketching, applies a horizontal or vertical constraint before any other possible constraints that affect the geometry being created.

Display

- Grid Lines

Toggles both minor and major grid lines on the screen on and off. To set the grid distance, click Tools > Document Settings, and on the Sketch tab of the Document Settings dialog box, change the Snap Spacing and Grid Display.

- Minor Grid Lines

Toggles the minor grid lines on the screen on and off.

- Axes

Toggles the lines that represent the X- and Y-axis of the current sketch on and off.

- Coordinate System Indicator

Toggles the icon on and off that represents the X-, Y-, and Z-axis at the 0,0,0 coordinates of the current sketch.

Overconstrained Dimensions

- Apply Driven Dimensions

When checked, and you add dimensions that would overconstrain the sketch, adds the dimension as a driven (reference) dimension.

- Warn of Overconstrained Condition

When checked, and you add dimensions that would overconstrain the sketch, a dialog box appears warning of the condition. This allows you to accept the placement of a driven dimension or cancel the dimension placement.

Snap to Grid

When checked, endpoints of sketched objects snap to the intersections of the grid as the cursor moves over them.

Edit dimensions when created

When checked, edits the values of a dimension in the Edit Dimension dialog box immediately after you position the dimension.

Autoproject edges during curve creation

When checked, and while sketching objects that are in a different plane, *scrubbing* an object automatically projects it onto the current sketch. You can also toggle AutoProject on and off while sketching by right-clicking and selecting AutoProject from the menu.

Autoproject edges for sketch creation and edit

When checked, automatically projects all of the edges that define that plane onto the sketch plane as reference geometry when you create a new sketch.

3D Sketch

- Auto-Bend with 3D Line Creation

When checked, automatically places a tangent between two 3D lines as they are sketched. To set the radius of the arc, click Tools > Document Settings, and change the Auto-Bend Radius on the Sketch tab of the Document Settings dialog box.

PART OPTIONS

You can customize Autodesk Inventor Part options to your preferences. Click Tools > Application Options, and click on the Part tab, as shown in the following image. Descriptions of the Part options follow. These settings are global–they will affect all active and new Autodesk Inventor documents.

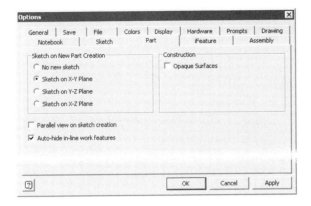

Figure 2-2

Sketch on New Part Creation

- No New Sketch

When checked, does not set sketch plane when you create a new part.

- Sketch on X-Y Plane

When checked, sets the X-Y plane as the current sketch plane when you create a new part.

- Sketch on Y-Z Plane

When checked, sets the Y-Z plane as the current sketch plane when you create a new part.

- Sketch on X-Z Plane

When checked, sets the X-Z plane as the current sketch plane when you create a new part.

Parallel View on Sketch Creation

When checked, automatically changes the view orientation to look directly at the sketch plane.

Auto-Hide In-Line Work Features

When checked, automatically hides work features that are consumed by another work feature.

Construction

- Opaque Surfaces

When checked, and as you create a surface, makes the surface opaque–otherwise, it will be translucent. After you create a surface, its translucency can be controlled from a menu.

UNITS

Autodesk Inventor uses a default unit of measurement for every part and assembly file. The default unit is set from the template file from which you created the part or assembly file. When specifying numbers in dialog boxes with no unit, the default unit will be used. You can change the default unit in the active part or assembly document by clicking Tools > Document Settings and on the Units tab, as displayed in the following image. The unit system values change for all of the existing values in that file.

 NOTE In a drawing file, the appearance of dimensions is controlled by dimension styles. Drawing settings will be covered in Chapter 5.

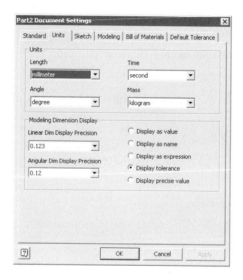

Figure 2-3

You can override the default unit for any value by entering the desired unit. If you were working in a mm file, for example, and placed a horizontal dimension whose default value was 50 mm, you could enter 2 in. Dimensions appear on the screen in the default units.

For the previous example, 50.8 mm would appear on the screen. When you edit a dimension, the overridden unit appears in the Edit Dimension dialog box as shown in the following image.

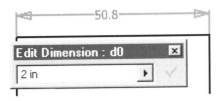

Figure 2-4

TEMPLATES

As you create new files, they are each created from a template. You can modify existing templates or add your own templates. As you work, make note of changes that you make to each file. You then create a new template file or modify an existing file that contains all of the changes, and save that file to your template directory–which by default is located at \Autodesk\Inventor (version number)\templates directory–or create a new subdirectory under the templates folder and place the file there. After you create the new template subdirectory and add a template file to it, a new tab will appear with the directory name. You can place any Autodesk Inventor file in this new directory and it will be available as a template.

There are two methods used to share template files among many users. You can modify the location of templates by clicking Tools > Application Options, clicking on the File tab, and modifying the Templates location as shown in the following image. The Templates location will need to be modified for each user who needs access to these common templates.

Figure 2-5

Another method is to set the Templates location in each project file. This method is useful for companies that need different templates files for each project. While editing a project file, change the Templates location in the Folder options area as shown in the following image. The Template location in the project file takes precedence over the Templates option in the Application Options, File tab.

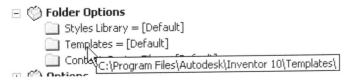

Figure 2-6

NOTE Template files have file extensions that are identical to other files of the same type, but they are located in the template directory. Template files should not be used as production files.

CREATING A PART

The first step in creating a part is to start a new part file or create a new part file in an assembly. The first four chapters in this book deal with creating parts in a new part file, and Chapter 6 covers creating and documenting assemblies. You can use the following methods to create a new part file:

- Click the New icon in What To Do area of the Open dialog box and then click the *Standard.ipt* icon on the Default tab, as shown in the following image, or click *Standard (unit).ipt* on one of the other tabs.

- Click New on the File menu, and then click the *Standard.ipt* icon on the Default tab, as shown in the following image. You can also click *Standard (unit).ipt* on one of the other tabs.

- Use the shortcut key, CTRL-N, and then click the *Standard.ipt* icon on the Default tab, as shown in the following image. You can also click *Standard (unit).ipt* on one of the other tabs.

- Click the down arrow of the New icon and select Part from the left side of the standard toolbar. This creates a new part file based on the units that were selected when Autodesk Inventor was installed or the *Standard.ipt* that that existing in the Templates folder.

After starting a new part file using one of the previous methods, Autodesk Inventor's screen will change to reflect the part environment.

NOTE The units for the files located in the Default tab are based upon the units you selected when you installed Autodesk Inventor. The following image shows the Default template files that are created if millimeters was selected as the default unit when Autodesk Inventor was installed.

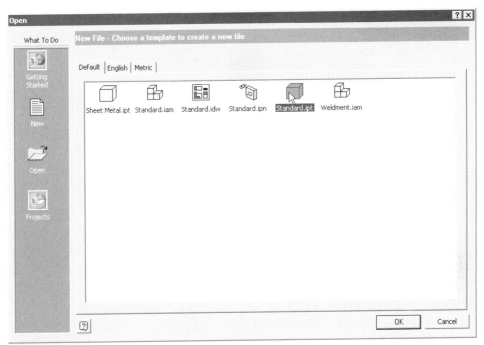

Figure 2-7

SKETCHES AND DEFAULT PLANES

Before you start sketching, you must have an active sketch on which to draw. A sketch is a plane on which 2D objects are sketched. You can use any planar part face or work plane to create a sketch. A sketch is automatically set by default when you create a new part file. You can change the default plane on which you create the sketch by selecting Tools > Application Options, and clicking on the Part tab. Choose the sketch plane to which new parts should default.

Each time you create a new Autodesk Inventor, there are three planes (XY, YZ, and XZ), three axes (X, Y, and Z), and the center (origin) point at the intersection of the three planes. You can use these default planes to create an active sketch. By default, visibility is turned off to the planes, axes, and center point. To see the planes, axes, or center point, expand the Origin entry in the Browser by clicking on the + to the left side of the text. You can then move the cursor over the names and they will appear in the graphics area. the following image illustrates the default planes, axes, and center point in the graphics area shown in an isometric view. The Browser shows the Origin menu expanded. To leave the visibility of the planes or axes on, right-click in the Browser while the cursor is over the name and select Visibility from the menu.

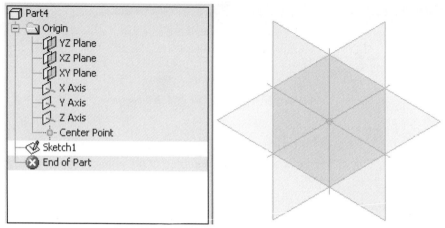

Figure 2-8

NEW SKETCH

Issue the Sketch tool to create a new sketch on a planar part face, a work plane, or to activate a non-active sketch in the active part. To create a new sketch or make an existing sketch active, follow one of these methods:

- Click the Sketch tool on the standard toolbar as shown in the following image. Then click a face, a work plane, or an existing sketch in the Browser.

- Click a face, a work plane, or an existing sketch in the Browser. Then click the Sketch tool on the standard toolbar.

- Press the hot key s and click a face of a part, a work plane, or an existing sketch in the Browser.

- Click a face of a part, a work plane, or an existing sketch in the Browser and then press the hot key s.

- While not in the middle of an operation, right-click in the graphics area and select New Sketch from the menu. Then click a face, a work plane, or an existing sketch in the Browser.

- While not in the middle of an operation, click a face of a part, a work plane, or an existing sketch in the Browser. Then right-click in the graphics area and select New Sketch from the menu.

Figure 2-9

After you have activated the sketch, the X- and Y-axes will align automatically to this plane and you can begin to sketch.

NOTE This book assumes that when you installed Autodesk Inventor you selected mm as the default unit. If you selected inch as the default unit, then select the *Standard(mm).ipt* template file from the Metric tab. This book will use the XY plane as the default sketch plane.

STEP 1–SKETCH THE OUTLINE OF THE PART

As stated at the beginning of the chapter, 3D parts usually start with a 2D sketch of the outline shape of the part. You can create a sketch with lines, arcs, circles, splines, or with any combination of these elements. The next section will cover sketching strategies, tools, and techniques.

SKETCHING OVERVIEW

When deciding what outline to start with, analyze how the finished shape will look. Look for the shape that best describes the part. When looking for this outline, try to look for a flat face. It is usually easier to work on a flat face than on a curved edge. As you gain modeling experience, you can reflect on how you created the model and think about other ways that you could have built it. There is usually more than one way to generate a given part.

When sketching, draw the geometry so that it is close to the desired shape and size—you do not need to be concerned about exact dimensional values. Autodesk Inventor allows islands in the sketch (closed objects that lie within another closed object). An example would be a circle that is drawn inside of a rectangle. When you extrude the sketch, the island may become a void in the solid. A sketch can consist of multiple closed objects that are coincident.

The following guidelines will help you generate good sketches:

- Select an outline that best represents the part. It is usually easier to work from a flat face.
- Draw the geometry close to the finished size. If you want a 20 mm square, for example, do not draw a 200 mm square.
- Create the sketch proportional in size to the finished shape. When drawing the first line segment, use the Precise Input dialog box to sketch the first line to size. This will give you a guide on how to sketch the rest.
- While sketching, notice that the distance and angle of the object being sketched will appear in the lower-right corner of the screen. Use this information as a guide.
- Draw the sketch so that it does not overlap. The geometry should start and end at the same point.
- Do not allow the sketch to have a gap—all of the connecting endpoints should be coincident.

- Keep the sketches simple. Leave out fillets and chamfers when possible. You can easily place them as features after the sketch turns into a solid. The simpler the sketch, the fewer the number of constraints and dimensions that will be required to constrain the model.

If you want to create a solid, the sketch must form a closed shape. If it is open, it can only be turned into a surface.

SKETCHING TOOLS

Before you start sketching the part, examine the 2D sketching tools that are available. By default, the 2D sketch tools appear on the Panel Bar with Expert mode turned off (text descriptions shown). As you become more proficient with Autodesk Inventor, you can turn off the text, as shown in the following image. Do this by clicking on the title area of the menu or by right-clicking on the Panel Bar and selecting Expert from the menu. You can also use the 2D Sketch toolbar to access the 2D sketch tools. Descriptions of these tools follow.

Figure 2-10

Sketch Tools		
Button	Tool	Function
	Line	Creates line segments. You can also draw an arc segment from the endpoint of a line using the line command. The next section will cover sketching techniques.
	Spline	Draws a spline by clicking points that lie on the spline.
	Center Point Circle	Creates a circle by clicking a center point for the circle and then a point on the circumference of the circle.
	Tangent Circle	Creates a circle that will be tangent to three lines or edges by clicking the lines or edges.
	Ellipse	Creates an ellipse by clicking a center point for the ellipse, a point on the first axis, and another point that will lie on the ellipse.
	Three Point Arc	Creates an arc by clicking a start and end point and then a point that will lie on the arc.
	Tangent Arc	Creates an arc that is tangent to an existing line or arc by clicking the endpoint of a line or arc and then clicking a point for the other endpoint of the arc.

Sketch Tools (continued)

Button	Tool	Function
	Center Point Arc	Creates an arc by clicking a center point for the arc and then clicking a start and endpoint.
	Two Point Rectangle	Creates a rectangle by clicking a point and then clicking another point to define the opposite side of the rectangle. The edges of the rectangle will be horizontal and vertical.
	Three Point Rectangle	Creates a rectangle by clicking two points that will define an edge and then clicking a point to define the third corner.
	Fillet	Creates a fillet between two nonparallel lines, two arcs, or a line and an arc at a specified radius. If you select two parallel lines, a fillet is created between them without specifying a radius.
	Chamfer	Creates a chamfer between lines. There are three options to create a chamfer: both sides equal distances, two defined distances, or a distance and an angle.
	Point, Hole Center	Creates a hole center that will be used to place a hole or points that can be used as vertices for other objects, like splines.
	Polygon	Creates an inscribed or a circumscribed polygon with the number of faces you specify.
	Mirror	Mirrors the selected objects about a centerline. A symmetry constraint will be applied to the mirrored objects.
	Rectangular Pattern	Creates a rectangular array of a sketch with a number of rows and columns you specify.
	Circular Pattern	Creates a circular array of a sketch with a number of copies and spacing you specify.
	Offset	Creates a duplicate of the selected objects that are a given distance away. By default, an equal distance constraint is applied to the offset objects.
	General Dimension	Creates a parametric or driven dimension.
	Auto Dimension	Automatically places dimensions and constraints on a selected profile.
	Extend	Extends the selected object to the next object it finds. Click near the end of the object that you want extended. While using the Extend tool, hold down the SHIFT key to trim objects.
	Trim	Trims the selected object to the next object it finds. Click near the end of the object that you want trimmed. While using the Trim tool, hold down the SHIFT key to extend objects.
	Move	Moves the selected profile from one point to another point. You can make a copy of the sketch by selecting Copy in the Move dialog box. If the moved sketch has objects that are constrained to it, they will also be moved.

Sketch Tools (continued)		
Button	Tool	Function
	Rotate	Rotates the selected sketch about a specified point. You can make a copy of the sketch by selecting Copy in the Rotate dialog box. If the rotated sketch has objects that are constrained to it, they will also be rotated.
	Constraints	Place geometric constraints on the geometry. Access constraints tools from the context menu (see the arrow by the Constraints icon in the previous image.) Constraints will be covered later in this chapter.
	Show Constraints	Shows constraints for object(s) you click. When the constraints are visible, you can delete them by right-clicking on a constraint and selecting Delete from the menu. Showing constraints will be covered later in this chapter.
	Project Geometry	Projects selected edges, vertices, and work features onto the current sketch. You can use the projected edges, vertices, and work features as part of the sketch or place dimensions on them. The projected edges, vertices, and work features are constrained to the original edge, vertex, or work feature. If the parent object changes or moves, so will the projected object.
	Project Cut Edges	Projects edges that lie on the section plane onto the sketch of the new part. Geometry is only projected if the uncut part would intersect the sketch.
	Project Flat Pattern	Projects selected areas of a flat pattern of a sheet metal onto the part. You can then dimension sketches to both the part and projected flat.
f_x	Parameters	Creates or modifies a parameter.
	Insert AutoCAD File	Inserts an AutoCAD 2D file onto the active sketch.
	Create Text	Adds text to the active sketch.
	Insert Image	Inserts a file to the active sketch.
	Edit Coordinate System	Realigns the sketch coordinate system to existing objects. Arrows on the screen will indicate the directions of the X-axis and the Y-axis.

Using the Sketch Tools

After starting a new part, use the sketch tools to draw the shape of the part. As you are sketching, Autodesk Inventor gives you visual feedback about what is happening on the screen and there are text messages in the lower-left area of the Status Bar. The following image shows the message displayed when you activate the Line tool. To start sketching, issue the sketch tool that you need, click a point in the graphics area, and follow the prompt on the status bar. The following sections will introduce techniques that you can use to create a sketch.

Figure 2-11

Line Tool

The Line tool is one of the most powerful tools that you will use to sketch. Not only can you draw lines with the Line tool, but you can also draw an arc from the endpoint of a line segment. After issuing the Line tool, you will be prompted to click a first point, select a point in the graphics window, and then click a second point. You can continue drawing line segments, or move the cursor over the endpoint of a line segment or arc, and a small gray circle will appear at that endpoint. The following image shows how the endpoint of a line segment looks when the cursor moves over it.

Figure 2-12

Click on the small circle, and with the left mouse button pressed down, move the cursor in the direction that you want the arc to go. Up to eight different arcs can be drawn, depending upon how you move the mouse. The arc will be tangent to the horizontal or vertical edges that are displayed from the selected endpoint. The following image shows an arc being drawn that is normal to the sketched line.

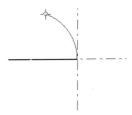

Figure 2-13

TIP When sketching, look at the bottom-right corner of the status bar (bottom of the screen) to see the coordinates, length, and angle of the objects that you are drawing. The following image shows the status bar when a line is being drawn.

| -12.917 mm, 33.313 mm | Length=17.834 mm | Angle=43.88 deg |

Figure 2-14

Inferred Points

As you sketch, dashed lines will appear on the screen. These dashed lines represent the endpoints of lines and arcs that represent their horizontal, vertical, or perpendicular positions. As the cursor gets close to these inferred points, it will snap to that location. If that is the point that you want, click that point; otherwise, continue to move the cursor until it reaches the desired location. When you select inferred points, no constraints (geometric rules such as horizontal, vertical, collinear, etc.) are applied from them. Using inferred points helps create more accurate sketches. The following image shows the inferred points from two endpoints that represent their horizontal and vertical position.

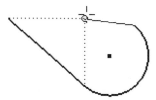

Figure 2-15

Automatic Constraints

As you sketch, small constraint symbols appear that represent geometric constraint(s) that will be applied to the object. If you do not want a constraint to be applied, hold down the CTRL key when you select the point. The following image shows a line being drawn from the arc, tangent to the arc and parallel to the angled line. The symbol appears near the object from which the constraint is coming. Constraints will be covered in the next section.

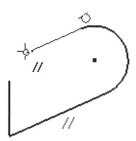

Figure 2-16

Scrubbing

As you sketch, you may prefer to have a different constraint applied than the one that automatically appears on the screen. You may want a line to be perpendicular to a given line, for example, instead of being parallel to a different line. The technique to change the constraint is called *scrubbing*. To place a different constraint while sketching, move the cursor so it touches (scrubs) the other object to which the constraint should be related. Move the cursor back to its original location and the constraint symbol changes to reflect the new constraint. The same constraint symbol will also appear near the scrubbed object, representing that it is the object to which the constraint is matched. Continue sketching as normal. The following image shows the top horizontal line being drawn with a perpendicular constraint that was scrubbed from the left vertical line. Without scrubbing the left vertical line, the applied constraint would have been parallel to the bottom line.

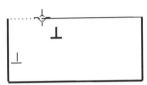

Figure 2-17

PRECISE INPUT

It may be easier to draw objects at a specified length or angle. You can enter these values in the Precise Input dialog box. To open the Inventor Precise Input dialog box, click View > Toolbars and click Inventor Precise Input, or right-click on either the standard toolbar or menu bar and select Inventor Precise Input from the menu. Before entering values into the dialog box, start a sketch tool, like Line. Click the method of input, enter the values into the dialog box, and press ENTER. The following image shows the Inventor Precise Input dialog box with Expert display mode on and with the input options displayed.

Figure 2-18

SELECTING OBJECTS

After sketching objects, you may need to move, rotate, or delete some or all of the objects. To edit an object, it must be part of a selection set. There are two methods to place objects into a selection set.

1. You can select objects individually by clicking on them. To select multiple individual objects, hold down the CTRL or SHIFT key while clicking the objects. You can remove selected objects from a selection set by holding down the CTRL or SHIFT key and reselecting them. As you select objects, their color will change to represent that they have been selected.

2. You can select multiple objects by defining a selection window. To define the window, click a starting point. With the left mouse button depressed, move the cursor to define the box. If you draw the window from left to right, only the objects that are fully enclosed in the window will be selected. If you draw

the window from right to left, as shown in the following image, all of the objects that are fully enclosed in the window *and* the objects that are touched by the window will be selected.

3. You may also use a combination of the methods to create a selection set.

When you select an object, its color will change according to the color style that you are using. To remove all of the objects from the selection set, click in a blank section of the graphics area.

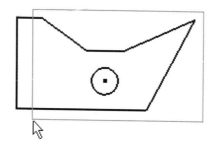

Figure 2-19

DELETING OBJECTS

To delete objects, select them and then either press the DELETE key or right-click and select Delete from the menu, as shown in the following image.

Figure 2-20

MEASURE TOOLS

Various measure tools are available that assist in analyzing sketch, part, and assembly models. You can measure distances, angles, and loops, and perform area calculations.

The Measure tools are on the Tools menu, as shown in the following image. The following sections discuss these tools in greater detail.

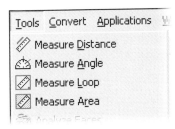

Figure 2-21

Measure Distance Click to measure the length of a line or edge, length of an arc, distance between points, radius and diameter of a circle, or the position of elements relative to the active coordinate system. A temporary line designating the measured distance appears and the Measure Distance dialog box displays the measurement for the selected length, as shown in the following image.

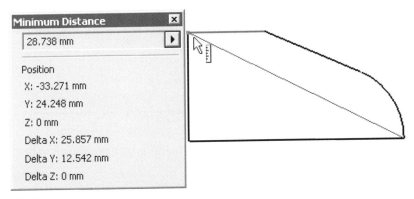

Figure 2-22

Measure Angle Click to measure the angle between two lines, edges, or points. The measurement box (see the following image) displays the angle based on the selection of two lines or edges.

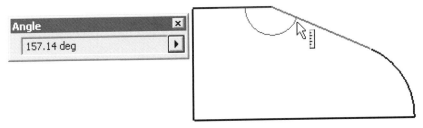

Figure 2-23

Measure Loop Click to measure the length of closed loops defined by face boundaries or other geometry. When moving your cursor over an edge, all edges that form a closed loop will become highlighted. Clicking on this edge, as shown in the following image, will calculate the closed loop or perimeter of the shape.

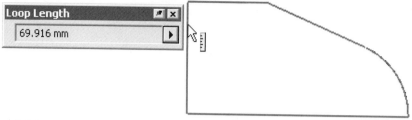

Figure 2-24

Measure Area Click to measure the area of enclosed regions or faces, as shown in the following image. Moving your cursor inside the closed outer shape will cause the outer shape and all holes (referred to as islands) to also become highlighted. Clicking inside this shape will calculate the area of the shape.

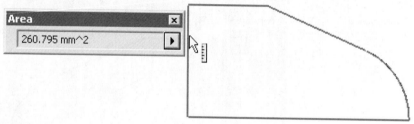

Figure 2-25

When you click the arrow beside the measurement box, a menu will appear similar to the following image.

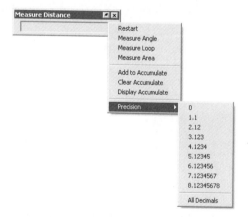

Figure 2-26

The following are brief explanations of each option.

Restart Click to clear the measurement from the measurement box so that you can make another measurement.

Measure Angle Click to change the measurement mode to Measure Angle.

Measure Loop Click to change the measurement mode to Measure Loop.

Measure Area Click to change the measurement mode to Measure Area.

Add to Accumulate Click to add the measurement in the measurement box to accumulate a total measurement.

Clear Accumulate Click to clear all measurements from the accumulated sum, resetting the sum to zero.

Display Accumulate Click to display the sum of all measurements you have added to the accumulated sum.

Precision Click to change the decimal display between showing all decimal places and showing the number of decimal places specified in the document settings for the active part or assembly.

EXERCISE 2-1 Creating A Sketch With Lines

In this exercise, you create a new part file, and then you create sketch geometry using basic construction techniques. You also learn how you can use the Autodesk Inventor Design Support System to assist in the design process.

1. Click the New tool and click the Metric tab and then double-click **Standard(mm).ipt**.

2. Click the Line tool in the Panel Bar.

3. Click near the left side of the graphics window, move the cursor to the right approximately 100 units, and, when the horizontal constraint symbol displays, click to specify a second point as shown in the following image.

Figure 2-27

 Note Symbols indicate the geometric constraint. In the figure above, the symbol indicates that the line is horizontal. When you create the first entity in a sketch, make it close to final size. The length and angle of the line are displayed in the lower right corner of the window to assist you.

4. To learn how to create a perpendicular line, use the Autodesk Inventor Design Support System. With the Line tool still active, right-click in the graphics window and click How To as shown in the following image.

Figure 2-28

5. In the Lines section, click Perpendicular, and then watch the animation.

6. Move the cursor up until the perpendicular constraint symbol displays beside the first line then click to create a perpendicular line that is approximately 100 units as shown in the following image.

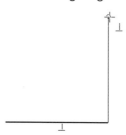

Figure 2-29

7. Move the cursor to the left and create a horizontal line of approximately 30 units. The parallel constraint symbol is displayed.

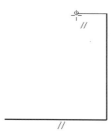

Figure 2-30

8. Move the cursor down and create a vertical line of approximately 60 units.

9. Move the cursor left to create a horizontal line of approximately 20 units.

10. Move the cursor up until the parallel constraint symbol is displayed and a dotted alignment line appears as shown in the following image.

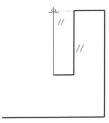

Figure 2-31

11. Click to specify a point.

12. Move the cursor left until the parallel constraint symbol is displayed and a dotted alignment line appears as shown in the following image, and then click to specify a point.

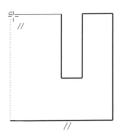

Figure 2-32

13. Move the cursor down until it touches the first point of the sketch. When the green circle (coincident constraint symbol) is displayed, click to place the line. Your screen should resemble the following image.

14. Right click in the graphics screen and click Done to end the Line tool.

15. Right click in the graphics screen again and click Finish Sketch.

16. Close the file. Do not save changes.

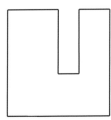

Figure 2-33

EXERCISE 2-2 Creating a Sketch With Tangencies

In this exercise, you create a new part file, and then you create a simple profile consisting of lines and tangential arcs. You also use the Precise Input toolbar.

1. Click the New tool and click the Metric tab and then double-click *Standard(mm).ipt*.

2. From the menu bar, select View > Toolbar > Inventor Precise Input to display the Precise Input toolbar.

3. Click the Line tool in the Panel Bar.

4. Click near the left side of the graphics window, and then enter **125** in the X field of the Inventor Precise Input toolbar as shown in the following image (the Precise Input toolbar is displayed in expert mode). If the Delta input button is off, click to turn it on.

Figure 2-34

5. When the horizontal constraint symbol is displayed, click to create a 125 mm horizontal line.

Figure 2-35

6. Click in the Y field, and then enter 90. Click a second point when the perpendicular constraint symbol is displayed as shown in the following image. If the line is not visible use the Zoom tool to view the entire line.

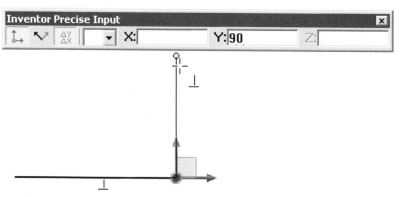

Figure 2-36

7. Click in the X field, and then enter **-40**. Click a second point when the parallel constraint symbol is displayed.

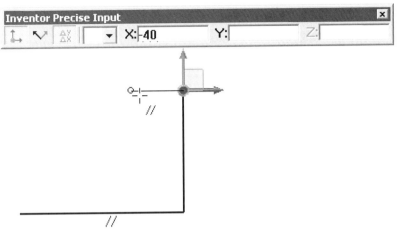

Figure 2-37

8. In the next two steps you infer points (no sketch constraint is applied). Move the cursor over the midpoint of the vertical line until the coincident constraint displays. Do not click.

9. Move the cursor over the midpoint of the bottom line until the coincident constraint displays. Do not click.

10. Move the cursor to the intersection of the dotted alignment lines (inferred points) as shown in the following image, then click to create the line.

Figure 2-38

11. Move the cursor to the left until the vertical alignment line and the parallel constraints displays as shown in the following image, then click to place the line.

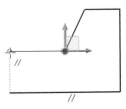

Figure 2-39

12. Right-click the graphics window, and then select How To.
13. In the Arcs section of the dialog box click Tangent. Watch the animation.
14. Click the gray dot at the end of the line, hold and drag the endpoint to create a tangent arc. **Do not release the mouse button.**
15. Move the cursor over the endpoint of the first line segment until a coincident constraint displays as shown in the following image.

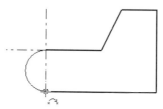

Figure 2-40

16. Release the mouse button to place the endpoint of the arc.
17. Right-click in the graphics window then click Done.
18. Close the Precise Input toolbar.
19. Close the file without saving the changes.
20. Repeat this exercise and sketch the objects close in size; do NOT use Precise Input.

STEP 2–CONSTRAINING THE SKETCH

After you draw the sketch, you may want to add geometric constraints to it. Geometric constraints apply behavior to a specific object or create a relationship between two objects. An example of using a constraint is applying a vertical constraint to a line so that it will always be vertical. You could apply a parallel constraint between two lines to make them parallel to one another; then, as one of the line's angles changes, so will the other's. You can apply a tangent constraint to a line and an arc or to two arcs.

When you add a constraint, the number of constraints or dimensions that are required to fully constrain the sketch will decrease. A fully constrained sketch is a sketch whose objects cannot move.

To help you see which objects are constrained, Autodesk Inventor will change the color of constrained objects if you have applied a fix constraint to the sketch. If you have not applied a fix constraint, the color of the constrained objects will not change. The fix constraint also prevents a sketch from moving in the sketch plane.

Points and edges can be fixed using the fix constraint. If you have not applied a fix constraint to the sketch, objects are free to move in their sketch plane.

 NOTE Autodesk Inventor does not force you to fully constrain a sketch. It is recommended to fully constrain a sketch, however, as this will allow you to better predict how a part will react when you change dimensions values.

CONSTRAINT TYPES

Autodesk Inventor has 11 geometric constraints that you can apply to a sketch. The following image shows the constraint types and the symbols that represent them. Descriptions of the constraints follow.

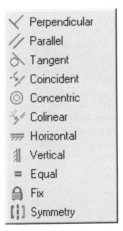

Figure 2-41 Constraints menu

Constraint Tools

Button	Tool	Function
	Perpendicular	Lines will be repositioned at 90° angles to one another—the first line sketched will stay in its position and the second will rotate until the angle between them is 90°.
	Parallel	Lines will be repositioned so they are parallel to one another—the first line sketched will stay in its position and the second will move to become parallel to the first.
	Tangent	An arc or circle and a line will become tangent to another arc or circle.
	Coincident	A gap between two endpoints of arcs and/or lines will be closed.
	Concentric	Arcs and/or circles will share the same center point.

Constraint Tools (continued)

Button	Tool	Function
	Collinear	Two selected lines will line up along a single line—if the first line moves, so will the second. The two lines do not have to be touching.
	Horizontal	Lines are positioned parallel to the X-axis, or a horizontal constraint can be applied between the center points of arcs or circles. The center points will share the same horizontal axis.
	Vertical	Lines are positioned parallel to the Y-axis, or a vertical constraint can be applied between the center points of arcs or circles. The center points will then share the same vertical axis.
	Equal	If two arcs or circles are selected, they will then have the same radius or diameter. If two lines are selected, they will become the same length. If one of the objects changes, so will the other object to which the Equal constraint has been applied. If the Equal constraint is applied after one of the arcs, circles, or lines have been dimensioned, the second arc, circle, or line will take on the size of the first one.
	Fix	Applying a fixed point or points will prevent the endpoints or edges of objects from moving. The fixed point overrides any other constraint. Any endpoint or segment of a line, arc, circle, spline segment, or ellipse can be fixed. Multiple points in a sketch can be fixed. If you select near the endpoint of an object, the endpoint will be locked from moving. If you select near the midpoint of a segment, the entire segment will be locked from moving. If applying constraints, and the profile is moving in directions that are undesirable, you can apply Fix constraints to hold the endpoints of the objects in place. You can remove a fix constraint as needed. Deleting constraints will be covered later in this chapter.
	Symmetric	Selected geometry will be symmetric about another line, centerline, or edge.

NOTE To fully constrain a base sketch, fix a point in the sketch or constrain or dimension the sketch to a projected center point.

ADDING CONSTRAINTS

As stated previously in this chapter, you can apply constraints while you sketch the objects. You may also apply additional constraints after the sketch is drawn. Autodesk Inventor will not allow you to overconstrain the sketch or add duplicate constraints. If you add a constraint that would conflict with another, you will be warned with the message, "Adding this constraint will overconstrain the sketch." If you try to add a vertical constraint to a line that already has a horizontal constraint, for example, you will be alerted. To apply a constraint, follow these steps:

1. Click a constraint from the Constraint pop-up menu in the Sketch Panel Bar or Sketch toolbar, or right-click in the graphics window, select Create

Constraint from the menu, and choose the specific constraint from the menu as shown in the following image.

2. Click the object or objects to apply the constraints.

SNAPS

Another method to place geometry with a coincident constraint is to use the snaps—midpoint, center, and intersection. After using a snap, a coincident constraint will be applied. The coincident constraint will maintain the relationship that you selected. If you use a midpoint snap, for example, the sketched point will always be in the middle of the selected object, even if the selected object's length changes.

To use snaps, follow these steps:

1. Select a sketching tool, and right-click in the graphics window when you need one of the snaps.

2. From the menu, select the specific snap.

3. Click on the object to which the sketched object will be constrained. For the intersection snap select two objects.

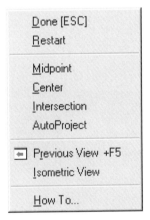

Figure 2-42

DRAGGING A SKETCH

To help determine if an object is constrained, you can drag it to a new location. While not in a command, click a point or an edge on the sketch. With the left mouse button depressed, drag it to a new location. If the geometry stretches, it is underconstrained. For example, if you draw a rectangle that has two horizontal and vertical constraints applied to it and drag a point on one of the corners, the size of the rectangle will change, but the lines will maintain their horizontal and vertical behavior. If dimensions are on the object, they too will prevent the object from stretching.

SHOWING AND DELETING CONSTRAINTS

To see the constraints that are applied to an object, use the Show Constraints tool from the Sketch Panel Bar, as shown on the left side of the following image. After issuing the Show Constraints tool, select an object and a row of constraint icons will appear similar to what is shown on the right side of the following image.

Figure 2-43

As you move the cursor over a constraint icon, the objects that are linked to that constraint will change color.

To delete the constraint, either click on it and then right-click, or right-click while the cursor is over the constraint and select Delete from the menu. You can also click on it and press the DELETE key. The following image shows the parallel constraint being deleted.

To close constraint icons, click the X on the right side of the constraint symbols.

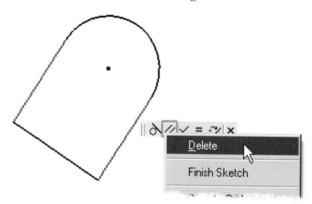

Figure 2-44

To show all the constraints for all of the objects in the sketch, do the following:

- While not in an operation, right-click in the graphics window and from the menu click Show All Constraints, as shown in the following image. To hide all the constraints, right-click in the graphics window and click Hide All Constraints from the menu.

- While not in an operation, right-click in the graphics window and select Show All Constraints from the menu, as shown in the following image. You can also press the F8 key. To hide all the constraints, right-click in the graphics window and select Hide All Constraints from the menu or press the F9 key.

Figure 2-45

EXERCISE 2-3 Adding and Displaying Constraints

In this exercise, you add geometric constraints to sketch geometry to control the shape of the sketch.

1. Click the New tool and click the Metric tab and then double-click *Standard(mm).ipt*.

2. Sketch the geometry as shown, approximate size 35mm in the X and 20mm in the Y. Right click in the graphics window then click Done.

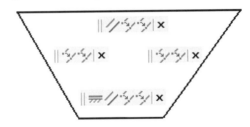

Figure 2-46

3. Right-click in the graphics window and click Show All Constraints or press the F8 key. Your screen should resemble the previous image.

4. If other constraints appear, move the cursor over it, right-click and click delete from the menu.

5. On the 2D Sketch panel click the down arrow beside the Constraint tool and click Fix.

6. Click the endpoint on the lower left line.

7. On the 2D Sketch panel click the down arrow beside the Constraint tool and click Parallel.

8. Click the two angled lines and your screen should resemble the following image. Note: depending upon the order in which you sketched the lines, the angles may be opposite of the image.

Figure 2-47

9. Press the Esc key to stop adding constraints.

10. Click on a line in the sketch and drag the line. Notice how the sketch changes shape.

11. Click on an endpoint in the sketch and drag the endpoint. The lines remain parallel due to the parallel constraint.

12. Move the cursor over a parallel constraint, right-click and click delete from the menu. Both parallel constraints are deleted.

13. On the 2D Sketch panel click the down arrow beside the Constraint tool and click Perpendicular.

14. Click the top horizontal line and the angled line on the right side.

15. Click the top horizontal line and the left vertical line and your screen should resemble the following image.

Figure 2-48

16. Right-click in the graphics window and click Hide All Constraints or press the F9 key.

17. Close the file without saving the changes.

18. Click the New tool and click the Metric tab and then double-click *Standard(mm).ipt*.

19. Sketch the geometry as shown, approximate size 50mm in the X and 35mm in the Y. Right click in the graphics window then click Done.

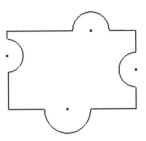

Figure 2-49

20. On the 2D Sketch panel click the down arrow beside the Constraint tool and click Fix.

21. Click the lower left point where the two lines touch.

22. On the 2D Sketch panel click the down arrow beside the Constraint tool and click Equal.

23. Click the bottom arc and then the arc on the left vertical line.

24. Add an equal constraint between the bottom arc and the two other arcs.

25. On the 2D Sketch panel click the down arrow beside the Constraint tool and click Collinear o right-click in the graphics window, click Create Constraint and the click Collinear. Note: In the following steps if the constraint cannot be placed you have a constraint that prevents it from being placed. Delete the constraint that is preventing the collinear constraint from being placed.

26. Click the two bottom horizontal lines.

27. Click the two top horizontal lines.

28. Click the two left vertical lines.

29. Click the two right vertical lines.

30. Click on a line in the sketch and drag the line. Notice how the sketch changes shape.

31. Click on an endpoint in the sketch and drag the endpoint. Notice how the sketch changes shape.

32. On the 2D Sketch panel, click the down arrow beside the Constraint tool and click Vertical.

33. Click the center point of the bottom arc and then the center point of the top arc.

34. On the 2D Sketch panel click the down arrow beside the Constraint tool and click Horizontal.

35. Click the center point of the left arc and then the center point of the right arc.

36. Align the center point of the arc with the adjacent lines add a horizontal or vertical constraint between the center point of an arc and an end point of an adjacent line.

37. Right-click in the graphics window then click Done and your screen should resemble the following image.

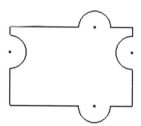

Figure 2-50

38. Examine the constraints, click the Show Constraints tool and click a few lines and arcs.

39. Click on a line in the sketch and drag the line. Notice how the sketch changes shape.

40. Click on an endpoint in the sketch and drag the endpoint. Notice how the sketch changes shape.

41. Close the file without saving the changes.

STEP 3–ADDING DIMENSIONS

The last step to constrain a sketch is to add dimensions. The dimensions you place will control the size of the sketch and will also appear in the part drawing views when they are generated. When placing dimensions, try to avoid having extension lines go through the sketch, as this will require more clean-up when drawing views are generated. Click near the side from which you anticipate the dimensions will originate in the drawing views.

All dimensions you create are parametric, which means that they will change the size of the geometry. All parametric dimensions are created with either the General Dimension or Auto Dimension tools.

GENERAL DIMENSIONING

The General Dimension tool can create linear, angle, radial, or diameter dimensions one at a time. To start the General Dimension tool, follow one of these techniques:

- Click the Create Dimension tool from the Sketch toolbar as shown in the following image.
- Right-click in the graphics area and select Create Dimension from the menu.
- Press the hot key D.

Figure 2-51

When you place a linear dimension, the extension line of the dimension will snap automatically to the nearest endpoint of a selected line; when an arc or circle is selected, it will snap to its center point. To dimension to a quadrant of an arc or circle, see "Dimensioning to a Tangent of an Arc or Circle" later in this chapter.

After you select the General Dimension tool, follow these steps to place a dimension:

1. Click a point or points to locate where the dimension is to start and end.
2. After selecting the point(s) to dimension, a preview image will appear attached to your cursor showing the type of dimension. If the dimension type is not what you want, right-click and then select the correct style from the menu. After changing the dimension type, the dimension preview will change to reflect the new style.
3. Click a point on the screen to place the dimension.

The next sections cover how to dimension specific objects and how to create specific types of dimensioning with the General Dimension tool.

Dimensioning Lines

There are two techniques for dimensioning a line. Issue the General Dimension tool and do one of the following:

- Click near two endpoints, move the cursor until the dimension is in the correct location, and click.
- To dimension the length of a line, click anywhere on the line (the two endpoints will be selected automatically), move the cursor until the dimension is in the correct location, and click.

Dimensioning Angles

To create an angular dimension, issue the General Dimension tool, click near the midpoint of two lines between which you want the angle dimension to be, move the cursor until the dimension is in the correct location, and click.

Dimensioning Arcs and Circles

To dimension an arc or a circle, issue the General Dimension tool, click on the circle's circumference, move the cursor until the dimension is in the correct location, and click. By default, when you dimension an arc, the result is a radius dimension. When you dimension a circle, the default is a diameter dimension. To change the radial dimension to a diameter or a diameter to radial, right-click before you place the dimension and select the other style from the menu.

Linear Diametric Dimensions

Diametric dimensions are used to create diameter dimensions for sketches that represent a quarter outline of a revolved part. To create a diametric dimension, follow these steps:

1. Draw a sketch that represents a quarter section of the finished part.
2. Draw a line, if needed, around which the sketch will be revolved. This line can be on the closed profile of the sketch.
3. Issue the General Dimension tool.
4. Click the line (not an endpoint) that will be the axis of rotation.
5. Click the other point to be dimensioned.
6. Right-click and select Linear Diameter from the menu.
7. Move the cursor until the diameter dimension is in the correct location and click.

The left side of the following image shows a sketch with the menu for changing the dimension to Linear Diameter. The right side shows the placed diametric dimensions—the left vertical line will be the axis of rotation. More options for creating diametric dimensions will be covered in the "Revolve" section of Chapter 3.

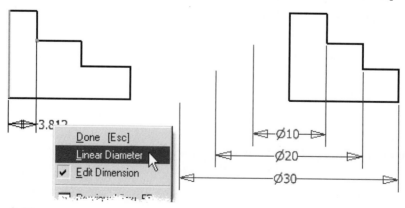

Figure 2-52

Dimensioning to a Tangent of an Arc or Circle

To dimension to a tangent of an arc or circle, follow these steps:

1. Issue the General Dimension tool.
2. Click a line that is parallel to the tangent.
3. Move the cursor over the tangent that should be dimensioned.
4. Move the cursor over the tangent until the constraint symbol changes to reflect a tangent, as shown on the left side of the following image.
5. Click and move the cursor until the dimension is in the correct location, and click as shown on the right side of the following image.

Figure 2-53

To dimension to two tangents, follow these steps:

1. Issue the General Dimension tool.
2. Click an arc or circle that includes one of the tangents to which it will be dimensioned.
3. Move the cursor over the tangent edge of the second arc or circle to which it will be dimensioned.
4. Move the cursor over the tangency until the constraint symbol changes to *tangent*.
5. Click and then move the cursor until the dimension is in the correct location, and click as shown in the following image.

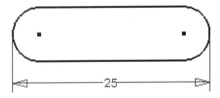

Figure 2-54

ENTERING AND EDITING A DIMENSION VALUE

After placing the dimension, you can change its value. Depending on your setting for editing dimensions when you created them, the Edit Dimension dialog box may or may not appear automatically after you place the dimension. To set the Edit Dimension option, do the following:

1. Click Tools > Application Options.
2. On the Sketch tab of the Options dialog box, select or deselect Edit dimension when created as shown on the left side of the following image.

If the Edit dimension when created option is selected, the Edit Dimension dialog box will appear automatically after you place the dimension. Otherwise, the dimension will be placed with the default value, and you will not be prompted for a different value. You can also set this option by right-clicking in the graphics area while placing a dimension and selecting or deselecting Edit Dimension from the menu, as shown on the right side of the following image.

Figure 2-55

To edit a dimension that has already been created, double-click on the value of the dimension and the Edit Dimension dialog box will appear, as shown in the following image. Enter the new value and unit for the dimension; then either press ENTER or click the check mark in the Edit Dimension dialog box. If no unit is entered, the units that the file was created with will be used. When inputting values, enter the exact value—do not round up or down. The accuracy shown in the dimension is from the current dimension style. Autodesk Inventor parts are accurate to six decimal places; for example, 1.0625 is more accurate than 1.06.

 NOTE When placing dimensions, it is recommended that you place the smallest dimensions first—this will help prevent the geometry from flipping in the opposite direction.

Figure 2-56

Repositioning a Dimension

Once you place a dimension, you can reposition it, but the origin points cannot be moved. Follow these steps to reposition a dimension:

1. Exit the current operation by either pressing ESC twice, or right-clicking and then selecting Done from the menu.

2. Move the cursor over the dimension until the move symbol appears as shown in the following image.

3. With the left mouse button depressed, move the dimension or value to a new location and release the button.

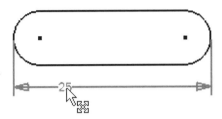

Figure 2-57

Overconstrained Sketches

As explained in the "Adding Constraints" section, Autodesk Inventor will not allow you to overconstrain a sketch or add duplicate constraints. The same is true when adding dimensions. If you add a dimension that will conflict with another constraint or dimension, you will be warned that this dimension will overconstrain the sketch or that it already exists. You will then have an option to either not place the dimension or place it as a driven dimension.

A driven dimension is a reference dimension. It is not a parametric dimension–it just reflects the size of the points to which it is dimensioned. A driven dimension will appear with parentheses around the dimensions value, for example, as (30). When you place a dimension that will overconstrain a sketch, a dialog box will appear similar to the one in the following image. You can either cancel the operation (and no dimension will be placed) or accept the warning (and a driven dimension will be created).

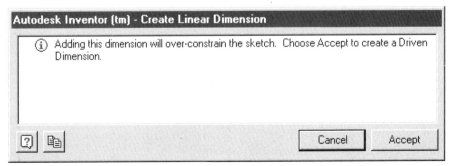

Figure 2-58

Autodesk Inventor gives you an option for handling overconstrained dimensions. To set the overconstrained dimensions option, click **Tools > Application Options**, and on the Sketch tab of the Options dialog box change the Overconstrained Dimensions option as shown in the following image. You have the following two options:

- Apply Driven Dimension

When checked, this option automatically creates a driven dimension without warning you of the condition.

- Warn of Overconstrained Condition

When checked, this option causes a dialog box to appear stating that the dimension will overconstrain the sketch. Click Cancel to not place a driven dimension and click Accept to place a driven dimension.

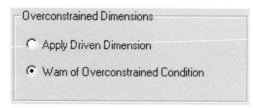

Figure 2-59

Another option for controlling the type of dimension that you create is to use the Driven Dimension tool on the standard toolbar. If you issue the Driven Dimension tool, any dimension you create will be a driven dimension. If you do not issue the Driven Dimension tool, you create a regular dimension. The following image shows the Driven Dimension tool on the standard toolbar in its normal condition. The same Driven Dimension tool can be used to change an existing dimension to either a normal or driven dimension by selecting the dimension and clicking the Driven Dimension tool.

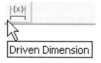

Figure 2-60

AUTO DIMENSION

Adding constraints and dimensions to a sketch or removing dimensions from a sketch can be a time-consuming task. To automate this process, you can use the Auto Dimension tool to create or remove dimensions or add constraints to selected geometry automatically. Before using the Auto Dimension tool, you should apply critical constraints and dimensions using the appropriate sketch constraint or Dimension command. The Auto Dimension tool will not override or replace any existing constraints or dimensions. Click the Auto Dimension tool on the Sketch

Panel Bar, as shown on the left in the following image. The Auto Dimension dialog box will appear as shown on the right.

Figure 2-61

To use the Auto Dimension tool, follow these steps:

1. Click the Auto Dimension tool on the Sketch Panel Bar.

2. The number of constraints and dimensions required to fully constrain the sketch appear in the lower left corner of the dialog box.

3. Determine if you want to create dimensions/constraints or remove the dimensions/constraints that you previously added using the Auto Dimension tool.

4. Click the Dimensions box and/or Constraints box.

5. Click the Curves option and then select the objects with which to work in the graphics window.

6. Click the Apply button to create the dimensions and/or apply constraints to the selected curves, or click the Remove button to delete the selected dimensions and/or constraints.

7. If you clicked Apply, you can change the values of the dimensions you placed by double-clicking on the dimension text and entering in a new value in the Edit Dimension dialog box.

8. After the dimensions are placed, you can change their values by double-clicking on the numbers and entering in a new value.

 NOTE If you use the Auto Dimension tool on the first sketch in the part, two dimensions or constraints will be required to fully constrain the sketch. Use the Fix or add dimensions or a Coincident constraint to a projected point to add these two dimensions.

EXERCISE 2-4 Dimensioning A Sketch

In this exercise, you add dimensional constraints to a sketch. Note: this exercise assumes the "Edit dimension when created" option is checked in the Options dialog box under the Sketch tab.

1. Click the New tool and click the Metric tab and then double-click *Standard(mm).ipt*.
2. Sketch the geometry as shown, approximate size 90mm in the X and 40mm in the Y. Right click in the graphics window then click Done.

Figure 2-62

3. Fix the lower right end point.
4. If needed add a tangent constraint between the arc and the two lines.
5. Add a horizontal constraint between the midpoint of the right vertical line and the center of the arc as shown in the following image.

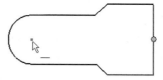

Figure 2-63

6. Add a vertical constraint between the lower endpoint of the lower angled line and the endpoint of the top angled line as shown in the following image.

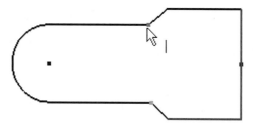

Figure 2-64

7. Add an equal constraint between the two angled lines.
8. Click the General Dimension tool in the Panel Bar.
9. Click the top horizontal line, drag the dimension up, click a point to locate it and Enter **30**, and then click the check mark as shown in the following image.

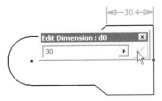

Figure 2-65

10. If the Edit Dimension dialog box did not appear, double-click on the dimension.

11. Add an angle dimension by clicking bottom horizontal line and the lower angled line, drag the dimension to the left, click a point to locate it and Enter **30**, and then click the check mark.

12. Add a radial dimension by clicking arc, drag the dimension to the left, click a point to locate it and Enter **15**, and then click the check mark.

13. Add a vertical dimension by clicking the vertical line, drag the dimension to the right, click a point to locate it and Enter **40**, and then click the check mark. When complete your sketch should resemble the following image.

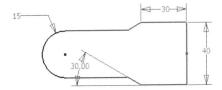

Figure 2-66

14. Click the Auto Dimension tool in the Panel Bar. The Auto Dimension dialog box displays and states that 1 dimension is required. This is a useful method to check how many additional dimensions are required.

15. Click Done to close the dialog box without adding any dimensions. Although the Auto Dimension tool can be used to apply dimensions automatically to the sketch, you will apply them one at a time in this exercise.

16. Add an overall horizontal dimension by clicking the vertical line (not an endpoint). Move the cursor near the left tangent point of the arc until the glyph of dimension to a circle appears then click, drag the dimension down, click a point to locate it and Enter **90**, and then click the check mark.

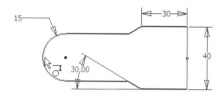

Figure 2-67

17. When complete your sketch should resemble the following image.

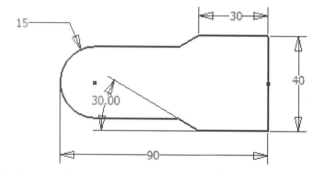

Figure 2-68

18. Edit the value of the dimensions and examine how the sketch changes. The arc should always be in the middle of the vertical line.

19. Delete the horizontal constraint between the center of the arc and the midpoint of the vertical line.

20. Delete the vertical constraint between the left endpoints of the angled lines.

21. Practice adding other type of constrains and dimensions.

22. Close the file without saving the changes.

OPENING/IMPORTING AUTOCAD FILES

Many Autodesk Inventor users have a large amount of existing data in the AutoCAD DWG format. Instead of redrawing this data, you can import it into Autodesk Inventor drawings or into a part feature sketch. You can also import AutoCAD files containing 3D solids. When you import an AutoCAD file with 3D solids, each solid is translated to an Autodesk Inventor part file. If multiple solids exist in the AutoCAD file, an Autodesk Inventor assembly file is created and all translated parts are inserted into the assembly. Parts are placed in the same position and orientation as in the AutoCAD file and by default are given the same name as the AutoCAD file, with a sequenced number at the end of the part name. An AutoCAD file named *Piston.dwg* that contained two solids, for example, would be imported into an Autodesk Inventor assembly named *Piston.iam*, and the parts would be named *Piston1.ipt* and *Piston2.ipt*. When importing a DWG file into Autodesk Inventor, there is an import wizard that guides you through the process. The following sections will introduce you to the options available when importing AutoCAD data into Autodesk Inventor.

OPENING AN AUTOCAD FILE OR AUTOCAD MECHANICAL FILE

One method for importing DWG data is to open a drawing file. The DWG File Import Options wizard guides you through a series of dialog boxes and enables you to import the data into the following destinations:

New Drawing   Use to create a new Autodesk Inventor drawing file and import the 2D geometry into a draft view. Dimensions and annotations are imported; blocks are translated to sketched symbols and placed in the Drawing Resources folder in the Browser. The geometry can be copied to the clipboard and pasted onto a sketch in a part's *.ipt* file.

New Part   Use to create a new part file and import geometry and associative dimensions to Autodesk Inventor sketch geometry. Text, non-associative dimensions and other annotations are not imported.

Title Block   Use to create a new Autodesk Inventor drawing file and import geometry, including annotations, into a new title block.

Border   Use to create a new Autodesk Inventor drawing file and import geometry, including annotations, into a new border.

Symbol   Use to create a new Autodesk Inventor drawing file and import geometry, including annotations, into a new sketched symbol. The sketched symbol is stored in the drawing and the symbol is placed on the drawing sheet.

When you open an AutoCAD file, a wizard guides you through the import process. The next time you open the same file type, the previous wizard settings are used automatically to import the file. To open an AutoCAD file into a new Autodesk Inventor file, follow these steps:

1. Click File > Open on the main menu or on the standard toolbar.

2. In the Open dialog box, click DWG Files (*.dwg) in the Files of type list, as shown in the following image.

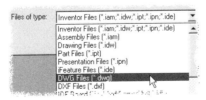

Figure 2-69

3. Browse to and pick the desired DWG file. A preview image of the DWG file will appear in the lower-left corner of the dialog box, as shown in the following image.

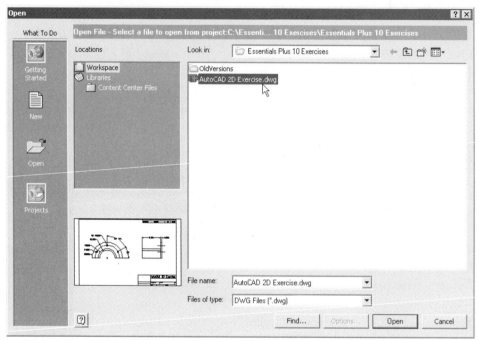

Figure 2-70

4. Click Open to start the DWG File Import Options wizard, whose first screen is shown in the following image.

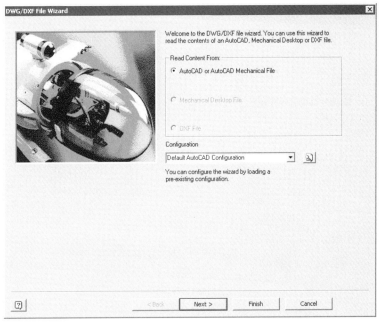

Figure 2-71

The type of DWG file—AutoCAD, AutoCAD Mechanical, or Mechanical Desktop—is selected automatically depending upon the data found in the DWG file. You can select a configuration file that will fill in the prompts based on previous responses to the DWG File Import Options wizard.

   5. Click Next. The Layers and Objects Import Options dialog box appears, as shown in the following image.

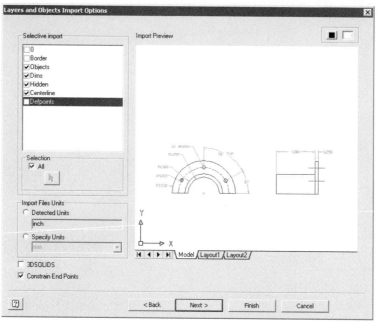

Figure 2-72

6. Select the units with which you created the file from the drop list in the Specify Units area.

7. To change the background color of the preview image, click the black or white icon at the top-right of the dialog box.

8. Click 3D SOLIDS if you want 3D solids to be imported. If the option's box is clear, no solids will be imported. 3D solids must exist in the file for this option to be valid.

9. Import objects from Model Space or from a layout within the DWG file by clicking the different tabs at the bottom of the screen. The names of the tabs are identical to the tab names in the DWG file.

10. In the Selective import area, click the layers that you will import from the preview. As you select the layers, the preview image will update to reflect the change.

11. To select objects in the Import Preview window, clear the All option in the Selection area of the dialog box and select objects in the Import Preview window by clicking or using window selection methods.

12. Click Next. The Import Destination Options dialog box appears, as shown in the following image.

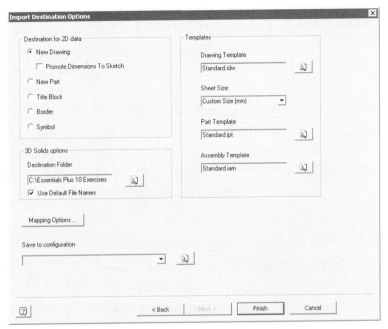

Figure 2-73

13. Select the destination for the imported data. You can import 2D data into a draft view in a new drawing or as the contents of a sketch in a new part file. You can also import the data into a title block, border, symbol, or an existing sketch.

When importing 3D solids, you can change the Destination Folder option (in which you will save the new Autodesk Inventor part and/or assembly files) by clicking the Browse button and then navigating to and selecting a folder. The area is grayed out if you are not importing 3D Solids.

Also, when you import 3D solids, you can generate the file names automatically based on the AutoCAD file name by clicking Use Default File Names. If you leave this option's box clear, you will be prompted for the file names for each 3D solid.

14. Click the Mapping Options button to specify mapping options for how you want to import the contents of each AutoCAD layer, or choose how you want to map AutoCAD fonts to Autodesk Inventor Fonts.

15. In the Templates area, specify the template that Autodesk Inventor will use for the geometry that you will import.

16. To save these configuration settings for future use, click the Save the Configuration button and enter a name and location to store the created *.ini* file.

17. To complete the operation, click the Finish button.

IMPORTING/INSERTING AUTOCAD DATA INTO A SKETCH

Another method that utilizes existing AutoCAD 2D data inserts this AutoCAD 2D data into the active sketch (in a part or drawing). To insert AutoCAD data into the active sketch, follow these steps:

1. Make a sketch active in a part file or a draft view active in a drawing file.

2. Click the Insert AutoCAD file icon on the 2D Sketch Panel Bar, as shown in the following image. Alternatively, in a drawing, click the Insert AutoCAD file tool on the Drawing Sketch Panel Bar.

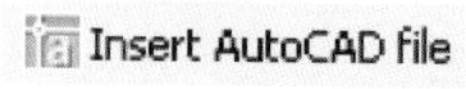

Insert AutoCAD file

Figure 2-74

3. Browse to and select the desired DWG file.

The Open dialog box will appear, and a preview image of the DWG file will appear on the right side of the dialog box, as shown in the following image.

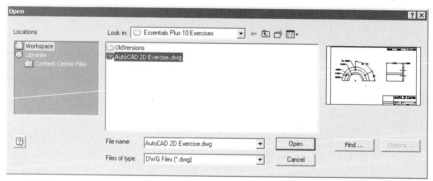

Figure 2-75

4. Click the Open button. The Layers and Objects Import Options dialog box will appear, as shown in step 5 of the previous section.

5. Select the units with which you created the file from the drop list in the Specify Units area.

6. You can change the background color of the preview image by clicking the black or white icon at the top of the dialog box.

7. Check the 3D SOLIDS area if you want to import 3D Solids. If the option's box is clear, no solids will be imported. 3D solids must exist in the file for this option to be valid.

8. Import objects from Model Space or from a layout within the DWG file by clicking the different tabs at the bottom of the screen. The names of the tabs are identical to the tab names in the AutoCAD file.

9. In the Selective import area, click the layers that you will import from the preview. As you select the layers, the Import Preview window will update to reflect the change.

10. To select objects in the preview window, clear the All option in the Selection area of the dialog box, and then select objects in the Import Preview window.

11. To complete the operation, click the Finish button.

IMPORTING OTHER FILE TYPES

Autodesk Inventor can also import parts and assemblies exported from other CAD systems. Autodesk Inventor models created from these formats are base solids or surface models, and no feature histories or assembly constraints are generated when you import a file in any of these formats. You can add features to imported parts, edit the base solids using Autodesk Inventor's solids editing tools, and add assembly constraints to the imported components. To open file types such as SAT, STEP, PRO/E, DXF, and IGES, click File > Open on the main menu or click Open on the standard toolbar.

In the Open dialog box, click the desired file format in the Files of type list. The following sections describe the available formats.

SAT

SAT is a file format available from most CAD systems based on the ACIS modeling kernel. CAD systems based on other kernels can often export parts and assemblies in this format. You can import SAT files either as solid bodies or as surface bodies using the Insert > Import menu option in the part environment.

STEP

The STEP format is used widely to translate 3D solid models between CAD systems. Autodesk Inventor can import models saved in AP-203 or AP-214 STEP formats using the Open command.

PRO/E

Autodesk Inventor can translate and import native Pro/E files up to Version 20 automatically. Complex solid models and Pro/E surfaces can cause import problems. Pro/E files exported in STEP format may be imported more reliably.

You must export Pro/E 2000i and newer files to another format for import into Autodesk Inventor.

DXF

The Drawing Interchange File format, DXF, is a neutral file format supported by many CAD systems. Import a DXF file into Autodesk Inventor using the same techniques that you would use to import a DWG file.

IGES

Early data translation between CAD systems almost always used the IGES file format, and it is still commonly used today. IGES files can contain many different types of data, but typically consist of a set of wireframes and surfaces representing the boundaries of the part. The contents of an IGES file that you import while the Promote option is not enabled will be placed in a Construction folder. You can then select surfaces and stitch them into a surface quilt. If the surfaces can be combined into a surface quilt with no gaps, the quilt can be promoted to an Autodesk Inventor base solid. You can promote individual surfaces and surface quilts to grounded Autodesk Inventor surfaces. Surface healing of IGES data is not done automatically when you import it. Gaps between surfaces due to file imprecision or other factors will prevent an imported IGES file from being promoted to a base solid. You can view the source of the IGES file by selecting the Custom tab, which is made available when you access the iProperties of the file after you have imported the data.

EXERCISE 2-5 Inserting an AutoCAD File

In this exercise, you insert an AutoCAD drawing into a part drawing.

1. Click the New tool and click the Metric tab and then double-click *Standard(mm).ipt*.
2. Click the Insert AutoCAD file tool in the 2D Sketch Panel Bar.
3. In the Open dialog box, select the file *AutoCAD 2D Exercise.dwg*.
4. Click the Open button at the bottom of the dialog box.
5. In the upper left corner of the dialog box uncheck the layers 0, Border, Hidden and Centerline boxes.
6. To select only portions of the displayed geometry click the All button to uncheck it.
7. In the preview window, select all the objects in the front view as shown in the following image.

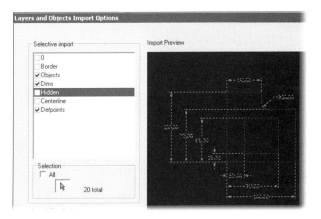

Figure 2-76

8. Click the Specify Units option on the lower left side of the dialog box and verify that mm is selected.

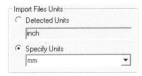

Figure 2-77

9. Verify that Constrain End Points option on the lower left corner of the dialog box is checked.
10. Click the Finish button on the bottom of the dialog box.
11. Click the F8 key to see all the constraints. Notice only coincident constraints were applied.
12. Click the F9 key to hide all the constraints.
13. Apply a Fix constraint to the lower left corner of the sketch.
14. Apply a Horizontal constraint to the lower horizontal line.
15. Click the Auto Dimension tool in the Panel Bar. The Auto Dimension dialog box displays and states that 11 dimension are required.
16. With the Curves option selected, window the entire sketch.
17. Click the Dimensions option to uncheck it.

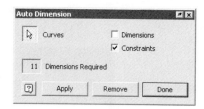

Figure 2-78

18. Click the Apply button, and now there should be 0 dimensions required.
19. Practice editing the values of the dimensions.
20. Close the file without saving the changes

CHAPTER SUMMARY

To	Do This	Tool
Modify the Sketch options of Autodesk Inventor	Click Tools > Application Options and click on the Sketch tab.	
Modify the Part options of Autodesk Inventor	Click Tools > Application Options and click the Part tab.	
Create a new part file	Click the New icon in What To Do and then click the *Standard.ipt* icon on the Default tab, or click *Standard (unit).ipt* on one of the other tabs.	Standard.ipt
Make a planar face, a work plane, or a nonactive sketch in the active part the active sketch	Click the Sketch tool on the standard toolbar and click a face, a work plane, or an existing sketch in the Browser. You can also click a face, a work plane, or an existing sketch in the Browser and then click the Sketch tool from the standard toolbar or press S.	
Sketch the outline of the part	Use the 2D Sketch tools on either the Panel Bar or the Sketch toolbar.	Sketch
Add geometric constraints to a sketch	Click a constraint from the constraint pop-up menu on the Sketch Panel Bar or Sketch toolbar, or right-click in the graphics area, select Create Constraint from the menu, and select the specific constraint.	
Add parametric dimensions to a sketch	Click the General Dimension tool on the Sketch toolbar, or right-click in the graphics area and select Create Dimension from the menu or press D.	
Import AutoCAD data	Either open the file using the Open tool or click the Insert AutoCAD file tool on the 2D Sketch Panel Bar or Drawing Sketch Panel	Insert AutoCAD file

Chapter 2 Sketching, Constraining, and Dimensioning

Applying Your Skills

SKILL EXERCISE 2-1

In this exercise, you create a sketch and then you add geometric and dimensional constraints to control the size and shape of the sketch. Start a new part file based on the *Standard(mm).ipt* and create the fully constrained sketch as shown in the following image. Assume the horizontal lines are collinear, the center points of the arcs are aligned vertically, the endpoints of the arcs are aligned horizontally and the angled lines are equal in length.

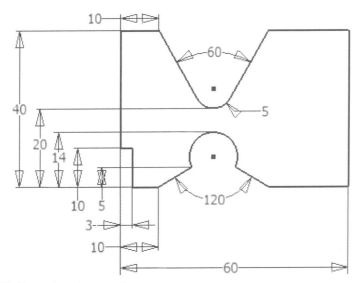

Figure 2-79 Completed exercise

SKILL EXERCISE 2-2

- In this exercise, you create a sketch with linear and arc shapes, and then you add geometric and dimensional constraints to fully constrain the sketch. Start a new part file based on the *Standard(mm).ipt* and create the fully constrained sketch as shown in the following image. Create the two outer circles first, create the two lines, trim the two circles and place a vertical constraint between the line endpoints on both ends.

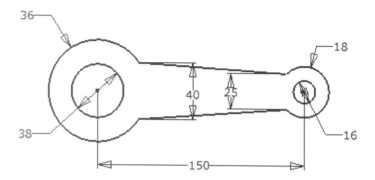

Figure 2-80 Completed exercise

CHECKING YOUR SKILLS

Use these questions to test your knowledge of the material in this chapter.

1. True__ False__ When you sketch, constraints are not applied to the sketch by default.

2. True__ False__ When you sketch and a point is inferred, a constraint is applied to represent that relationship.

3. True__ False__ A sketch does not need to be fully constrained.

4. True__ False__ When working on an mm part, you cannot use English units.

5. True__ False__ After a sketch is constrained fully, you cannot change a dimension's value.

6. True__ False__ A driven dimension is another name for a parametric dimension.

7. True__ False__ If you use the Auto Dimension tool on the first sketch in the part, the sketch will be constrained fully.

8. True__ False__ You can only import 2D AutoCAD data into Autodesk Inventor.

9. Explain how to draw an arc while still in the Line command.

10. Explain how to remove a geometric constraint from a sketch.

    _____

    _____

11. Explain how to change a vertical dimension to an aligned dimension while you create it.

    _____

    _____

12. Explain how to create a dimension between two quadrants of two arcs.

    _____

    _____

CHAPTER 3

Creating and Editing Sketched Features

After you have drawn, constrained, and dimensioned a sketch, your next step is to turn the sketch into a 3D part. This chapter takes you through the process to create and edit sketched features.

CHAPTER OBJECTIVES

- After completing this chapter, you will be able to
- Understand what a feature is
- Use the Autodesk Inventor Browser to edit parts
- Extrude a sketch into a part
- Revolve a sketch into a part
- Edit features of a part
- Edit the sketch of a feature
- Make an active sketch on a plane
- Create sketched features using one of three operations: cut, join, or intersect

UNDERSTANDING FEATURES

After creating, constraining, and dimensioning a sketch, the next step in creating a model is to turn the sketch into a 3D feature. The first sketch of a part that is used to create a 3D feature is referred to as the base feature. In addition to the base feature, you can create sketched features, in which you draw a sketch on a planar face or work plane and either add or subtract material to or from existing features in a part. Use the Extrude, Revolve, Sweep, or Loft tools to create sketched features in a part. You can also create placed features such as fillets, chamfers, and holes by applying them to features that have been created. Placed features will be covered in Chapter 4. Features are the building blocks that help create a part.

A plate with a hole in it, for example, would have a base feature representing the plate and a hole feature representing the hole. As features are added to the part, they appear in the Browser, showing the history of the part or assembly (the order in which the features are created or the parts are assembled). Features can be edited, deleted from, or reordered in the part as required.

CONSUMED AND UNCONSUMED SKETCHES

You can use any sketch as a profile in feature creation. A sketch that has not yet been used in a feature is called an unconsumed sketch. When you turn a 2D sketched profile into a 3D feature, the feature consumes the sketch. The following image shows an unconsumed sketch in the Browser on the left and a consumed sketch in the Browser on the right.

Figure 3-1

Although a consumed sketch is not visible as you view the 3D feature, you may need to access sketches and change their geometric or dimensional constraints in order to modify their associated features. A consumed sketch can be accessed from the Browser by either double-clicking on its name or icon or right-clicking and selecting Edit Sketch from the menu. You may also access the sketch by starting the Sketch command and selecting the sketch from the Browser.

Figure 3-2

USING THE BROWSER FOR CREATING AND EDITING

The Autodesk Inventor Browser, by default, is docked along the left side of the screen and displays the history of the file. In the Browser you can create, edit, rename, copy, delete, and reorder features or parts. You can expand or collapse the Browser to display the history of the part(s) (the order in which the features were created) by clicking the + and - on the left side of the part or feature name in the Browser. An alternate method of expanding the Browser is to place your cursor so that it is on top of a feature icon (do not click). After a couple of seconds, the item in the Browser automatically expands. You can also expand or collapse all part children by right-clicking an item in the Browser and selecting either Expand All Children or Collapse All Children from the menu.

The following image shows the Browser with all features of the part expanded.

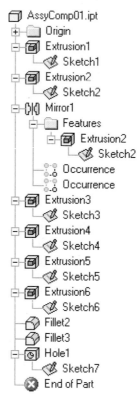

Figure 3-3

As parts grow in complexity, so will the information found in the Browser. Dependent features are indented to show that they relate to the item listed above it. This is referred to as a *parent-child relationship*. If a hole is created in an extruded rectangle, for example, and the extrusion is then deleted, the hole will also be deleted. To help filter out some of the object types that appear in the Browser, you can click the icon that looks like a funnel from the top of the Browser. After clicking the funnel, you can select objects to hide from the drop-down list, as shown in the following image.

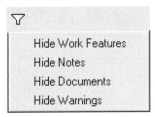

Figure 3-4

Each feature in the Browser is given a default name. The first extrusion, for example, will be named Extrusion1, and the number in the name will sequence as you add similar features. The Browser can also help you locate parts and features in the graphics area. To highlight a feature or part in the graphics window, simply move your cursor over the feature or part name in the Browser.

To zoom in on a selected feature, right-click on the feature's name in the Browser and select Find in Window on the menu or press the END key on your keyboard. The Browser itself functions similarly to a toolbar, except that you can resize it while it is docked. To close the Browser, click the X in its upper-right corner. If the Browser is not visible on the screen, you can display it by clicking Browser Bar on the Toolbar pop-up menu on the View menu or right-clicking while over a toolbar and selecting Browser Bar from the list.

Specific functionality of the Browser will be covered throughout the book in the sections where it pertains. A basic rule, however, is to either right-click or double-click on the feature's name or icon to edit or perform a function on the feature. A sample Browser shown in the following image illustrates some of the topics described above.

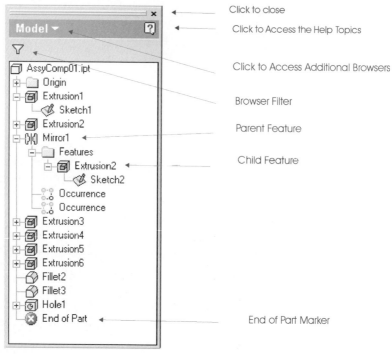

Figure 3-5

SWITCHING ENVIRONMENTS

Up to this point, you have been working in the sketch environment where the work is done in 2D. The next step is to turn the sketch into a feature. To turn a sketch into a feature, you need to exit the *sketch environment* and enter the *part environment*. There are a number of methods that can be used to accomplish this:

- Click the Return tool on the left side of the standard toolbar as shown in the following image.

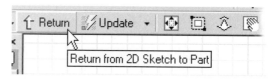

Figure 3-6

- Click the arrow on the Panel Bar near 2D Sketch Panel and click Part Features from the drop-down list as shown in the following image.

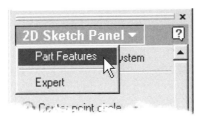

Figure 3-7

 NOTE If you exit the sketch environment with the Return tool, this is done automatically.

- Right-click in the graphics area and select Finish Sketch from the menu. You can also right-click in the graphics area, select Create Feature from the menu (as shown in the following image), and then click the tool you need to create the feature. Only the tools that are applicable to the current situation will be available.

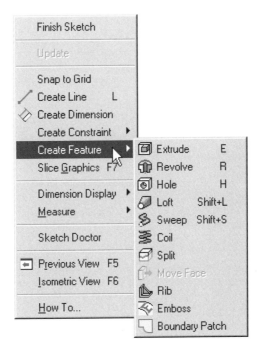

Figure 3-8

- Enter a shortcut key to initiate one of the feature tools.

FEATURE TOOLS

Once you are in the part environment, the Panel Bar icons will change to the part feature tools. The following image shows the Part Features Panel Bar with the Expert mode turned on. The following list includes descriptions of the part feature tools:

Figure 3-9

Feature Tools

Button	Tool	Function
	Extrude	Extrudes a sketch in the positive or negative Z-axis. This extrusion can form a base feature or add or remove material from a part.
	Revolve	Rotates a sketch around a straight edge or axis at a specified angle. This revolution can form a base feature or add or remove material from a part.
	Hole	Creates a drilled, tapped, countersunk, or counterbored hole feature in a part.
	Shell	Removes material from the part, leaving a specified wall thickness.
	Rib	Creates a thin-walled extrusion from a 2D sketch.
	Loft	Creates a lofted base feature or loft feature by blending between sketches that lie on different planes.
	Sweep	Creates a swept feature by sweeping a sketch about a defined path.
	Coil	Creates a 3D helical part or feature by revolving a sketch around a centerline. The specifics of the 3D helical are determined by data entered into a dialog box.
	Thread	Creates a thread feature in a hole or on a cylinder. The thread appears in the graphic area as well as in drawing views.
	Fillet	Creates a fillet feature on an edge or edges. The feature can be a fillet or a round.
	Chamfer	Creates a chamfer feature on a selected edge or edges.

Feature Tools (continued)

Button	Tool	Function
	Move Face	Moves a face a specified distance or a planar move to specific coordinates.
	Face Draft	Creates a face draft feature that angles a selected planar face in or out.
	Split	Allows a face to be split into two faces or a part to be split into two parts.
	Delete Face	Creates a delete face feature to delete faces and includes a *Healing* option to delete lumps or voids from a part. When you delete a face, the parametric model is converted to a surface model.
	Boundary Patch	Creates a boundary patch surface feature using edges that lie on a planar face.
	Stitch Surface	Stitches two or more surfaces together to form a single surface. When this tool is activated in the construction environment it can be used to analyze edge conditions of surfaces. You can knit surfaces together to form a quilt or you can knit a combination of faces and surfaces to promote the model from a surface model to a solid model.
	Replace Face	Replaces existing faces of features in a part by merging them with a surface boundary that is selected.
	Thicken/Offset	Creates a thicken feature by adding material to a selected set of faces or surfaces. You can also choose to offset a face or surface. Depending on whether you choose to thicken or offset, you can enhance the part to describe surfaces or change a surface model to a solid part.
	Emboss	Creates a profile that physically alters a face by raising or lowering regions of the face.
	Decal	Creates a decal-type, or silkscreen-type, feature. Decals are usually separate parts, such as stickers.
	Rectangular Pattern	Arrays selected feature(s) in a rectangular pattern or along a selected path.
	Circular Pattern	Arrays selected feature(s) in a circular pattern around a centerline.
	Mirror Feature	Mirrors selected feature(s) or solid bodies about a plane.
	Content Center	Opens the Content Center where you can access the Content Center Library which can contain custom features, fasteners, shafts, steel shapes, etc.
	Work Plane	Creates a work plane feature that is based on user input. You can use a work plane as a sketch plane or as a plane for the mirror feature tool or for extrusion terminations.
	Work Axis	Creates a work axis feature that can be used as an axis for circular patterns or to define other work features.

Feature Tools (continued)

Button	Tool	Function
	Work Points	Creates a work point feature that can be used to help create work planes, work axes, and other tools that require points to be selected.
	Grounded Work Points	Creates a work point feature that initially is locked into position. You can modify its position using the 3D Move/Rotate tool.
	Promote	Stitches an imported surface model into a single surface and then promotes it so the surface can be used for creating features. You can also promote a part as a surface that can be used in another part file. This is typically done when working in the assembly environment.
	Derived Component	Creates a part based on another part or assembly. Any changes to the source will also appear in the derived part. You can also derive sketches, work features, surfaces, parameters, or iMates from another file.
	Parameters	Displays and defines parameters used in the model. You can create user-defined parameters, rename, add equations, and link parameters to a Microsoft Excel file.
	Create iMate	Creates a predefined constraint or group of constraints (Composite iMates) on a component to specify how parts will connect when they are inserted into an assembly.
	Insert iFeature	Places an iFeature on the current part.
	View Catalog	Opens Windows Explorer to let you browse any iFeatures or punches that have been created.

EXTRUDING A SKETCH

The most common method for creating a feature is to extrude a sketch and give it depth along the Z-axis. Before extruding, it is helpful to view the part in an isometric view. Autodesk Inventor previews the extrusion depth and direction in the graphics window.

To extrude a sketch, click the Extrude tool on the Part Features Panel Bar, press the hot key E, or right-click in the graphics screen and select Create Feature > Extrude on the menu. After you issue the tool, the Extrude dialog box appears as shown in the following image.

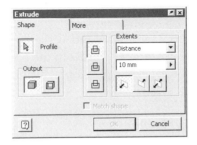

Figure 3-10

The Extrude dialog box has two tabs: Shape and More. When you make changes in the dialog box, the shape of the sketch will change in the graphics area to represent these values and options. When you have entered the values and the options you need, click the OK button to create the extruded feature.

SHAPE

The Shape tab is where you specify the profile to use, the operation (middle column), extents, and output type. It contains several options.

   Profile   Click this button to choose the sketch to extrude. If there are multiple closed profiles, you will need to select which sketch area you want to extrude. If there is only one possible profile, Autodesk Inventor will select it for you and you can skip this step. If you select the wrong profile or sketch area, click the Profile button again and choose the desired profile or sketch area. To remove a selected profile, hold the CTRL key and click the area you wish to remove.

Operation

This is the middle column of buttons that is not labeled. If this is the first sketch with which you are working, it is referred to as a base feature and only the top button is available. The operation defaults to Join (top button). Once the base feature has been established, you can then extrude a sketch, adding or removing material from the part using the Join or Cut options, or you can retain the common volume between the existing part and the newly defined extrude operation using the Intersect option.

	Join	Adds material to the part.
	Cut	Removes material from the part.
	Intersect	Removes material, keeping what is common to the existing part feature(s) and the new feature.

Extents

Extents determines the type, distance, and direction for an extrusion. This section of the dialog box has the following three areas:

Termination

The termination determines how the extruded sketch will be terminated. There are five options from which to choose. Similar to the operation section, this section has options that may not be available until a base feature exists.

Distance   This option determines that the sketch will be extruded a specified distance.

To Next   This option determines that the sketch will be extruded until it reaches a plane or face. The sketch must be fully enclosed in the area to which it is projecting. Otherwise use the To termination with the Extend to Surface option. Click the Direction button to determine the extrusion direction.

To   This option determines that the sketch will be extruded until it reaches a selected face or plane. To select a plane or face to end the extrusion, click the Select Surface button, as shown in the following image, and then click a face or plane at which the extrusion should terminate.

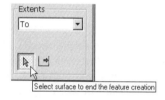

Figure 3-11

From To   This option determines that the extrusion will start at a selected plane or face and stop at another plane or face. Click the Select the surface to start feature creation button, and then click the face or plane where the extrusion will start. You should then click the Select the surface to end the feature creation button, and then click the face or plane where the extrusion will terminate.

All   This option determines that the sketch will be extruded all the way through the part in one or both directions.

Distance

If Distance is selected as the termination, do one of the following: enter a value at which the sketch will be extruded, click the arrow to the right and measure two points to determine a value, display dimensions of previously created features to select from, or select from the list of the most recent values used. After you enter a value, a preview image appears in the graphics area representing how the extrusion will look. Another method is to click the edge of the extrusion preview shown in the graphics area and drag it.

A preview image appears in the graphics area, and the corresponding value appears in the distance area.

If values and units appear red when you enter them, it means that the defined distance is incorrect and needs to be corrected.

For instance, if you entered too many decimal places (i.e., 2.12.5) or incorrect units for the dimension value, as shown in the following image, the values appear in red. You need to correct them before the extrusion can be created.

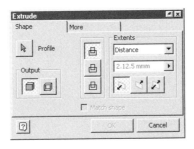

Figure 3-12

 NOTE When extruding a sketch a given distance, drag the sketch to get a preview of how it will look at different distances.

Direction

There are three buttons from which to choose for determining the direction. Choose the first two to flip the extrusion direction, or click the last button (mid-plane) to have the extrusion go equal distances in the negative and positive directions. If the extrusion distance is 2", for example, the extrusion will go 1" in both the negative and positive Z directions when using the mid-plane option.

Output

There are two options available to define the type of output that the Extrude tool will generate.

Solid   Extrudes the sketch and the result is a solid body.

Surface Extrudes the sketch and the result is a surface.

MATCH SHAPE

You can use the match shape option when working with an open profile that you want to extrude. The edges of the open profile are extended until they intersect geometry of the model. This provides a "flood-fill"-type effect for the extrude feature

When you convert a sketch into a part or feature, the dimensions on the sketch are consumed (or disappear). When you edit the feature, the dimensions reappear. The dimensions can also be displayed when drawing views are made. For more information on editing parts or features, see the Editing 3D Parts section later in this chapter.

MORE

The More tab, as shown in the following image, contains additional options to refine the feature that is being created:

Taper Taper extrudes the sketch and applies a taper angle to the feature.

Alternate Solution Alternate Solution terminates the feature on the most distant solution for the selected surface. An example is shown in the following image.

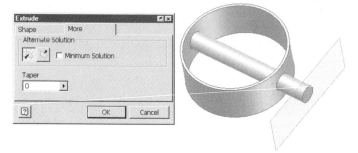

Figure 3-13

Minimum Solution Minimum Solution terminates the feature on the first possible solution for the selected surface. An example is shown in the following image.

 TIP To extend the taper angle out from the part, give the taper angle a negative number. This reduces/increases the volume of the resulting extruded feature. This is also known as reverse draft.

Figure 3-14

EXERCISE 3-1 Extruding a Sketch

In this exercise, you create a base feature by extruding an existing profile. You will examine the direction options available in the Extrude dialog box

1. Open *ESS_E03_01.ipt*.

Chapter 3 Creating and Editing Sketched Features 109

Figure 3-15

2. Click the Extrude tool. Because there is only one possible sketch, the profile is automatically selected.

3. In the Extrude dialog box, set the Distance to 15 mm.

4. Select the preview edge of the extrusion in the graphics window.

5. Drag the edge until a distance of 25 mm is displayed in the Distance field of the Extrude dialog box.

Figure 3-16

6. Click the More tab in the Extrude dialog box.

7. Adjust the value of the Taper to 10.

8. Press the F4 key and rotate the part until you can see the arrow previewing that the taper will add mass to the part.

9. Release the F4 key.

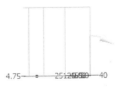

Figure 3-17

10. Return to an Isometric View of the model.

11. Click the Shape tab in the Extrude dialog box.

12. Change the direction of the extrusion by selecting the middle button on the direction area. This will flip the direction of the extrusion.

13. Change the direction to Midplane by selecting the right button on the direction area. The model should resemble the following image.

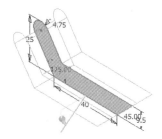

Figure 3-18

14. Click OK. The extrusion is created.

Figure 3-19

15. Close the file. Do not save changes.

REVOLVING A SKETCH

Another method for creating a part is to revolve a sketch around a straight edge or axis (centerline). You can use revolved sketches to create cylindrical parts or features. To revolve a sketch, you follow the same steps you did to extrude a sketch (create the sketch and add constraints and dimensions), then click the Revolve tool on the Part Features Panel Bar, press R or right-click in the graphics window and select Create Feature > Revolve from the menu. The Revolve dialog box appears, as shown in the following image.

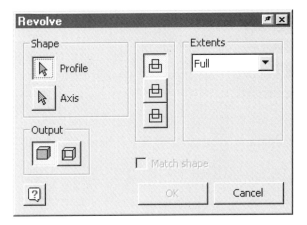

Figure 3-20

The Revolve dialog box has five sections: Shape, Operation (middle column), Extents, Output, and Match Shape. When you make changes in the dialog box, the preview for the revolved feature changes in the graphics area to represent the values and options selected. When you have entered the values and the options you need, click the OK button to create the revolve feature.

SHAPE

This section has two options: Profile and Axis.

Profile Click this button to choose the profile to revolve. If the Profile button is shown depressed, this is telling you that a profile or sketch needs to be selected. If there are multiple closed profiles, you will need to select the profile you want to revolve. If there is only one possible profile, Autodesk Inventor will select it for you and you can skip this step. If the wrong profile or sketch area is selected, click the Profile button and choose the new profile or sketch area. To remove a selected profile, hold the CTRL key and click the profile to remove.

Axis Click a straight edge or centerline about which the sketch should be revolved. If a centerline is used, it needs to be part of the sketch. See the section below on how to create a centerline and create diametric dimensions.

OPERATION

This is the middle column of buttons that is not labeled. If this is the first sketch with which you are working, it is referred to as a base feature and only the top button is available. The operation defaults to Join (top button). Once the base feature has been established, you can then revolve a sketch, adding or removing material from the part using the Join or Cut options, or you can keep what is common between the existing part and the completed revolve operation using the Intersect option.

	Join	Adds material to the part.
	Cut	Removes material from the part.
	Intersect	Removes material, keeping what is common to the existing part feature(s) and the new feature.

EXTENTS

The Extents area determines if the sketch will be revolved 360° or another specified angle.

Full Full is the default option and it will revolve the sketch 360° about a specified edge or axis.

Angle Click this option from the drop-down list and the Revolve dialog box displays additional options, as shown in the following image. Enter an angle for the sketch to be revolved. There are three buttons below the degree area that will determine the direction of the revolution. Choose the first two to flip the revolution direction or click the last button to have the revolution go an equal distance in the negative and positive directions. If the angle was set to 90°, for example, the revolution will go 45° in both the negative and positive directions.

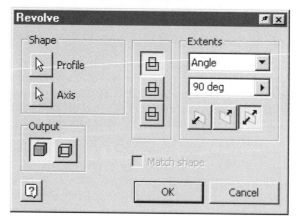

Figure 3-21

OUTPUT

There are two options available to select the type of output that the Revolve tool will generate.

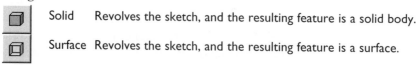

Solid   Revolves the sketch, and the resulting feature is a solid body.

Surface Revolves the sketch, and the resulting feature is a surface.

MATCH SHAPE

You can use the Match shape option when working with an open profile. The edges of the open profile are extended until they intersect the geometry of the model. This provides a "flood-fill"-type effect for the revolve feature.

CENTERLINES AND DIAMETRIC DIMENSIONS

When revolving a sketch, you may want to specify diametric linear dimensions instead of radial dimensions. Sketches you revolve are usually a quarter section of the completed part. The following image shows a sketch that represents a quarter section of the completed part with a centerline and the diametric linear dimensions.

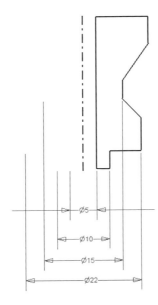

Figure 3-22

The dimensions placed on a sketch are used for drawing views. If you want to place a diametric linear dimension on the sketch, you should select a centerline. To create a centerline, you activate the Centerline tool on the standard toolbar, as shown in the following image. You can activate the Centerline tool before sketching the line to be displayed as a centerline, or you can select an existing sketched entity and then select the Centerline tool.

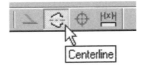

Figure 3-23

To create a diametric linear dimension, use the General Dimension tool and select either the centerline and the other point or line to be dimensioned, or click a point or edge and then the centerline to place the diametric dimension. When selecting the centerline, be certain to select the entire centerline, not just an endpoint of the centerline. If you do not use a centerline as part of the dimension, you can right-click after selecting the geometry to be dimensioned and select Linear Diameter from the menu to place a diametric dimension.

EXERCISE 3-2 Revolving a Sketch

In this exercise, you create a sketch and then create a revolved feature to complete a part. This exercise demonstrates how to revolve sketched geometry about an axis to create a revolved feature.

1. Click the New tool.

2. Select the Metric tab and then double-click *Standard(mm).ipt*.

3. Create the sketch geometry as shown.

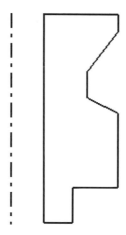

Figure 3-24

4. Click the General Dimension tool.

5. Add the linear diameter dimensions shown.

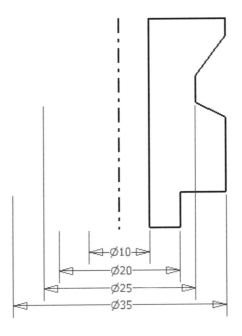

Figure 3-25

    6. Right-click in the graphics window.

    7. Select Isometric View from the menu.

    8. Finish the sketch and click the Revolve tool.

    9. Change the extents to Angle.

   10. Enter **45** degrees. The preview of the model updates to reflect the change.

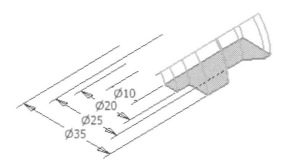

Figure 3-26

   11. Change the direction of the revolve by picking the middle flip button. The preview image will reverse the direction.

12. Set the revolve direction to Midplane.
13. Enter **90** degrees for the angle. The preview should resemble the following image.

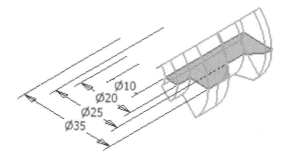

Figure 3-27

14. Change the extent type to Full.
15. Select the OK button to create the feature.
16. Rotate the model as shown. Your part should resemble the following image.

Figure 3-28

17. Close the file. Do not save changes.

EDITING A FEATURE

After you create a feature, the feature consumes all of the dimensions that were visible in the sketch. If you need to change the dimensions' values, taper, operation, or termination that was entered in the dialog box during the feature creation, you will need to edit the feature. To edit it, follow these steps:

1. Right-click on the feature's name in the Browser.
2. Select Edit Feature from the menu.

The following image shows the menu that appears after right-clicking on Extrusion1.

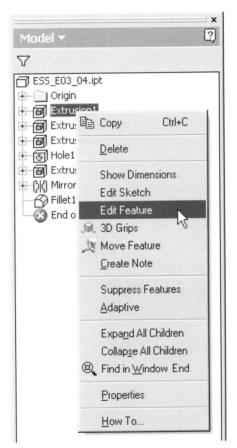

Figure 3-29

3. In the dialog box, enter new values or change the settings.
4. Click the OK button to complete the edit.

Everything can be changed in the dialog box except the join operation on a base feature and the output.

EDITING FEATURE SIZE

To change the dimensional values of a feature, you need to edit the feature so the dimensions are visible. There are multiple methods you can use to edit the dimensions. There is no preferred method; use the method that works best for your workflow.

- In the Browser, double-click on the feature's name or icon.
- In the Browser, right-click on the feature's name and select Edit Sketch from the menu.

In the Browser, expand the children of the feature, right-click on the name of the sketch, and choose Edit Sketch from the menu as shown in the following image.

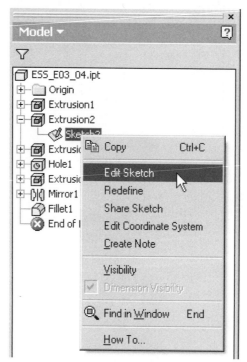

Figure 3-30

On the command bar, click the down arrow by the Select button and choose Feature Priority from the drop-down menu as shown in the following image. Then double-click on the feature that you want to edit, and the dimensions will appear on the part. You can also access the Select tools by holding down the SHIFT key and right-clicking in the graphics area.

Figure 3-31

When the dimensions are visible on the screen, double-click on the dimension text that you want to edit. The Edit Dimension dialog box appears. Enter a new value and then click the check mark in the dialog box, or press the ENTER key. Continue to edit the dimensions and, when finished, click the Update button on the command bar as shown in the following image. The dimensions will disappear and the new values will be used to regenerate the part.

Figure 3-32

3D GRIPS

An alternate method for editing a feature size is to use 3D Grips. The 3D Grips tool can be used to push or pull the faces of an extruded, revolved, or a sweep feature. The 3D Grips tool is accessed from the shortcut menu after you have selected a feature, face, or sketch as shown in the following image.

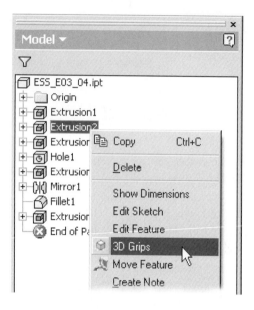

Figure 3-33

After the 3D Grips are displayed, click and drag the grip that is associated with the geometry that you want to modify, then right-click and click Done from the menu. Arrows are displayed as you move the cursor over existing faces of the feature that are being modified as shown in the following image.

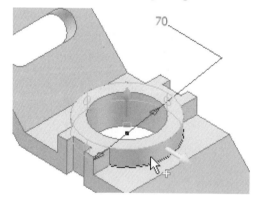

Figure 3-34

You can also use the grip edit capabilities to modify the feature by a specific distance or angle. This is done by right-clicking the arrow and choosing the appropriate option from the menu as shown in the following image. Depending on the available geometry of the feature, the menu will contain different options such as, Edit Angle, Edit Offset, Edit Radius, Edit Extent, etc. After selecting a menu option, enter the new value in the edit dialog box and click Done from the menu.

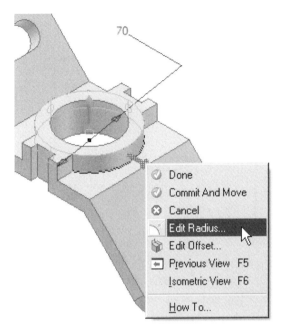

Figure 3-35

When using the grip edit functionality, parametric dimensions are ignored and are modified as if they were reference dimensions. When the grip edit is complete, the dimension values are updated to reflect the new values.

MOVE FEATURE

Another method that can be used to edit a feature is the Move Feature tool. This tool provides the ability to click and drag a feature and move it from one face of a part to another. Similar to the 3D Grips tool, the Move Feature tool is accessed by right-clicking after selecting a feature as shown in the following image.

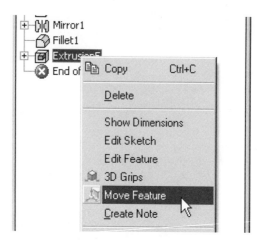

Figure 3-36

After dragging the feature to the new face, right-click and select Done from the menu. If you choose Commit and 3D Grip Edit, the feature is relocated and is ready for modification using grips as described above and shown in the following image.

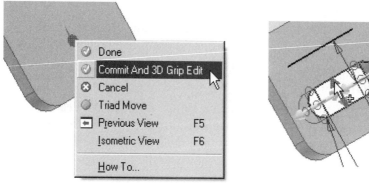
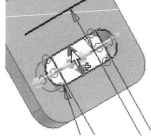

Figure 3-37

You can also use the 3D Move/Rotate tool to move the feature if the Triad Move option is selected from the menu. The 3D Move/Rotate tool is discussed in Chapter 4.

RENAMING FEATURES AND SKETCHES

By default, each feature is given a name. These feature names may not help you when trying to locate a specific feature of a complex part. The feature names will not be descriptive to your design intent. The first extrusion, for example, is given the name Extrusion1 by default, whereas the design intent may be that the extrusion is the thickness of a plate. You can edit any feature names, however. To rename a feature, slowly double-click the feature name. Enter a new name (spaces are allowed).

FEATURE COLOR

When parts become complex, you may want to change colors of specific features. This is also useful when you want to differentiate between a cast and a machined surface. To change a feature color, right-click on the feature name in the Browser and select Properties from the menu. In the Feature Properties dialog box, select a new feature color from the drop-down menu, as shown in the following image, and then click OK to complete the operation. You can also change a feature's name in the top area of the Feature Properties dialog box.

You can change the color of model faces using a similar method. Right-click a selected face in the graphics area and select Properties from the menu. You can then select a new face color from the drop-down menu in the Feature Properties dialog box.

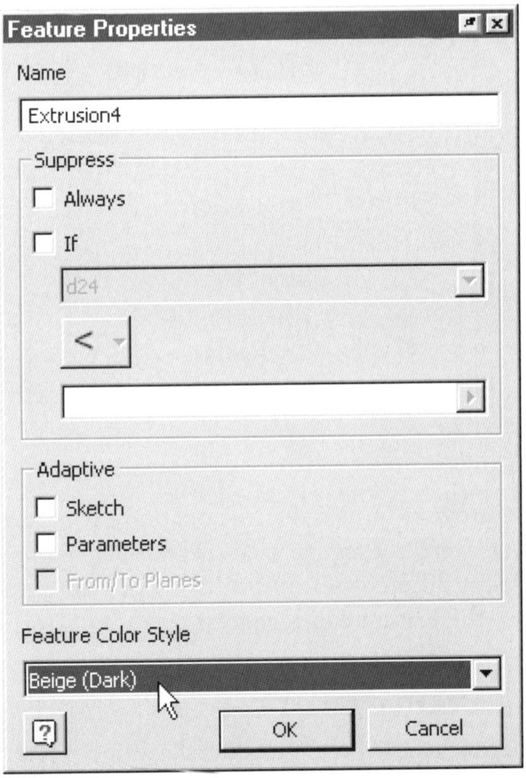

Figure 3-38

DELETING A FEATURE

You may choose to delete a feature after it has been placed. To delete a feature, right-click on the feature name in the Browser and select Delete from the menu, as shown in the following image. The Delete Features dialog box will then appear, and you should choose what you want to delete from the list. You can delete multiple features by holding down the CTRL or SHIFT key, clicking their names in the Browser, right-clicking on one of the names, and then selecting Delete from the menu.

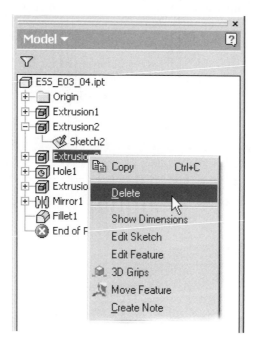

Figure 3-39

FAILED FEATURES

If the feature in the Browser turns red after updating the part, it is an alert that the new values or settings were not regenerated successfully. You can then edit and enter new values or select different settings to define a valid solution. Once you define a valid solution, the feature will regenerate without error.

EDITING A FEATURE SKETCH

In the last section, you learned how to edit the dimensions and the settings in which the feature was created. In this section, you will learn how to add and delete constraints, dimensions, or geometry in the original 2D sketch. To edit the 2D sketch of a feature, do the following:

1. In the Browser, right-click on the name or icon of the feature or sketch that you want to edit and select Edit Sketch from the menu, as shown in the following image.

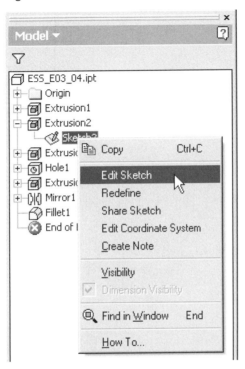

Figure 3-40

2. Add or remove geometry from the sketch and add or delete constraints and dimensions, as described in Chapter 2.

While editing the sketch, you can both add and remove objects. You can add geometry-lines, arcs, circles, and splines to the sketch. To delete an object, right-click on it and select Delete from the menu or click on it and press the DELETE key. If you erase an object from the sketch that has dimensions associated with it, the dimensions are no longer valid for the sketch and they will be deleted. You can also delete the entire sketch and replace it with an entirely new sketch. When replacing entire sketches, you should delete other features that would be consumed by the new objects first and re-create them. Once you modify the sketch, update the part by clicking the Return or the Update button on the command bar.

 NOTE If you get an error after updating the part, make sure that the sketch forms a closed profile. If the appended or edited sketch forms multiple closed profiles, you will need to reselect the profile area.

EXERCISE 3-3 Editing Features and Sketches

In this exercise, you edit a consumed sketch in an extrusion and update the part.

1. Open *ESS_E03_03.ipt*.

Figure 3-41

2. In the Browser, right-click Extrusion1.
3. Click Edit Sketch. The consumed feature sketch is displayed.

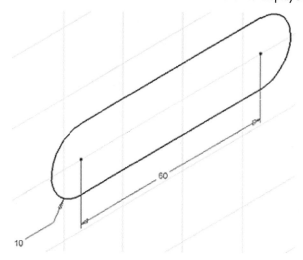

Figure 3-42

4. Double-click the 10 dimension.
5. In the edit box, enter **8** then click the checkmark.
6. Double-click the 60 dimension then edit the value to 80.
7. Click Return on the Standard toolbar.
8. Click the Zoom All tool. The feature is updated with the new values.

Figure 3-43

Edit the termination method for an extrusion.

9. In the Browser, right-click Extrusion2.

10. Click Edit Feature. The Extrude dialog box is displayed.

11. Under Extents, select To.

12. Select the bottom face.

Figure 3-44

13. Place a checkmark beside the right button to terminate on an extended face.

14. Click OK. The feature is updated.

15. Close the file. Do not save changes.

SKETCHED FEATURES

A sketched feature is a feature that you draw on a plane or face and add material to or remove material from. The basic steps to create a sketched feature are:

1. Create or make an existing sketch active.

2. Draw the geometry that defines the sketch.

3. Add constraints and dimensions.

4. Perform a Boolean operation that will either add material to or remove material from the part or will keep whatever is common between the part and the completed feature.

There are no limits to the number of sketched features that can be added to a part. Each sketched feature, however, needs to be on its own plane. In the following section, you will learn how to assign a plane to the active sketch and then how to work with sketched features.

DEFINING THE ACTIVE SKETCH PLANE

As stated already, each sketch must exist on its own plane. The active sketch has a plane on which the sketch is drawn. To assign a plane to the active sketch, there are three requirements:

1. The part in which the plane will be placed must be an Autodesk Inventor part.
2. The part must be active.
3. A plane on which the sketch will be created must be a planar face or a work plane. The planar face does not need to have a straight edge. A cylinder has two faces; for example, one on the top and the other on the bottom of the part. Neither has a straight edge, but a sketch can be placed on either face.

To make a sketch active, use one of the following methods:

- Issue the Sketch tool from the command bar, shown in the following image, and then click the plane where you want to place the sketch.

Figure 3-45

- Press the hot key S and click the plane where you want to place the sketch.
- Click a plane that will contain the active sketch and issue the Sketch tool from the command bar.
- Click a plane that will contain the active sketch, and press the hot key S.
- Click a plane that will contain the active sketch, right-click in the graphics area, and select New Sketch from the menu.
- Issue the Sketch tool from the command bar, expand the Origin folder in the Browser, and click one of the default work planes.
- Expand the Origin folder in the Browser, right-click on one of the default work planes, and select New Sketch on the menu.
- To make a previously created sketch active, issue the Sketch tool and click the sketch name in the Browser.

Once you have created a sketch, it appears in the Browser with the name Sketch#, and sketch tools appear in the 2D Sketch Panel Bar. The number will sequence for each new sketch that is created. In the Browser, slowly double-click the existing name and enter a new one to rename a sketch. When you have created a new sketch on a plane, it is sometimes easier to work in a plan view (looking straight at, or normal to, the current plane). This can be done by using the Look At tool and clicking the plane or the sketch in the Browser. You can place sketch curves, apply constraints, and apply dimensions, exactly as you did with the first sketch. In addition to constraining and dimensioning the new sketch, you can also constrain the new sketch to the existing part. You can place dimensions to geometry that do not lie on the current plane; the dimensions, however, will be placed on the current plane. When you look at a part from different viewpoints, you will see arcs and circular edges appearing as lines. Remember that they are still circular edges. When you constrain or dimension them, the constraints and dimensions will be placed on their center points. After the sketch has been constrained and dimensioned, it can be extruded, revolved, swept, or lofted. Exit the sketch environment by clicking the sketch name in the Browser and either right-clicking in the graphics area and selecting Finish Sketch or clicking the Return tool in the command bar. There can only be one active sketch at a time.

TIP Rename the sketches to better explain the purpose for which you use them.

FACE CYCLING

Autodesk Inventor has dynamic face highlighting to help you select the correct face to activate, and to aid in selecting objects. As you move the cursor over a given face, it highlights the edges of the face. If you continue to move the cursor, different faces are highlighted as the mouse passes over them.

To cycle to a face that is behind another one, move the cursor over the face that is in front of the one that you want to select and hold the cursor still. The Select Other tool appears, as shown in the following image. Select the left or right arrow to cycle through the faces until the correct face is highlighted, and then press the left mouse button or click the green rectangle in the middle of the Select Other tool. You can also access the Select Other tool by right-clicking the desired location in the graphics area and selecting Select Other from the menu.

Figure 3-46

You can specify the amount of time before the Select Other tool will appear automatically. Click Application Options on the Tools menu and click on the General tab, as shown in the following image. The "Select Other" delay (sec) can be specified in tenths of a second. If you do not want the Select Other tool to open automatically, specify OFF in the edit box. The default value is 1.0 seconds.

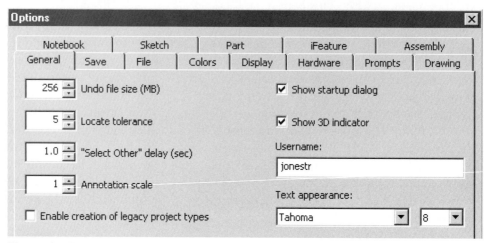

Figure 3-47

EXERCISE 3-4 Sketch Planes

In this exercise, you create a sketch plane on the angled face of a part then create a slot using the new sketch plane.

1. Open *ESS_E03_04.ipt*.

2. Click the Sketch tool on the command bar then click the angled face.

Figure 3-48

3. Click the Look At tool, then select the sketch in the Browser.

Create a sketch for a slot in the part, making sure the slot is centered vertically.

4. Click the 2 Point Rectangle tool. Draw a rectangle similar to the following image.

Chapter 3 Creating and Editing Sketched Features   131

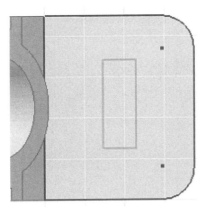

Figure 3-49

5. Click the 3 Point Arc tool. Draw tangent arcs on the two ends of the rectangle as shown in the following image.

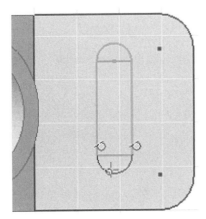

Figure 3-50

6. Click the Horizontal constraint tool.
7. Select the midpoint of the vertical rectangle edge and the midpoint of the projected edge of the part as shown in the following image.

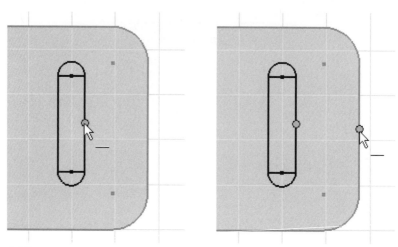

Figure 3-51

The rectangle is centered vertically using just geometric constraints.

8. Click the General Dimension tool. Place three dimensions as shown in the following image.

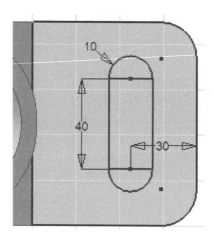

Figure 3-52

9. Press E to start the Extrude tool.
10. Press the F6 key to return to an isometric view.
11. Click inside the two arcs and inside the rectangle to select all 3 sketch loops.
12. Select the Cut operation.
13. In Extents, select All and ensure the direction is into the part as shown in the following image.

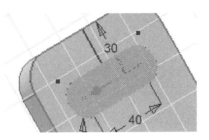

Figure 3-53

14. Click OK.

Figure 3-54

15. Close the file. Do not save changes.

PROJECTING PART EDGES

Building parts based partially on existing geometry is done often, and you will frequently need to reference faces, edges, or loops of features or parts that have been created.

While in a part file, you can project an edge, face, or loop onto a sketch. Projected geometry can maintain an associative link to the original geometry that is projected. If you project the face of a feature onto another sketch, for example, and the parent sketch is modified, the projected geometry will update to reflect the changes.

DIRECT MODEL EDGE REFERENCING

While you sketch, you can use direct model edge referencing to

- Automatically project edges of the part to the sketch plane as you sketch a curve
- Create dimensions and constraints to edges of the part that do not lie on the sketch plane
- Control the automatic projection of part edges to the sketch plane

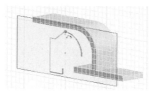

Figure 3-55

Creating Reference Geometry

The following are two ways to automatically project part edges to the sketch plane:

1. Rub the cursor on an edge of the part while sketching a curve.
2. Click an edge of the part while creating a dimension or constraint.

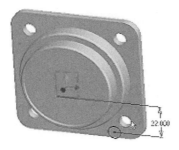

Figure 3-56

On the Sketch tab in the Application Options dialog box you can use

- Autoproject edges during curve creation. This controls the ability to rub and project edges while sketching a curve.
- Autoproject edges for sketch creation and edit. This controls the automatic projection of part edges coplanar with the sketch plane.

 NOTE Neither of these options disables the ability to reference part edges when creating dimensions and constraints.

PROJECT EDGES

In this section you learn about using the Project Geometry tool that can project selected edges, vertices, work features, curves, or the silhouette edges of another part in an assembly or other features in the same part to the active sketch. There are three project tools available on the Sketch Panel Bar, as shown in the following image. The three tools are Project Geometry, Project Cut Edges, and Project Flat Pattern.

Project Geometry Use to project geometry from a sketch or feature onto the active sketch.

Project Cut Edges Use to project edges that lie on the section plane of a part on the active sketch. The geometry is only projected if the uncut part would intersect the sketch plane.

Project Flat Pattern Use to project a selected face or faces of a sheet metal part flat pattern onto the active sheet metal part sketch plane (Chapter 9 will cover creating sheet metal parts).

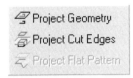

Figure 3-57

To project geometry, follow these steps:

1. Make a plane the active sketch.
2. Click the Project Geometry tool on the 2D Sketch Panel Bar.
3. Select the geometry to be projected onto the active sketch. Click a point in the middle of a face and all the edges of the face will be projected onto the active sketch plane. If you want to project all edges that are tangent to an edge (a loop), use the Select Other tool to cycle through until they all appear highlighted, as shown in the following image.
4. To exit the operation, press the ESC key or click another tool.

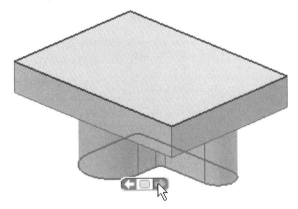

Figure 3-58

If you clicked a loop for the projection, the sketch is updated to reflect the modification when any part of the profile changes. If a face is projected, the internal islands that are defined on the face are also projected and will update accordingly.

For example, if the face of the following image is projected, the outer loop of the face and all of the circles that define the hole pattern on the face will be projected and updated if modified.

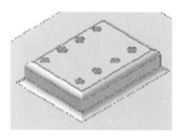

Figure 3-59

CHAPTER SUMMARY

To	Do This	Tool
Place a sketch plane	Click the Sketch tool on the command bar and click an existing face or work plane.	Sketch
Edit sketch features	Right-click the sketch icon in the Browser, and then select Edit Sketch from the menu.	
Define relationships to existing features	Click the Join, Cut, or Intersect operation.	
Create an extruded feature	Click the Extrude tool on the Panel Bar or on the Features toolbar.	
Create a revolved feature	Click the Revolve tool on the Panel Bar or on the Features toolbar.	
Display sketch geometry	Right-click the Sketch icon in the Browser, and then select Visibility from the menu.	
Project faces, loops, or edges from existing features	Click the Project Geometry tool on the Panel Bar or on the Sketch toolbar.	

Applying Your Skills

SKILL EXERCISE 3-1

In this exercise, you create a bracket from a number of extruded features.

1. Start a new part based on the metric *Standard(mm).ipt* template.
2. Create the sketch geometry for the base feature and slot.

3. Add geometric constraints and dimensions. Make sure the sketch is fully constrained.

4. Extrude the base feature.

5. Create the other two features to complete the part.

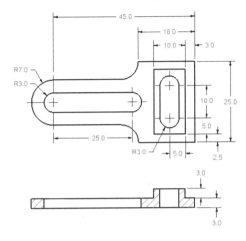

Figure 3-60

The completed part should resemble the following image.

Figure 3-61

SKILL EXERCISE 3-2

In this exercise, you create a connecting rod, and add draft to extrusions during feature creation.

1. Start a new part based on the metric *Standard(mm).ipt* template.

2. Create the sketch geometry for the outside of the connecting rod.

3. Add geometric constraints and dimensions to fully constrain the sketch.

4. Extrude the base feature using the midplane option, adding a -3 degree taper.
5. Create a sketch for the two holes and extrude (cut) them.

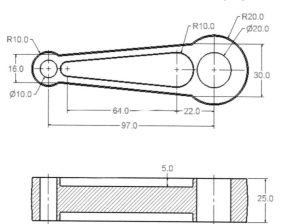

Figure 3-62

6. Create separate sketches for each pocket then constrain the sketch geometry and add dimensions. The sides of the pocket are parallel to the sides of the connecting rod.
7. Extrude (cut) each pocket using a draft of -3 degrees

The completed part should resemble the following image.

Figure 3-63

SKILL EXERCISE 3-3

In this exercise, you create a pulley using a revolved feature.

1. Start a new part based on the metric *Standard(mm).ipt* template.
2. Create the sketch geometry for the cross-section of the pulley.

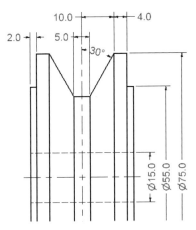

Figure 3-64

 Tip To create a centerline, draw a line, select it, and then select the Centerline tool on the Standard toolbar.

3. Apply appropriate geometric constraints.
4. Add dimensions.

 Tip To create a linear diameter dimension, select the centerline then the line to dimension to. When you select the centerline, do not select the endpoint of the centerline.

5. Revolve the sketch.
6. Close the file. Do not save changes.

The completed part should resemble the following image.

Figure 3-65

CHECKING YOUR SKILLS

Use these questions to test your knowledge of the material covered in this chapter.

1. What is a base feature?

   _____

   _____

2. **True__ False__** When creating a feature with the Extrude or Revolve tool, you can drag the sketch to define the distance or angle.

3. Which objects can be used as an axis of revolution?

   _____

   _____

4. Explain how to create a diametric dimension on a sketch.

   _____

   _____

5. Name two ways to edit an existing feature.

   _____

6. True___ False___   Once a sketch becomes a base feature, you cannot delete or add constraints, dimensions, or objects to the sketch.

7. Name three operation types used to create sketched features.

   _____

   _____

8. True___ False___   A cut operation cannot be performed before a base feature is created.

9. True___ False___   Once a sketched feature exists, its termination cannot be changed.

10. True___ False___   Geometry that is projected from one feature to a sketch that defines another feature will update automatically based on changes to the original projected geometry.

CHAPTER 4

Creating Placed Features

In Chapter 3 you learned how to create and edit base and sketched features. In this chapter you will learn how to create *placed* features. Placed features are features that are predefined except for specific values and only need to be located. You can edit placed features in the Browser like sketched features. When you edit a placed feature, either the dialog box that you used to create it will open or feature values will appear on the part.

When creating a part, it is usually better to use placed features instead of sketched features wherever possible. To make a through hole as a sketched feature, for example, you can draw a circle profile, dimension it, and then extrude it with the cut operation, using the All extension. You can also create a hole as a placed feature—you can select the type of hole, size it, and then place it using a dialog box. When drawing views are generated, the type and size of the hole are easy to annotate, and they automatically update if the hole type or values change.

CHAPTER OBJECTIVES

After completing this chapter, you will be able to

- Create fillets
- Create chamfers
- Create holes
- Create internal and external threads
- Shell a part
- Add face draft to a part
- Create work axes
- Create work points
- Create work planes
- Pattern features

FILLETS

Fillet features consist of fillets and rounds. Fillets add material to interior edges to create a smooth transition from one face to another. Rounds remove material from exterior edges. The following image shows a part without fillets on the left and a part with fillets on the right.

Figure 4-1

When creating fillets in 3D, you select the edge that needs to be filleted and the fillet is created between the two faces that share the edge. This is different from placing a fillet in 2D. In the 2D environment, you click two objects and a fillet is created between them. When creating a part, it is good practice to create fillets and chamfers as some of the last features in the part. Fillets add complexity to the part, which in turn add to the size of the file. They also remove edges that you may need to place other features.

To create a fillet feature, click the Fillet tool from the Part Features Panel Bar, as shown in the following image, or press the hot key SHIFT+F.

Figure 4-2

After you click the tool, the Fillet dialog box appears as shown in the following image.

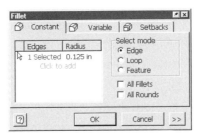

Figure 4-3

The Fillet dialog box has three tabs: Constant, Variable, and Setbacks. Each tab creates a different type (or style) of fillet. The options for each of the tabs are described in the following sections.

Before we look at the tabs, let's look at the methodology that you will use to create fillets. You can either click an edge or edges to fillet or select the type of fillet to create. If you click an edge before you issue the Fillet feature tool, it is placed in the first selection set. Selection sets contain the edges that will be filleted when the OK button is clicked to create the feature.

Each fillet feature can contain multiple selection sets, each having its own unique fillet value. There is no limit to the number of selection sets that can exist in a single instance of the feature. An edge, however, can only exist in one selection set. All of the selection sets included in an individual fillet command appear as a single fillet feature in the Browser. Click the type of fillet that you want to create, and then to add edges to the first selection set, select the edges that you want to have the same radius. To create another selection set, select Click to add and then click the edges that will be part of the next selection set. To remove an edge that you have selected, click the selection set that includes the edge, and the edges will be highlighted. Hold down the CTRL key and click the edge(s) to be removed from the selection set. Enter the desired values for the fillet. As changes are made in the dialog box, a representation of the fillet is previewed in the graphics area. When the fillet type and value are correct, click the OK button to create the fillet.

To edit a fillet's type and radius, follow these steps:

1. Issue the Edit Feature tool by right-clicking on the fillet's name in the Browser and selecting Edit Feature from the menu. The Fillet dialog box appears with all of the settings that you used to create the fillet.
2. Change the fillet settings as needed.
3. To edit only the dimensional value of a fillet, you can double-click on the fillet's name or icon in the Browser or change the select priority to Feature Priority and double-click on the fillet that is on the part.
4. The dimension(s) appear on the part. Double-click the dimension to change it and enter new values in the Edit Dimension dialog box. Then click the check mark in the dialog box or click Enter.
5. Click the Update tool to update the part.

CONSTANT TAB

With the options on the Constant tab, shown in the previous image, you can create fillets that have the same radius from beginning to end. There is no limit to the number of part edges that you can fillet with a constant fillet. You can select the edges as a single set or as multiple sets—each set can have its own radius value. The order in which you select the edges of a selection set is not important. You need to select the edges that are to be filleted individually, though, and the use of the window or crossing selection method is not allowed. If you change the value of a selection set, all of the fillets in that group will change. To remove an edge from a group, choose the group in the select area and then hold down the CTRL key and click the edge to be removed. Descriptions of the options that are available on the Constant tab follow.

Select Edge and Radius

By default, after issuing the Fillet tool, you can click edges, and they appear in the first selection set. To create another selection set, select Click to add and then click the edges that will be part of the next selection set. After clicking an edge, a preview image of the fillet appears on the edge that reflects the current values, as shown in the following image.

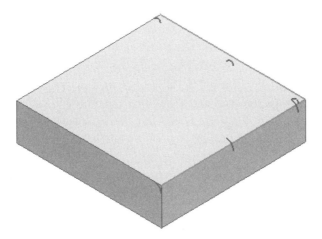

Figure 4-4

Select Mode

Edge Click the Edge mode to select individual edges to fillet. By default, any edge that is tangent to the clicked edges is also selected. If you do not want to have tangent edges automatically selected, uncheck the Automatic Edge Chain option in the More (>>) section.

Loop Click the Loop mode to have all of the edges filleted that form a closed loop with the selected edge.

Feature Click the Feature mode to select all of the edges of a selected feature.

All Fillets Click the All Fillets option to select all concave edges of a part that you have not filleted already (see the following image). The All Fillets option adds material to the part and requires a separate edge selection set from All Rounds.

Figure 4-5

All Rounds   Click the All Rounds option to select all convex edges of a part that you have not filleted already (see the following image). The All Rounds option removes material from the part and requires a separate edge selection set from All Fillets.

Figure 4-6

More (>>) Options

Click the More (>>) button to access other options, as shown in the following image.

Roll along sharp edges   Click this option to adjust the specified radius when necessary to preserve the edges of adjacent faces.

Rolling ball where possible   Click this option to create a fillet around a corner that looks like a ball has been rolled along the edges that define the corners, as shown on the left side in the following image. When the rolling ball solution is possible but you have not selected it, a blended solution is used, as shown on the right side in the following image.

Figure 4-7

Automatic Edge Chain Click this option to select tangent edges automatically when you click an edge.

Preserve All Features Click to check all features that intersect with the fillet and to calculate their intersections during the fillet operation. If the option's check box is clear, only the edges that are part of the fillet operation are calculated during the operation.

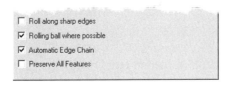

Figure 4-8

VARIABLE TAB

With the options under the Variable tab, shown in the following image, you can create a fillet that has a different start and end radius.

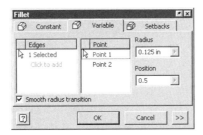

Figure 4-9

Edges By default, after clicking the fillet tool, you can click an edge and it appears in the first selection set. For a variable fillet, only one edge can exist per selection set. After clicking an edge, a preview image of the fillet appears on the edge that reflects the current values. To add another point, click on the edge and it appears in the Point column and a preview of the points location and radius appear on the display.

Point To identify where a point is on the selected edge, select the point text and a filled cyan circle appears at the point to highlight it.

Radius After clicking a point's name, enter a radius.

Position If you add another point, enter a value between 0.000001 and 1. This number represents a percentage of the distance between the start and end points relative to the start point.

Smooth Radius Transition Click this option to blend the fillet from the start to the end radius as a smooth transition, similar to a cubic spline, as shown on the left side in the following image. If this option's box is clear, the fillet blends from the start to the end radius in a straight line, as shown on the right side in the following image.

Figure 4-10

SETBACKS TAB

You can specify the distance at which a fillet starts its transition from a vertex with the options on the Setbacks tab, as shown in the following image.

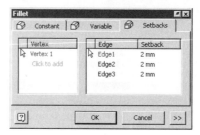

Figure 4-11

Using these options, you can model special fillet applications where three or more edges converge, as shown in the the following image. You can choose a different radius for each converging edge if needed. You can only use setbacks where three or more filleted edges form a vertex.

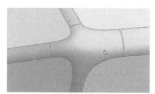

Figure 4-12

TIP If you get an error when creating or editing a fillet, try to create it with a smaller radius. If you still get an error after trying to use a smaller radius, you can try to create the fillets in a different sequence or create multiple fillets in the same operation.

CHAMFERS

Chamfers are similar to fillets except that their edges are beveled rather than rounded. When you create a chamfer on an interior edge, material is added to your model. When you create a chamfer on an exterior edge, material is cut away from your model (see the following image).

Figure 4-13

To create a chamfer feature, follow the same steps as you did to create the fillet features. Click the common edge and the chamfer is created between the two faces sharing the edge. To create a chamfer feature, click on the Chamfer tool on the Part Features Panel Bar, as shown in the following image, or press the hot key SHIFT+K.

Figure 4-14

After you click the tool, the Chamfer dialog box appears, as shown in the following image. As with fillet features, you can select multiple edges to be included in a single chamfer feature. From the dialog box, click a method, click the edge or edges to chamfer, enter a distance and/or angle, and then click OK to create the chamfer.

Figure 4-15

To edit the type of chamfer feature or distances, use one of the following methods:

- Double-click the feature's name or icon in the Browser to edit a distance.
- Right-click the chamfer's name in the Browser and select Edit Feature from the menu. The Chamfer dialog box appears with all the settings that you used to create the feature. Adjust the settings as desired.
- Change the select priority to Feature Priority and then double-click the chamfer on the part. The chamfer dimensions appear for editing.

METHOD

Distance   Click the Distance option to create a 45° chamfer on the selected edge. You determine the size of the chamfer by typing a distance in the dialog box. The value is offset from the common edge of the two adjacent faces. You can select a single edge, multiple edges, or a chain of edges. A preview image of the chamfer appears on the part. If you select the wrong edge, hold down the CTRL key and select the edge to remove.

Distance and Angle   Click the Distance and Angle option to create a chamfer offset from a selected edge on a specified face, at the defined angle. In the dialog box, enter an angle and distance for the chamfer, then click the face to which the angle is applied and specify an edge to be chamfered. You can select one edge or multiple edges. The edges must lie on the selected face. A preview image of the chamfer appears on the part. If you selected the wrong face or edge, click on the Edge or Face button and choose a new face or edge. The following image illustrates the use of the Distance and Angle option (on the right side).

Two Distances   Click the Two Distances option to create a chamfer offset from two faces, each being the amount that you specify. Click an edge first and enter a value for Distance1 and Distance2. A preview image of the chamfer appears. To reverse the direction of the distances, click the Flip button. When the correct information about the chamfer is in the dialog box, click the OK button to create the chamfer. You can only use a single edge or chained edges with the Two Distance option. The following image illustrates the use of the Two Distances option.

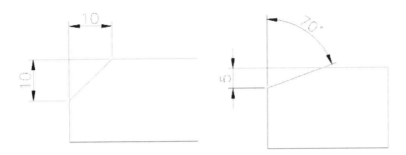

Figure 4-16 Distance for the image on the left. Distance and angle for the image on the right.

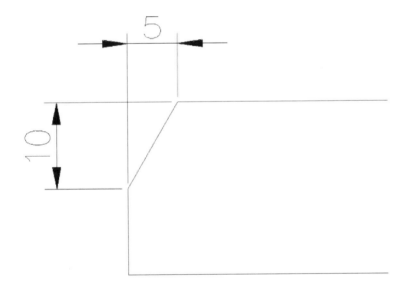

Figure 4-17 Distance and angle for the image on the right. Two distances

EDGE AND FACE

Edges   Click an edge or edges to be chamfered.

Face   Click a face on which the chamfer will be based.

Flip   Click the button to reverse the direction of the distances for a Two Distance chamfer.

DISTANCE AND ANGLE

Distance Enter a distance to be used for the offset.

Angle Enter a value that will be used for the angle if creating the Distance and Angle chamfer type.

EDGE CHAIN AND SETBACK

Edge Chain Click this option, shown in the following image, to include tangent edges in the selection set automatically.

Setback When the Distance method is used and three chamfers meet at a vertex, click this option to choose to have the intersection of the three chamfers form a flat edge (left button) or to have the intersection meet at a point as though the edges were milled (right button), as shown in the following image.

Preserve All Features Click this option, shown in the following image, to check all features that intersect with the chamfer and to calculate their intersections during the chamfer operation. If the option's check box is clear, only the edges that are part of the fillet operation are calculated during the operation.

Figure 4-18

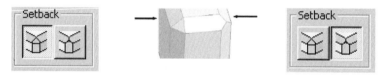

Figure 4-19

EXERCISE 4-1 Creating Fillets and Chamfers

In this exercise, you create constant radius fillets, variable radius fillets, and chamfers.

1. Open *ESS_E04_01.ipt*.
2. Click the Fillet tool.
3. Click the edge of the slot as shown in the following image.

Chapter 4 Creating Placed Features   153

Figure 4-20

4. In the Fillet dialog box, click on the first entry in the Radius column then type **2**.

5. Click OK to create the fillet.

6. Click the fillet tool. In the Fillet dialog box, click the Variable tab.

7. Click the edge of the model as shown in the following image.

Figure 4-21

8. In the Fillet dialog box, in the Point list, click on Start then type **12** in the Radius field.

9. In the Point list, click on End then type **12** in the Radius field.

10. In the Point list, click on End again.

11. In the model, move the cursor over the endpoint of the first line segment then click to specify a point.

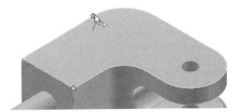

Figure 4-22

1. In the Fillet dialog box, type **2** in the Radius field.

2. In the Point list, click on Point1.

3. In the model, move the cursor over the endpoint of the last line segment then click to specify a point.

Figure 4-23

4. In the Fillet dialog box, click OK to create the fillet.

Figure 4-24

5. Repeat the procedure to create a variable radius fillet on the opposite side of the part.
6. Click the Fillet tool.
7. Click the bottom edge as shown in the following image.

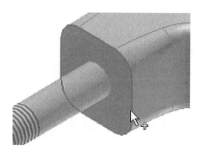

Figure 4-25

8. In the fillet dialog box, click on the first entry in the Radius column then type **2**.
9. Click OK to create the fillet.
10. Click the Chamfer tool.
11. In the Chamfer dialog box, click on Distance and Angle.
12. Click the face on the end of the threaded stud.

Figure 4-26

13. Click the edge of the same face.

Figure 4-27

14. In the Chamfer dialog box, enter **3** in the Distance field and **30** in the Angle field.

15. Click OK to create the chamfer.

Figure 4-28

16. Close the file. Do not save changes. End of exercise.

HOLES

The Hole tool lets you create drilled, counterbored, countersunk, clearance, and tapped holes, as shown in the following image. You can place holes using sketch geometry or existing planes, points, or edges of a part. You can also specify the type of drill point and thread parameters for a hole.

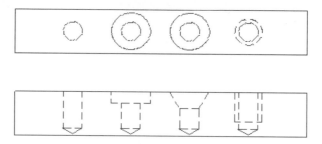

Figure 4-29

To create a hole feature, follow these steps:

1. Create a part that contains faces or planes where you want to place a hole.
2. Click the Hole tool from the Part Features Panel Bar (shown in the following image), or press the hot key H.

Figure 4-30

3. The Holes dialog box appears as shown in the following image. There are four placement options available: From Sketch, Linear, Concentric, and On Point. The Placement options are covered in the next section. After you have chosen the hole placement options, select the desired hole style options from the Holes dialog box. As you change the options, the preview image of the hole(s) is updated. When you are done making changes, click the OK button to create the hole(s).

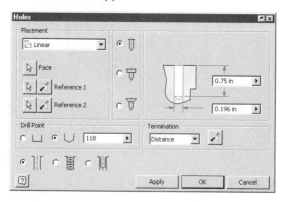

Figure 4-31

EDITING HOLE FEATURES

To edit the type of hole feature or distances, use one of the following methods:

- Double-click the feature's name or icon in the Browser to edit the hole feature dimensions.
- Right-click the hole's name or icon in the Browser and select Edit Feature from the menu to display the Holes dialog box with the dimensions and option you used to create the hole feature.
- Change the select priority to Feature Priority and then double-click the hole feature on the part in the graphics area to edit the hole feature dimensions.

HOLES DIALOG BOX

In the Holes dialog box, you establish the placement method, type of hole, its termination, and additional options such as type of drill point, angle, and tapped properties.

PLACEMENT

Select the appropriate placement method. If you select From Sketch, a sketch containing hole centers must exist on the part. Hole centers are described in the next section. The Linear, Concentric, and On Point options do not require an unconsumed or shared sketch to exist in the model and are based on previously created features. Depending on the placement option you select, the input parameters will change as shown in the following image and described below.

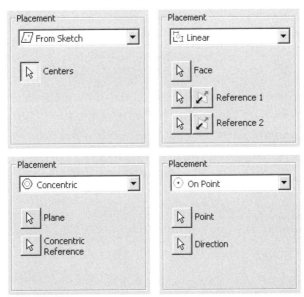

Figure 4-32

From Sketch

Select the From Sketch option to create holes that are based on a location defined within an unconsumed or shared sketch. You can base the center of the hole on a point/hole center or endpoints of sketched geometry. You can also use endpoints of projected geometry that reside in the unconsumed sketch.

Centers
Select the hole center point or sketch points where you want to create a hole.

Linear
Select the Linear option to place the hole relative to two selected face edges.

Face   Select the face on the part where the hole will be created.

Reference 1   Select a face edge as a positional reference for the center of the hole. When you select the edge, a dimension appears that can be edited to constrain the center of the hole dimensionally.

Reference 2   Select a face edge as a positional reference for the center of the hole. When you select the edge, a dimension appears that can be edited to constrain the center of the hole dimensionally.

Flip Side   Click this button to position the hole on the opposite side of the selected edge.

Concentric
Select the Concentric option to place the hole on a planar face and concentric to a circular edge or cylindrical face.

Plane   Select a planar face or work plane where you want to create the hole.

Concentric Reference   Select a circular model edge or face to constrain the center of the hole to be concentric with the selected entity.

On Point
Select the On Point option to place the center of the hole on a work point. The work point must exist on the model prior to selecting this option.

Point   Select a work point to position the center of the hole.

Direction   Select a plane, face, work axis, or model edge to specify the direction of the hole. When selecting a plane or face, the hole direction will be normal to the face or plane.

HOLE OPTION
Click the type of hole that you want to create: drilled, counterbore, or countersink, and enter the appropriate dimensions as shown in the following image.

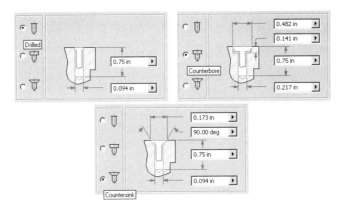

Figure 4-33

Termination   Select how the hole will terminate.

> *Distance*–Specify a distance for the depth of the hole.
> *Through All*–Choose to extend the hole through the entire part in one direction.
> *To*–Select a plane at which the hole will terminate.
> *Flip*–Reverse the direction in which the hole will travel.

Drill Point   Select either a flat or angle drill point. If you select an angle drill point, you can specify the angle of the drill point.

Dimensions   To change the diameter, depth, countersink, or counterbore diameter, or countersink angle or counterbore depth of the hole, click the dimension in the dialog box and enter a desired value.

HOLE TYPE

Click the type of hole you want to create. There are three options: Simple Hole, Tapped Hole, and Clearance Hole.

Simple Hole

Click the Simple Hole option to create a (drilled) hole feature with no thread features or properties.

Tapped Hole

Click the Tapped Hole option if the hole is threaded. Thread information appears in the dialog box area (shown in the following image) so you can specify the thread properties.

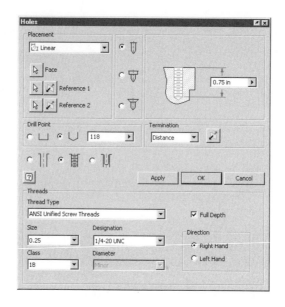

Figure 4-34

Thread Type   From the drop list, select the standard on which the hole is based: ANSI Unified Screw Threads, ANSI Metric M Profile, ISO Metric Profile, ISO Metric Trapezoidal Threads, ISO Pipe Threads, JIS Pipe Threads, DIN Pipe Threads, or BSI Pipe Threads.

Full Depth   Click this option if the threads extend the full depth of the hole.

Size   Select the nominal size for the thread.

Designation   Select the pitch size for the thread.

Class   Select the class of thread.

Diameter   The drop list displays the current setting for the Tapped Hole Diameter. You can change the nominal diameter type for the hole by clicking on the Tools > Document Settings > Modeling tab, and selecting one of the four available types. The available types are: Minor, Pitch, Major, or Tap Drill.

Right Hand   Click this option to create threads that wind clockwise and recede.

Left Hand   Click this option to create threads that wind counterclockwise and recede.

Clearance Hole

Click the Clearance Hole option to base the hole size on a selected fastener. No fastener is created with the hole, but the hole is sized so that a fastener can be placed in the hole when in an assembly. Clearance information appears in the dialog box area (shown in the following image) so you can specify the clearance properties

Standard   Select the fastener standard on which to base the hole.

Fastener Type Select the fastener type on which to base the hole. The options will change based on the selected standard.

Size Select the size for the fastener.

Fit Select the type of clearance fit to use. The available types are: Close, Normal, and Loose.

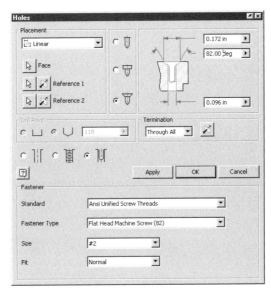

Figure 4-35

HOLE CENTERS

Hole centers are sketched entities that can be used to locate hole features. To create a hole center, follow these steps:

1. Make a sketch active.

2. Click the Point, Hole Center tool on the 2D Sketch Panel Bar, as shown in the following image.

3. Click a point to locate the hole center where you want to place the hole and constrain it as desired.

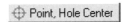

Figure 4-36

You can also switch between a hole center and a sketch point by using the Hole Center tool on the standard toolbar, as shown in the following image, when the sketch environment is active. When you press the button, you create a hole center. If you use a hole center style when you activate the From Sketch option of the Hole tool, all hole centers that reside in the sketch are selected as centers for the hole feature automatically. This can expedite the process of creating multiple holes in a single hole feature.

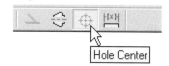

Figure 4-37

EXERCISE 4-2 Creating Holes

In this exercise, you add drilled, tapped, and counterbored holes to a cylinder head.

1. Open *ESS_E04_02.ipt*.
2. Click the Sketch tool on the command bar, and then click the rectangular face.

The edges of the face and arc centers are projected onto the new sketch, allowing you to position the hole features.

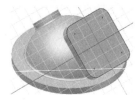

Figure 4-38

3. Right-click the graphics window and then click Finish Sketch.
4. Click the Hole tool in the panel bar.
5. Click the four arc centers.
6. Select To from the Termination drop-down list.

Notice that the Select surface to end feature creation button next to the Termination drop-down list is already depressed. You are now ready to select the termination face.

7. Pause the cursor over the side face of the flange to highlight the underside face as shown in the following figure. When the underside face is highlighted, click to select it.

 NOTE If you have difficulty selecting the correct face, use Select Other to cycle through the selections.

Chapter 4 Creating Placed Features   163

Figure 4-39

8. In the Holes dialog box, enter **6** for the hole diameter dimension.

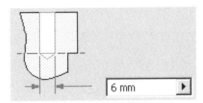

Figure 4-40

9. Click OK. The holes are created in the model and a hole feature icon is added to the Browser. Notice that one feature defines all four holes.

Figure 4-41

10. Click the Rotate tool and rotate the model to view the location for the tapped hole.

TIP You can also press and hold the F4 function key to activate the Rotate tool. When you release the F4 key, the Rotate tool ends.

11. Click the Hole tool in the panel bar.
12. Select Concentric from the Placement list.
13. Select the plane where the hole will be created.

Figure 4-42

14. Click the circular edge of the face.

15. In the Holes dialog box, select Through All from the Termination drop-down list.

16. Set the hole type to Tapped Hole.

17. Select ANSI Metric M Profile from the Thread Type drop-down list.

18. Select 16 from the Size drop-down list, and then select M16x1.5 from the Designation drop-down list.

19. Click OK to create the hole.

Figure 4-43

Notice that the threads are displayed on the part.

20. Click the Sketch tool on the command bar, and then click the top face of the large circular flange.

21. Click the Look At tool then click the same face.

Figure 4-44

22. Click the Center Point Circle tool in the panel bar.
23. Place a circle with the center point coincident with the center of the circular flange.

Figure 4-45

24. Click the General Dimension tool in the panel bar.
25. Dimension the circle with a diameter of **88 mm**.
26. Click the Point, Hole Center tool in the panel bar.
27. Place a hole center point coincident with the circle. See the animation below.
28. Click the Vertical constraint tool.
29. Click the center point of the circular flange then click the hole center point to constrain the point to the quadrant of the circle.

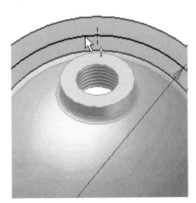

Figure 4-46

30. Press H to end the sketch and start the Hole tool.
31. The Hole dialog box displays and the point you placed is automatically selected as the hole center.
32. In the Hole dialog box:
    a. Click the Simple Hole type.

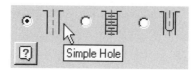

Figure 4-47

b. Select Counterbore.

Figure 4-48

c. Make sure Through All is selected in the Termination list.

33. In the Holes dialog box, enter **1.5** for the counterbore depth, **9** for the counterbore diameter, and **6** for the hole diameter.

34. Click OK to create the hole then rotate the part and examine the hole.

35. Close the file. Do not save changes. End of exercise.

THREADS

You use thread features to create both internal and external threads. The threads appear on the parts with a graphical representation, and when you create a drawing view, you can call out the thread per drafting standards. Since the threads are graphical representations, they do not physically exist on the part—if a model was to be cast directly from the part, no threads would exist on the finished model. You can add thread features to any cylindrical or conical face. If you are adding threads to a hole, you can do so using either the Hole or Thread tool. If you add it using the Thread tool, it is a separate feature in the Browser. For this reason, it is recommended that you use the Hole tool with the tapped option when creating a threaded hole.

To create external threads, follow these steps:

1. Create a cylinder and dimension it to the size that will represent the major diameter of the thread (this value must be between the maximum and minimum major diameter).

2. To create the thread feature, issue the Thread tool from the Part Features Panel Bar, as shown in the following image.

3. Enter the data into the Thread Feature dialog box as needed.

Figure 4-49

There are two tabs in the Thread Feature dialog box—Location and Specification. The thread data comes from an Excel spreadsheet named *Thread.xls*. By default, the spreadsheet is in the *Program Files\ Autodesk\ Inventor 10\ Design Data* directory. You can modify this spreadsheet to match your company's standards. The thread data is used when the thread appears on the part and when the thread is annotated in a drawing view. The thread data is not associative to the threads on existing parts. When you make changes to the spreadsheet, the new values are only used when new threads are created. If you make a dimensional change to the diameter of the cylinders where you placed a thread, a warning dialog box appears when you update the part. The dialog box notifies you that you used an inappropriate thread size. Accept the warning message and then modify the thread feature to a size that fits the corresponding diameter.

LOCATION TAB

The following image shows the options available on the Location tab, and the following sections describe them.

Face   Click this button, and then select a cylindrical or conical face on which or in which to place a thread.

Display in Model   Click this option to display a visual representation of the thread on the part. If the option's box is clear, the thread only appears when you create a drawing view.

Thread Length

Full Length   Click this option so the thread continues the entire length of the selected face.

Flip   Click this button to reverse the direction of the thread.

Length   When you do not click Full Length (the box is clear), enter a value for the length of the thread.

Offset   When you do not click Full Length enter a value to which the thread will be offset. The offset distance is from the closest plane relative to the selected cylindrical or conical face.

Figure 4-50

SPECIFICATION TAB

The following image shows the options available on the Specification tab, and the following sections describe them.

Thread Type   Select the thread type—the types are defined by the tabs in the *Thread.xls* file.

Size   Select the nominal size for the thread.

Designation   Select the pitch for the thread.

Class   Select the class of thread.

Right hand or Left hand   Select the direction for the thread, thread size, and its representation. The graphical representation of the thread will change based on the selected direction.

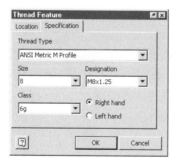

Figure 4-51

EXERCISE 4-3 Creating Threads

In this exercise, you use the Thread tool to create custom thread features on a cylinder.

   1. Open *ESS_E04_03.ipt*.

   2. Click the Thread tool.

   3. Select the inside face of the cylinder, then on the Specification tab set the Nominal Size to 25 and the Designation to M25 x 1 as shown in the following image.

Chapter 4 Creating Placed Features   169

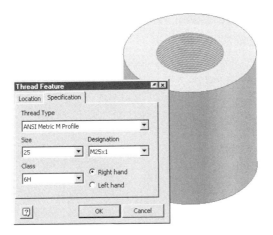

Figure 4-52

4. Click OK to create the thread.

5. Click the Thread tool.

6. Select the outside face of the cylinder, uncheck Full Length, and set the length to 20 mm and the offset to 15 mm as shown in the following image.

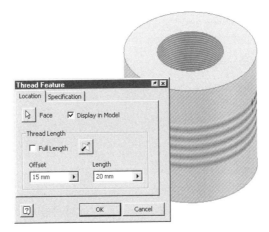

Figure 4-53

7. On the Specification tab set the Nominal Size to 50 and the Designation to M50x4.

8. Click OK to create the thread.

9. Edit the last Thread feature that you created and change the thread length to Full Length.

10. Click OK to update the thread feature.

11. Edit the first extrusion and change the diameter of the outside circle to 75 mm, then update the part.

12. A warning dialog box is displayed alerting you to a design problem as shown in the following image.

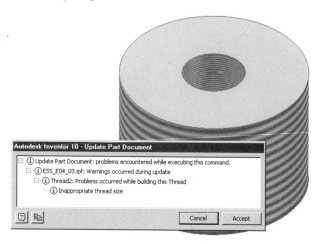

Figure 4-54

13. Accept the warning and then edit the Thread2 feature to change its Nominal size to 75 and the Designation to M75x4.

14. Click OK to complete the edit.

15. Close the file without saving changes. End of exercise.

SHELLING

As you design parts, you may need to create a model that is made up of thin walls. The easiest way to create a thin-walled part is to create the main shape and then use the Shell tool to remove material.

The term *shell* refers to giving a thickness (wall thickness) to the outside shape of a part and removing the remaining material–like scooping out the inside of a part, leaving the walls a specified thickness (see the following image). You can offset the wall thickness in, out, or evenly in both directions. If the part you shell contains a void, such as a hole, the feature will have the thickness built around it.

A part may contain more than one shell feature, and individual faces of the part can have different thicknesses. If a wall has a different thickness than the shell thickness, it is referred to as a unique face thickness. If a face that you select for a unique face thickness has faces that are tangent to it, those faces will also have the same thickness. You can remove faces from being shelled, and these faces remain open. If no face is removed, the part is hollow on the inside.

Figure 4-55

To create a shell feature, follow these steps:

1. Create a part that will be shelled.

2. Issue the Shell tool from the Part Features Panel Bar, as shown in the following image.

Figure 4-56

3. The Shell dialog box appears, as shown in the following image. Enter the data as needed.

4. After filling in the information in the dialog box, click OK and the part is shelled.

To edit a shell feature, use one of the following methods:

- Double-click the feature's name or icon in the Browser to edit wall thickness dimensions.

- Right-click the name of the shell feature in the Browser and select Edit Feature from the menu. The Shell dialog box appears with all of the settings you used to create the feature. Change the settings as needed.

- Change the select priority to Feature Priority and double-click the shell feature on the part to edit wall thickness dimensions.

Figure 4-57

The following sections explain the options that are available for the Shell tool.

REMOVE FACES

Click the Remove Faces button and then click the face or faces to be left open. To deselect a face, click the Remove Faces button and hold down the CTRL key while you click the face.

THICKNESS

Enter a value or select a previously used value from the droplist to be used for the shell thickness.

DIRECTION

Inside Click this button to offset the wall thickness into the part by the given value.

Outside Click this button to offset the wall thickness out of the part by the given value.

Both Click this button to offset the wall thickness evenly into and out of the part by the given value.

UNIQUE FACE THICKNESS

Unique face thickness is available by clicking the More (>>) button that is located on the lower-right corner of the dialog box, as shown in the following image.

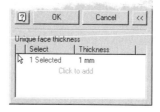

Figure 4-58

To give a specific face a thickness, select Click to add and then click the face and enter a value. A part may contain multiple faces that have a unique thickness (see the following image).

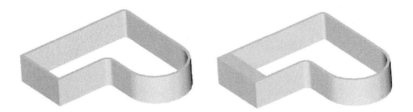

Figure 4-59

EXERCISE 4-4 Shelling a Part

In this exercise, you use the Shell tool to create a shell on a part.

1. Open *ESS_E04_04.ipt*.
2. Click the Shell tool.
3. Pick the Remove Faces button and select the left front face of the part.
4. Select a Thickness of 3 mm as shown in the following image.

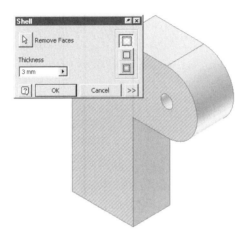

Figure 4-60

5. Change to an isometric view.
6. Edit the Shell feature that you just created.
7. Select the More button (>>), then click in the "Click to add" area and select the bottom face.
8. Enter a value of **10 mm** as shown in the following image.

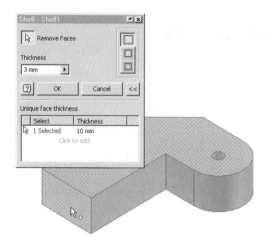

Figure 4-61

9. Click OK to create the shell feature.
10. Click the Shell tool.
11. Choose the Remove Face button and then select the same face (the one that you applied a unique thickness to) and type in a Thickness of 3 mm as shown in the following image.

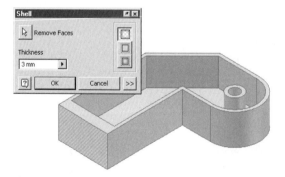

Figure 4-62

12. Click OK to create the shell.

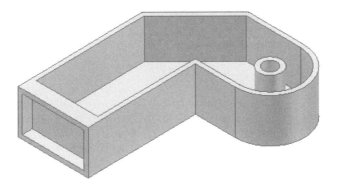

Figure 4-63

13. Close the file without saving changes. End of exercise.

FACE DRAFT

Face draft is a feature that applies an angle to a face. You can apply face draft to any specified internal or external face, including shelled parts. When you apply face draft, any tangent face also has the face draft applied to it.

To create a face draft feature, follow these steps:

1. Issue the Face Draft tool from the Part Features Panel Bar, as shown in the following image.

Figure 4-64

2. The Face Draft dialog box appears, as shown in the following image. The following section describes the options in the Face Draft dialog box.

3. Click the Pull Direction that shows how the mold will be pulled from the part.

4. Then click a face or faces to which to apply the face draft. If you select an incorrect face, you can deselect it by holding down the CTRL key and clicking the face.

Figure 4-65

DRAFT TYPE

Click the type of draft that you want to create. There are two types of drafts available.

Fixed Edge

Click the Fixed Edge button to allow the draft to be created from an edge or series of contiguous edges.

Fixed Plane

Click the Fixed Plane button to allow the draft to be created from a selected plane. The plane you select is used to specify both the pull direction and the fixed plane.

PULL DIRECTION AND FIXED PLANE

Click the direction that the mold will be pulled from the part. When you select the Fixed Plane draft type, click a planar face or work plane from which to draft the faces.

Direction Click the arrow and move the cursor around the part. A dashed line appears–this line points 90° from the highlighted face and shows the direction that the mold will be pulled. When the correct direction appears on the screen, left-click.

Flip Click the Flip button to reverse the pull direction 180°.

FACES

Click the face or faces to which to apply the face draft. As you move the cursor over the face, a symbol with an arrow appears that shows how the draft will be applied. As you move the cursor to different edges and faces, the arrow direction preview shows how the draft will be applied to that specific face.

DRAFT ANGLE

Enter a value for the draft angle, or select a previously used draft angle from the drop list.

EXERCISE 4-5 Creating Face Drafts

In this exercise, you apply draft angles to the faces of a part.

1. Start a new part file based on the *Standard (mm).ipt* template file.
2. Sketch and dimension a rectangle that measure 50 mm horizontally by 75 mm vertically.
3. Change to an isometric view.
4. Extrude the rectangle 25 mm.
5. Shell the part 3 mm and remove the top face.
6. Pick the Face Draft tool.
7. Select near the middle of the removed face, and a dashed line will appear as shown in the following image. When it is displayed, select the left mouse button and an arrow that points up will be shown.

 NOTE If the arrow is pointing in the wrong direction, pick the Flip button in the Face Draft dialog box.

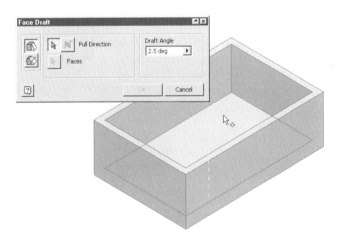

Figure 4-66

8. Change the draft angle to 3.
9. In the Face Draft dialog box, select the Faces button and pick near the inside top right edge of the box as shown in the following image. This is the edge that will be fixed.

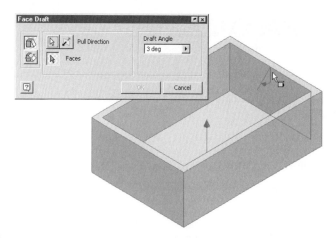

Figure 4-67

10. While still in the same operation, select near the inside back bottom edge of the box as shown in the following image. This is the edge that will be fixed.

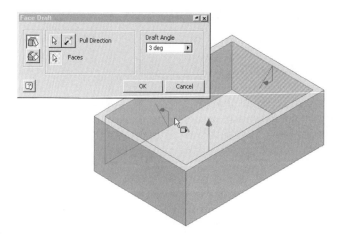

Figure 4-68

11. Then click the OK button to complete the Face Draft feature. Your part should resemble the following image, shown in wireframe display.

Chapter 4 Creating Placed Features   179

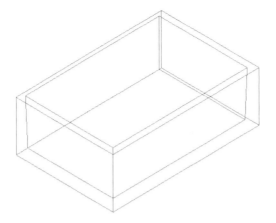

Figure 4-69

12. Delete the Face Draft feature that you just created.

13. Place a 5 mm fillet in each of the four inside vertical edges of the box.

14. Click the Face Draft tool.

15. Click near the middle of the removed face to select it and define the pull direction.

16. Change the draft angle to 5.

17. In the Face Draft dialog box, select the Faces button and pick near the inside top back edge of the box as shown in the following image. Since all the faces are tangent, all the top edges that are inside the part will be highlighted. These edges will be fixed.

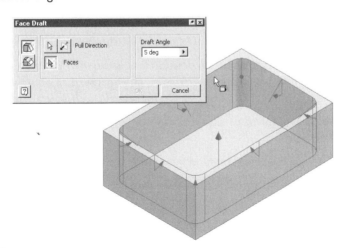

Figure 4-70

18. Click OK to complete the operation. The part should resemble the following image, shown in wireframe display.

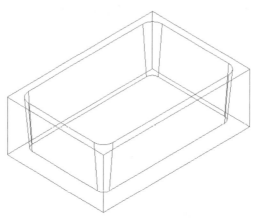

Figure 4-71

19. Close the file without saving changes. End of exercise.

WORK FEATURES

When you create a parametric part, you define how the features of the part relate to one another so that a change in one feature results in appropriate changes in all related features. Work features are special construction features that are attached parametrically to part geometry. You typically use work features to help you position and define new features in your model. There are three types of work features: work planes, work axes, and work points.

Use work features in the following situations:

- To position a sketch for new features when a part face is not available.
- To establish an intermediate position that is required to define other work features. You can create a work plane at an angle to an existing face, for example, and then create another work plane at an offset value from that plane.
- To establish a plane or edge from which you can place parametric dimensions and constraints.
- To provide an axis or point of rotation for revolved features and patterns.
- To provide an external feature termination plane off the part (i.e., a beveled extrusion edge) or an internal feature termination plane in cases where there are no existing surfaces.

CREATING A WORK AXIS

A work axis is a feature that acts like a construction line. It is infinite in length, and you can use it to help create work planes, work points, and subsequent part features. You can also use work axes as axes of rotation for polar arrays or to constrain parts in an assembly using assembly constraints. Their length always extends beyond the part—as the part changes size, the work axis also changes size. A work axis is tied parametrically to the part. As changes occur to the part, the work axis will maintain its relationship to the points, edge, or cylindrical face from which you created it. To create a work axis, use the Work Axis tool on the Part Features Panel Bar, as shown in the following image, or press the hot key / (forward slash).

Figure 4-72

Use one of the following methods to create a work axis:

- Click a cylindrical face to create a work axis along the axis of the face.
- Click two points on a part to create a work axis through both points.
- Click an edge on a part to create a work axis on the part edge.
- Click a work point or sketch point and a plane or face to create a work axis that is normal to the selected plane or face and passes through the point.
- Click two nonparallel planes or faces to create a work axis at their intersection.

CREATING WORK POINTS

A work point is a feature that you can create on the active part or in 3D space. You can create a work point any time a point is required. To create a work point, use the Work Point tool on the Part Features Panel Bar, as shown in the following image, or press the hot key (.) (period). You can also create a point when using the Work Plane or Work Axis tools—right-click and select Create Point when one of these work feature tools is active. If you issue the tool while you are creating a work feature, the tool will be exited after you create the work point. The created point will be indented as a child of the work axis or work plane in the Browser.

Figure 4-73

Use one of the following methods to create a work point:

- Click an endpoint or the midpoint of an edge and click on two intersecting edges or axes to create a work point at the intersection (or theoretical intersection) of the two.

- Click an edge and plane to create a work point at the intersection (or theoretical intersection) of the two.
- Click a spline that intersects a face or plane to create a work point at the intersection of the spline and face or plane.
- Click three nonparallel faces or planes to create a work point at their intersection (or theoretical intersection).

Grounded Work Points

You can create grounded work points that are positioned in 3D space. Grounded work points are not associated with the part or any other work features, including the original locating geometry. When you modify surrounding geometry, the grounded work point remains in the specified location. To create a grounded work point, use the Grounded Work Points tool on the Part Features Panel Bar, located by clicking the arrow next to the Work Points tool, as shown in the following image, or press the hot key ; semi-colon).

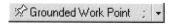

Figure 4-74

After clicking the Grounded Work Points tool, select a vertex, midpoint, sketch point, or a work point on the model. When you have selected the vertex or point, the 3D Move/Rotate dialog box and a triad appear, as shown in the following image. The initial orientation of the triad matches the principle axes of the part.

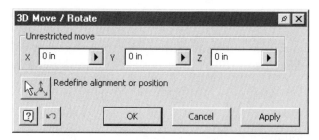

Figure 4-75

Enter values in the 3D Move/Rotate dialog box to position precisely the grounded work point relative to the selected point, or select areas of the triad to move the triad and locate the grounded work point in the desired direction, as shown in the following image and described in the following sections.

Figure 4-76

Arrowheads   Select an arrowhead to specify a position along a particular axis.

Legs   Select a leg to rotate about that axis.

Origin   Click to move the triad freely in 3D space.

Planes   Select a plane to move the triad in the two axes of that plane.

Once you position the triad properly, click Apply or OK in the 3D Move/Rotate dialog box to create the grounded work point. You can identify a grounded work point in the Browser by the thumbtack icon that is placed on the work point, as shown in the following image.

Figure 4-77

EXERCISE 4-6 Creating Work Axes

In this exercise, you create a work axis to position a circular pattern.

1. Open *ESS_E04_06.ipt*
2. Click the Sketch tool.
3. Select the face as shown.

Figure 4-78

    4. Click the Look At tool then select the new sketch .

    5. Click the Line tool.

    6. Create two line segments from the midpoint on the top line to the midpoint of the line on the right.

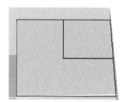

Figure 4-79

    7. Right-click in the graphics window then select Isometric View.

    8. Exit the Line tool.

    9. Right-click in the graphics window then select Finish Sketch.

    10. Click the Work Axis tool.

    11. Select the plane then select the point at the intersection of the two lines. The work axis is created through this point and normal to the plane.

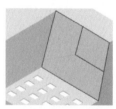

Figure 4-80

> **NOTE** To complete the model, a hole feature is created 20 mm from the work axis and a circular pattern of holes is created. The work axis is the center of the pattern. Circular patterns are covered in another exercise

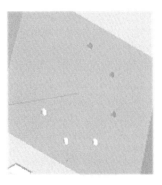

Figure 4-81

12. Close the file. Do not save changes. End of exercise.

CREATING WORK PLANES

Before introducing work planes, it is important that you understand when you need to create a work plane. You can use a work plane when you need to create a sketch and there is no planar face at the desired location, or if you want a feature to terminate at a plane and there is no existing face to select. If you want to apply an assembly constraint to a plane on a part and there is no existing part face, you will need to create a work plane. If a face exists in any of these scenarios, you should use it and not create a work plane. A new sketch can be created on a work plane.

A work plane is a feature that looks like a rectangular plane. It is tied parametrically to the part. Though extents of the plane will always appear slightly larger than the part, the plane is in fact infinite. If the part or related feature moves or resizes, the work plane will also move or resize. If a work plane is tangent to the outside face of a 1" diameter cylinder and the cylinder diameter changes to 2", for example, the work plane moves with the outside face of the cylinder. You can create as many work planes on a part as needed, and you can use any work plane to create a new sketch. A work plane is a feature and is modified like any other feature.

Before creating a work plane, ask yourself where this work plane needs to exist and what you know about its location. You might want a plane to be tangent to a given face and parallel to another plane, for example, or to go through the center of two arcs. Once you know what you want, select the appropriate options and create a work plane. There are times when you may need to create an intermediate (construction) work plane before creating the final work plane. You may need to create a work plane, for example, that is at 30\r¢ and tangent to a cylindrical face. You should first create a work plane that is at a 30\r¢ angle and located at the center of the cylinder, and then create a work plane parallel to the angled work plane that is also tangent to the cylinder.

To create a work plane, click the Work Plane tool on the Part Features Panel Bar, as shown in the following image, or press the hot key (]) (end-bracket). A work plane is created depending upon what, where, and how you select options. Work planes can also be associated to the default reference (origin) planes that exist in every part. These origin planes initially have their visibility turned off, but you can turn on the visibility of these planes by expanding the Origin folder in the Browser, right-clicking on a plane or planes, and selecting Visibility from the menu. You can use these default planes to create a new sketch or to create other work planes.

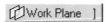

Figure 4-82

Use one of the following methods to create a work plane:

- Click three points.
- Click a plane and a point.
- Click a plane and an edge or axis.
- Click two edges, two axes, or an edge and axis.
- To create a work plane that is tangent to a face, click an axis, plane, or face and a cylindrical face, and the resulting work plane is created parallel or coincident with the selected axis, plane, or parallel to the face and tangent to the selected cylindrical face.
- To create an angled work plane, click a plane or face and an edge–a dialog box appears for you to enter the angle.
- To create an offset work plane from an existing face, click a plane and then drag the new work plane to a selected location. While dragging the work plane, an Offset dialog box appears that displays the offset distance. Click a point, enter a precise value for the offset distance, and click the check mark in the dialog box or press the ENTER key.
- To create a work plane at the midplane defined by the two selected planes or faces, click parallel planes or faces.

When you are creating a work plane, and more than one solution is possible, the Select Other tool appears. Click the forward or reverse arrows from the Select Other tool until you see the desired solution displayed. Click the check mark in the selection box. If you clicked a midpoint on an edge, the resulting work plane links to the midpoint. If the selected edge's length changes, the location of the work plane will adjust to the new midpoint.

NOTE The order in which points or planes are selected is irrelevant.

Use the Show Me animations located in the Visual Syllabus (shown in the following image) to view animations that display how to create certain types of work planes.

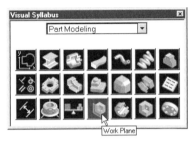

Figure 4-83

Types of Work Planes

The following is a list of work planes that you can use in the modeling process:

- Angled
- Edge and face normal
- Edge and tangent
- Offset
- Point and face normal
- Point and face parallel
- Sketch geometry
- Tangent and face parallel
- Tangent and edge or axis
- 3-point
- 2-edge or 2-axis
- Through line endpoint, perpendicular to line
- Normal to arc at a point on the arc
- Normal to a spline or work curve at a point on the spline or curve
- Midway between two parallel planes

FEATURE VISIBILITY

You can control the visibility of the origin planes, origin axes, origin point or user work planes, user work axes, user work points, and sketches by either right-clicking on them in the graphics area or on their name in the Browser, and selecting Object Visibility from the menu. You can also control the visibility for all origin planes, origin axes, origin point or user work planes, user work axes, user work points, and sketches from the Object Visibility option in the View menu. Visibility can be checked to turn visibility on or cleared to turn it off, as shown in the following image.

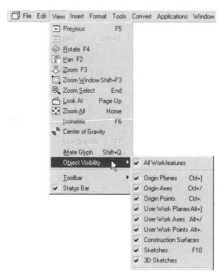

Figure 4-84

EXERCISE 4-7 Creating Work Planes

In this exercise, you create work planes in order to create a boss on a cylinder head and a slot in a shaft.

1. Open *ESS_E04_07.ipt*.
2. Click the Work Plane tool in the panel bar.
3. In the Browser, expand the Origin folder.
4. Click on X Axis.
5. Click on XY Plane.
6. In the Angle dialog box, type **45** and click the check mark.
7. A work plane is created.
8. Rotate the model so your view is similar to the following image.

Chapter 4 Creating Placed Features 189

Figure 4-85

9. You now create a work plane parallel to the new workplane and offset 42 mm.
10. Click the Work Plane tool in the panel bar.
11. Click on the work plane you just created and drag the new workplane up and to the left.

Notice the offset value changing as you drag the workplane.

12. In the Offset dialog box, type **-42** and click the check mark.
13. A second work plane is created.

Figure 4-86

14. In the Browser, right-click on the first workplane you created and turn off Visibility.
15. Click the Sketch tool on the command bar then click the second workplane you created.

 NOTE You can click the workplane in the graphics screen or in the Browser.

16. Rotate the model so you have a plan view of the sketch plane.
17. Click the Project Geometry tool in the panel bar.
18. In the Browser, click on Center Point in the Origin folder.
19. The center point is projected onto the sketch plane.
20. Click the Center Point Circle tool in the panel bar.

21. Place a circle with the center point coincident with the projected center point.
22. Click the General Dimension tool in the panel bar.
23. Place a 25 mm diameter dimension on the circle.

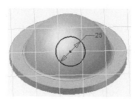

Figure 4-87

24. Press E to start the Extrude tool.
25. Right-click in the graphics window and click Isometric View.
26. Click in the circle to select it.
27. In the Extrude dialog box, select To from the Extents list.
28. Click on the spherical dome of the cylinder head.
29. Click OK.
30. The boss is created.
31. Turn off the visibility of the work plane

Figure 4-88

32. Close the file. Do not save changes. End of exercise.

PATTERNS

You can create two different types of patterns: feature patterns and body patterns. A feature pattern is a copy of an individual feature or features, and a body pattern is a pattern of a solid body (group of features). The type of pattern to create is selected in either the rectangular or circular pattern dialog box as shown in the following image.

Figure 4-89

Once you create a feature, you will find many instances when you need to duplicate a feature, group of features, or an entire solid multiple times with a set distance or angle between them. You can use the pattern tools to expedite this process.

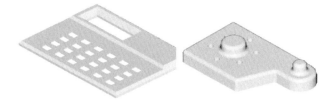

Figure 4-90

There are two types of available patterns: rectangular and circular (see the previous image). The pattern is held together as a single feature in the Browser, but the original feature and individual feature occurrences are listed under the pattern feature. You can suppress the entire pattern or individual occurrences, except for the first occurrence. Both rectangular and circular patterns have a child relationship to the parent feature(s) that you patterned. If the size of the parent feature changes, all of the child features will also change. If you patterned a hole, and the parent hole type changes, the child holes also change. Because a pattern is a feature, you can edit it like any other feature. You can pattern the base part or feature, as well as patterns. A rectangular pattern repeats the selected feature(s) along the direction set by two edges on the part or edges that reside in a sketch (these edges do not need to be horizontal or vertical as shown in the following image). A circular pattern repeats the feature(s) around an axis, a cylindrical or conical face, or an edge.

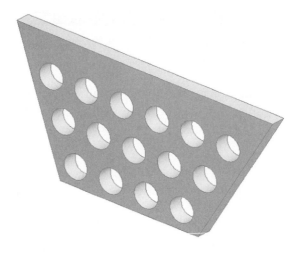

Figure 4-91

When creating a rectangular pattern, you define two directions by clicking an edge or sketch that defines alignment. Before creating a circular pattern, you must have a work axis, a part edge, or a cylindrical face about which the feature will rotate. After you click the Rectangular Pattern or Circular Pattern tool on the Part Features Panel Bar, as shown in the following images, or press the hot key SHIFT+O (circular pattern) or SHIFT+R (rectangular pattern), a Pattern dialog box appears. Click a feature to pattern and then enter the values, and click the edges and axis as needed. A preview image of the pattern appears in the graphics area.

Figure 4-92

Figure 4-93

The following are descriptions of the options available for rectangular and circular patterns.

RECTANGULAR PATTERNS

The following image shows the options in the Rectangular Pattern dialog box.

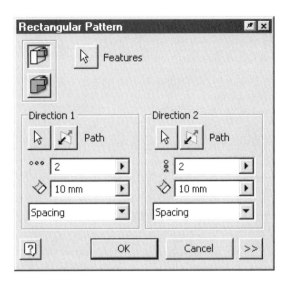

Figure 4-94

Pattern Individual Features   Click this button to pattern a feature or features. When you select this option, you activate a features button as described below.

Pattern the Entire Solid   Click this button to pattern a solid body. When you select this option, you select the entire part as the item to pattern. You also have the Include Work Features option when patterning an entire solid.

Features   Click this button and then click a feature or features to be patterned from either the graphics window or the Browser. You can add or remove features to or from the selection set by holding down the CTRL key and clicking them.

Include Work Features   Click this button and then click a work feature or work features to include in the pattern. You can add or remove work features to or from the selection set by holding down the CTRL key and clicking them.

Direction 1

In Direction 1, you define the first direction for the alignment of the pattern. It can be an edge, an axis, or a path.

Path   Click this button and then click an edge or sketch that defines the alignment along which you will pattern the feature.

Flip   If the preview image shows the pattern going in the wrong direction, click this button to reverse its direction.

Column Count   Enter a value or click the arrow to choose a previously used value that represents the number of feature(s) you will include in the pattern.

Column Spacing   Enter a value or click the arrow to choose a previously used value that represents the distance between the patterned features or the total overall distance for the patterned features.

Distance Define the occurrences of the pattern using the provided dimension as the spacing between the occurrences or the total overall distance for the patterned features.

Curve Length Create the occurrences of the pattern at an equal spacing along the length of the selected curve.

Direction 2

In Direction 2, you will define the second direction for the alignment of the pattern. It can be an edge, an axis, or path, but it cannot be parallel to Direction 1.

Path Click this button and then click an edge or sketch that defines the alignment along which you will pattern the feature.

Flip If the preview image shows the pattern going in the wrong direction, click this button to reverse its direction.

Row Count Enter a value or click the arrow to choose a previously used value that represents the number of feature(s) that you will include in the pattern.

Row Spacing Enter a value or click the arrow to choose a previously used value that represents the distance between the patterned features or the total overall distance for the patterned features.

Distance Define the occurrences of the pattern using the provided dimension as the spacing between the occurrences or the total overall distance for the patterned features.

Curve Length Create the occurrences of the pattern at an equal spacing along the length of the selected curve.

Options for the start point of the direction, compute type, and orientation method, as shown in the following image, are available by clicking the More (>>) button located in the bottom-right corner of the Rectangular Pattern dialog box.

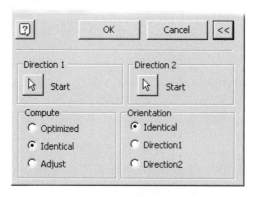

Figure 4-95

Start

Click the Start button to specify where the start point for the first occurrence of the pattern will be placed. The pattern can begin at any selectable point on the part. You can select the start points for both Direction 1 and Direction 2.

Compute

Optimized   Click this option to use faces instead of features to calculate all of the occurrences in the pattern. This option is ideal when the occurrences you are creating do not intersect and are all identical. It can improve the performance of pattern creation.

Identical   Click this option to use the same termination as that of the parent feature(s) for all of the occurrences in the pattern. This is the default option.

Adjust   Click this option to calculate the termination of each occurrence individually. Since each occurrence is calculated separately, the processing time can increase. You must use this option if a parent feature terminates to a face or plane.

Orientation

Identical   Click this option to orient all of the occurrences in the pattern the same as the parent feature(s). This is the default option.

DirectionI   Click this option to control the position of the patterned features by the selected direction. Each occurrence of the pattern is rotated to maintain proper orientation with the 2D tangent vector of the path.

Direction2   Click this option to control the position of the patterned features by the selected direction. Each occurrence of the pattern is rotated to maintain proper orientation with the 2D tangent vector of the path.

CIRCULAR PATTERNS

The following image shows the options in the Circular Pattern dialog box.

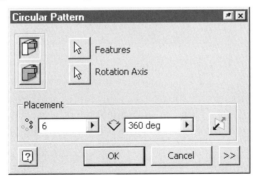

Figure 4-96

Pattern Individual Features   Click this button to pattern a feature or features. When you select this option, the features button is available and operates as described below.

Pattern the Entire Solid   Click this button to pattern a solid body. When you select this option, you select the entire part as the item to pattern. You also have the Include Work Features option when patterning an entire solid.

Features   Click this button and then click a feature or features to be patterned. You can add or remove features to or from the selection set by holding down the CTRL key and clicking them.

Include Work Features   Click this button and then click a work feature or work features to include in the pattern. You can add or remove work features to or from the selection set by holding down the CTRL key and clicking them.

Rotation Axis   Click the button and then click an edge, axis, or cylindrical face (center) that defines the axis about which the feature(s) will rotate.

Placement

Occurrence Count   Enter a value or click the arrow to choose a previously used value that represents the number of feature(s) that you will include in the pattern.

Occurrence Angle   Enter a value or click the arrow to choose a previously used value that represents the angle that you will use to calculate the spacing of the patterned features.

Flip   If the preview image shows the pattern going in the wrong direction, click this button to reverse its direction.

By clicking the More (>>) button located in the bottom-right corner of the Circular Pattern dialog box, you can access options for the creation method and positioning method of the feature, as shown in the following image.

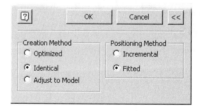

Figure 4-97

Creation Method

Optimized   Click this option to use faces instead of features to calculate all of the occurrences in the pattern. This option is ideal when the occurrences you are creating do not intersect and are all identical. It can improve the performance of pattern creation.

Identical   Click this option to use the same termination as that of the parent feature(s) for all of the occurrences in the pattern. This is the default option.

Adjust to Model   Click this option to calculate each occurrence termination individually. Because each occurrence is calculated separately, the processing time can increase. You must use this option if a parent feature terminates to a face or plane.

Positioning Method

Incremental Click this option to separate each occurrence by the number of degrees specified in Angle in the dialog box.

Fitted Click this option to space each occurrence evenly within the angle specified in Angle in the dialog box.

> **TIP** A work axis, an edge, or a cylindrical face about which the feature will rotate must exist before you create a circular pattern.

LINEAR PATTERNS–PATTERN ALONG A PATH

There are many modeling cases in which you need to create a pattern that follows a path. You can define a path by a complete or partial ellipse, an open or closed spline, or a series of curves (including lines, arcs, splines, etc.).

To pattern along a path, click the Path button and use the options described above for rectangular patterns. The path you use can be either 2D or 3D.

To create a pattern along 3D paths, follow these steps:

1. Create a 3D sketch.
2. Create a path (include model edges and work curves).
3. Create the pattern.

EXERCISE 4-8 Creating Rectangular Patterns

In this exercise, you add a rectangular pattern of holes to a plastic cover plate.

1. Open ESS_E04_08.ipt.
2. Rotate your view until the part looks like the following image.

Figure 4-98

3. Click the Sketch tool on the command bar, and then click the top surface of the part.
4. Click the Point, Hole Center tool in the panel bar.

5. Click a point anywhere in the lower left corner of the part.
6. Zoom in to the corner of the part, and then position the hole center using dimensions. The hole center is located 20 mm from the left edge and 10 mm above the bottom edge.

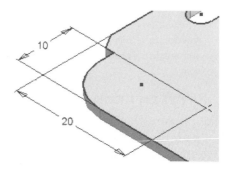

Figure 4-99

7. Click the Hole tool in the panel bar.
8. Add a hole with a diameter of 3 mm using the Through All termination.
9. You now add the hole pattern.
10. Click the Zoom All tool on the Standard toolbar to display the entire part.
11. Click the Rectangular Pattern tool in the panel bar.
12. Click the hole feature.
13. In Direction 1, click the Path button then click the bottom horizontal edge of the part.

A preview of the pattern is displayed.

14. Click the Flip button.
15. Enter **5** in Count and **17.5** in Spacing.
16. Click the Direction 2 Path button then click the vertical edge on the left of the part.
17. Enter **4** in Count and **17.5** in Spacing.

Figure 4-100

18. Click OK to create the pattern.

You now suppress three of the holes that are not required in the design.

19. Expand the rectangular pattern entry in the Browser to display the occurrences.

20. In the Browser, point to the occurrences. Each occurrence highlights in the graphics window as you point to it in the Browser.

21. Hold the CTRL key down and click the occurrences to suppress, as shown in the following image.

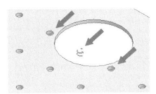

Figure 4-101

22. Right-click on any one of the highlighted occurrences in the Browser and choose Suppress.

The holes are suppressed in the model.

Figure 4-102

23. Close the file. Do not save changes. End of exercise.

EXERCISE 4-9 Creating Circular Patterns

In this exercise, you create a circular pattern of eight counterbored holes on a flange.

1. Open *ESS_E04_09.ipt*.
2. Click the Circular Pattern tool in the panel bar.
3. Click the counterbored hole feature.
4. Click the Rotation Axis button in the Circular Pattern dialog box.
5. Click the circular flange to specify the rotation axis.

Figure 4-103

A preview of the pattern is displayed. In this example, you will place 8 holes fitted over 360 degrees. You could also specify 8 holes with an incremental value of 45 degrees each.

   6. Type **8** in the Count field.
   7. Click the More button (>>).
   8. Under Positioning Method, make sure Fitted is selected.
   9. Enter **360** in the Angle field.
   10. Click OK to create the circular pattern.

Figure 4-104

   11. Close the file. Do not save changes. End of exercise.

EXERCISE 4-10 Creating Path Patterns

In this exercise, you pattern a boss and hole along a nonlinear path.

   1. Open *ESS_E04_10.ipt*.
   2. In the Browser, right-click Sketch and turn on Visibility.

Figure 4-105

   3. Click the Rectangular Pattern tool in the panel bar.

4. In the Browser, click the Extrusion feature then click the Hole feature to select both.

5. In Direction 1, click the Path arrow button then click the sketch line near the extrusion feature to select the entire path.

Figure 4-106

A preview of the pattern is displayed.

Figure 4-107

6. Click the More (>>) button.

7. In Direction 1, click the Start button then click the center point of the hole.

8. In Count, enter **40** then in Spacing, enter **36**.

The pattern preview updates.

Figure 4-108

9. In Orientation, click Direction 1.

10. Click OK to create the pattern feature.

11. In the Browser, right-click Sketch and turn off Visibility.

Figure 4-109

12. Close the file. Do not save changes. End of exercise.

CHAPTER SUMMARY

To	Do This	Tool
Create a hole feature	Click the Point, Hole Center tool on the 2D Sketch toolbar to sketch a hole center (if using the From Sketch option). Click the Hole tool on the Part Features toolbar or on the Panel Bar.	
Create a fillet feature	Click the Fillet tool on the Part Features toolbar or on the Panel Bar.	
Create a chamfer feature	Click the Chamfer tool on the Part Features toolbar or on the Panel Bar.	
Create a shell feature	Click the Shell tool on the Part Features toolbar or on the Panel Bar.	
Create a rectangular pattern	Click the Rectangular Pattern tool on the Part Features toolbar or on the Panel Bar.	
Create a circular pattern	Click the Circular Pattern tool on the Part Features toolbar or on the Panel Bar.	
Create a thread feature	Click the Thread tool on the Part Features toolbar or on the Panel Bar.	
Create a face draft	Click the Face Draft tool on the Part Features toolbar or on the Panel Bar.	
Create a work plane	Click the Work Plane tool on the Part Features toolbar or on the Panel Bar.	
Create a work axis	Click the Work Axis tool on the Part Features toolbar or on the Panel Bar.	
Create a work point	Click the Work Point tool on the Part Features toolbar or on the Panel Bar.	

To	Do This	Tool
Create a grounded work point	Click the Grounded Work Points tool on the Part Features toolbar or on the Panel Bar.	

Applying Your Skills

In this exercise, you create a drain plate cover.

1. Start a new part based on the metric *standard(mm).ipt* template.
2. Use the extrude, shell, face draft, hole, and rectangular pattern tools to complete the part.

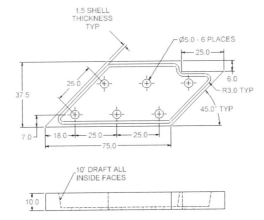

Figure 4-110

SKILL EXERCISE 4-2

In this exercise, you create a connector part.

1. Start a new part based on the metric *standard(mm).ipt* template.
2. Use the extrude, work plane, hole, thread, chamfer, fillet, and circular pattern tools to complete the part.

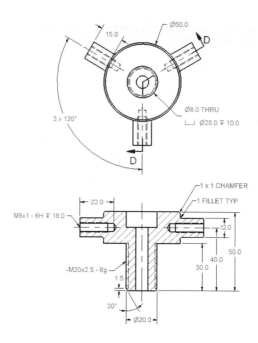

Figure 4-111

CHECKING YOUR SKILLS

Use these questions to test your knowledge of the material covered in this chapter.

1. **True__ False__** When creating a fillet feature that has more than one selection set, each selection set appears as an individual feature in the Browser.

2. In regards to creating a fillet feature, what is a smooth radius transition?

   _____

   _____

3. **True__ False__** When you are creating a fillet feature with the All Fillets option, material is removed from all concave edges.

4. **True__ False__** When you are creating a chamfer feature with the Distance and Angle option, you can only chamfer one edge at a time.

5. **True__ False__** When you are creating a hole feature, you do not need to have an active sketch.

6. For what reason is a hole center used?

_____

_____

7. True__ False__ Thread features are represented graphically on the part and will be annotated correctly when you generate drawing views.

8. True__ False__ A part may contain only one shell feature.

9. When you are creating a face draft feature, what is the definition of pull direction?

_____

_____

10. True__ False__ The only method to create a work axis is by clicking a cylindrical face.

11. True__ False__ You need to derive every new sketch from a work plane feature.

12. Explain the steps to create an offset work plane.

_____

_____

13. True__ False__ You cannot create work planes from the default work planes.

14. True__ False__ When you are creating a rectangular pattern, the directions along which the features are duplicated must be horizontal or vertical.

15. True__ False__ When you are creating a circular pattern, you can only use a work axis as the axis of rotation.

CHAPTER 5

Creating and Editing Drawing Views

After creating a part or assembly, the next step is to create 2D drawing views that represent that part or assembly. To create drawing views, you start a new drawing file, select a 3D part or assembly on which to base the drawing views, insert or create a drawing sheet with a border and title block, project orthographic views from the part or assembly, and then add annotations to the views. You can create drawing views at any point after a part or assembly exists. The part or assembly does not need to be complete because the part and drawing views are associative in both directions (bidirectional). This means that if the part or assembly changes, the drawing views will automatically be updated. If a parametric dimension changes in a drawing view, the part will get updated before the drawing views get updated. This chapter will guide you through the steps for setting up dimension styles, creating borders and title blocks, drawing views of a single part, editing dimensions, and adding annotations.

CHAPTER OBJECTIVES

- After completing this chapter, you will be able to
- Understand drawing options
- Create and edit drawing borders and title blocks
- Create base and projected drawing views from a part
- Create auxiliary, section, detail, broken, break-out, draft, and perspective views
- Edit the properties and location of drawing views
- Retrieve model dimensions to use in drawing views
- Edit, move, and hide dimensions
- Select drawing objects using a window or a crossing window
- Add automated centerlines
- Add general dimensions
- Create hole notes
- Add annotations such as GD&T, surface finish symbols, weld symbols, and datum identifiers

DRAWING OPTIONS, CREATING A DRAWING, AND DRAWING TOOLS

In this first section, you will learn about the drawing options that are available, learn how to create a new drawing, and learn about the tools that are used to create drawing views.

DRAWING OPTIONS

Before drawing views are created, the drawing options should be set to your preferences. To set the drawing options, click Application Options on the Tools menu. The Options dialog box will appear. Click the Drawing tab; your screen should resemble as shown in the following image. Make any changes to the options before creating the drawing views, or the changes may not affect drawing views that you have already created. The following sections describe the drawing options.

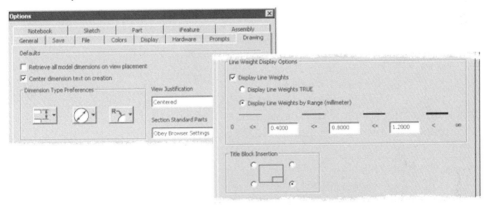

Figure 5-1

Retrieve All Model Dimensions on View Placement

Click this option to add applicable model dimensions to drawing views when they are placed. If the option's box is clear, no model dimensions will be placed automatically. You can override this option setting by manually selecting All Model Dimensions in the Drawing View dialog box when creating views.

Center Dimension Text on Creation

Click this option to have dimension text centered as you create the dimension.

Dimension Type Preferences

Use this area to set the preferred type of dimensions, referring to the following image as a guide.

Figure 5-2

View Justification

Use this option to set the default justification for drawing views. Two modes are available: Centered and Fixed.

Section Standard Parts

Use this option to control whether standard parts placed into an assembly from the supplied Inventor parts library, such as nuts, bolts, and washers, are sectioned. Three options are available in this area: Always, Never, and Obey Browser Settings.

Line Weight Display Options

Use this option to control the display of line weights in a drawing.

Display Line Weights

Click this option to allow for line weights to appear in drawings. Visible lines will appear in drawings with the line weights defined in the active drafting standard. Clear the option's box to display all lines without weights.

Display Line Weights TRUE   When clicked, line weights appear on the screen as they would appear when plotted on paper.

Display Line Weights by Range (millimeter)   When selected, line weights appear according to the values entered by the user. These values range from the smallest value on the left to the largest value on the right.

 NOTE This setting does not affect line weights when you print the views.

Title Block Insertion

Click to select the title block insertion point for the first and subsequent drawing sheets.

CREATING A DRAWING

The first step in creating a drawing from an existing part or assembly is to create a new drawing IDW file by either clicking the New icon in the What To Do section of the Getting Started page or by clicking New on the File menu, and then clicking the desired drawing template from one of the template tabs (as shown in the following image), or click the down arrow on the New icon on the left side of the standard toolbar and clicking Drawing.

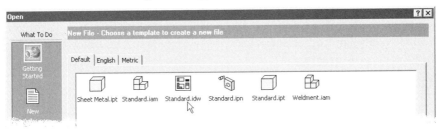

Figure 5-3

Inside the Metric tab are six drafting templates, as shown in the following image, that are included with Autodesk Inventor. They are

- ANSI (American National Standards Institute)
- BSI (British Standards Institute)
- DIN (Deutsche Industrie Norm-The German Institute for Standardization)
- GB (Guojia Biaozhun-The Chinese National Standard)
- ISO (International Organization for Standardization)
- JIS (Japan Industrial Standard)

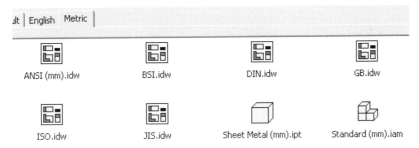

Figure 5-4

DRAWING TOOLS

After you start a new drawing, Autodesk Inventor's screen will change to reflect the new drawing environment. There are three toolbars available in the Panel Bar that you will use to create drawings: Drawing Views, Sketch, and Drawing Annotation. By default, when a new drawing is created, the Panel Bar will show the Drawing Views toolbar. These tools allow you to create drawing views, sheets, and draft views. The Sketch tools enable you to create custom symbols and add 2D geometry and dimensions to draft views. The Drawing Annotation tools enable you to add annotations to drawing sheets and existing drawing views. The three sets of tools will be explained throughout this chapter.

DRAWING VIEWS PANEL BAR

When you first enter the drawing environment, the Drawing Views Panel Bar appears as shown in the following image. Note that this image shows the Panel Bar with Expert mode turned on. Use this Panel Bar for creating base, orthographic, section, auxiliary, and other specialized drawing views. The following section describes the buttons briefly.

Figure 5-5

Drawing Views

Button	Tool	Function
	Base View	Creates the first view in a drawing.
	Projected View	Creates views projected from the base view. These views can be either orthogonal or isometric.
	Auxiliary View	Creates a view that is projected from a selected edge in a parent view.
	Section View	Creates full, half, offset, radial, and aligned views from a selected parent view.
	Detail View	Creates a detailed view of all or a portion of a specified parent view.
	Broken View	Creates a partial view by removing a section of the part with the ends still intact. This view is beneficial when documenting long shaft parts and structural steel members.
	Break Out View	Creates a view consisting of a specified area removed in order to expose internal parts or features.
	Overlay View	Overlay drawing views use positional representations to show an assembly in multiple positions in a single view. Creating overlay views is covered in Chapter 6.
	New Sheet	Adds a new drawing sheet to the current active drawing document.
	Draft View	Allows you to create a special view by sketching geometry.

DRAWING ANNOTATION PANEL BAR

A second Panel Bar is available in the drawing environment; namely the Drawing Annotation Panel Bar, as shown in the following image. Note that this image shows the Panel Bar with Expert mode turned on. Use this Panel Bar for adding drawing annotations such as dimensions, symbols, text, balloons, and parts list to your drawing. Each button is briefly described as follows.

Figure 5-6

Drawing Annotations

Button	Tool	Function
	General Dimension	Adds dimensions to document a part. These dimensions to not control part features or sizes.
	Baseline Dimension Set	Creates a series of dimensions that reference a common origin. All baseline dimensions are created as a set and are grouped together as a single object. You can edit the entire group as a single object.
	Baseline Dimension	Creates a series of dimensions that reference a common origin. Baseline dimensions are not grouped. Rather each dimension is considered a single object.
	Ordinate Dimension Set	Creates ordinate dimensions that reference a common origin. All ordinate dimensions you create as a set are grouped together and you can edit them as a group.
	Ordinate Dimension	Creates ordinate dimensions that reference a common origin. Ordinate dimensions consist of individual dimensions and are not grouped together as the ordinate dimension set.
	Hole/Thread Note	A special dimension that adds diameter, depth, thread, and hole type information to a hole or thread.
	Center Mark	Creates a center mark on a feature in a drawing view. Also found under this button are the capabilities of creating a centerline bisector, centerline, and centered pattern.
	Surface Texture Symbol	Adds a surface texture symbol to your drawing. A dialog box allows you to add extra information to the surface texture symbol.
	Welding Symbol	Places a welding symbol in the drawing. Also activates a welding symbol dialog box where various types of additional welding symbols and weld annotations are stored.
	Feature Control Frame	Places a feature control frame symbol and activates the Feature Control Frame dialog box where you can make changes to such items as tolerance type symbol, tolerance value, and datum references, in addition to material condition modifiers.
	Feature Identifier Symbol	Creates a feature identifier symbol along with a leader.
	Datum Identifier Symbol	Creates one or more datum identifier symbols. You can create the datum identifier with a leader or leave it as a standalone symbol.

Drawing Annotations (continued)

Button	Tool	Function
	Datum Target	Places a datum target in the drawing. While leader is the default mode, you can also place additional datum targets in circle, line, point, and rectangle modes.
	Text	Places text in the form of general notes in a drawing.
	Leader Text	Groups a leader to a general note in a drawing.
	Balloon	Allows you to place a single balloon in a drawing. A second button automatically applies balloons to all parts. Creating balloons is covered in Chapter 6.
	Parts List	Places a parts list consisting of Item, Quantity, Part Number, and Description by default. Additional data fields can be added to a parts list. Creating a parts list is covered in Chapter 6.
	Hole Table	Displays the X, Y, and Diameter values of holes in a table format (it is one of three buttons that does this). The other two modes allow you to create a hole table by View or Selection Type.
	Caterpillar	Creates a type of welding representation that displays a weld in a 2D drawing view. A second button is available to provide the end representation of a weld in the 2D drawing view.
	Revision Table	Places a revision table to document changes in a drawing. A second button allows you to place a revision tag on an area of the drawing where the change occurs.
	Symbols	Activates the Symbols dialog box used for inserting previously defined sketched symbols in a drawing.
	Retrieve Dimensions	Model dimensions that can be used in the dimensioning of drawing views. Making changes to a model dimension in a drawing will also change the associative part file.

DRAWING SHEET PREPARATION

When you create a new drawing file using one of the provided template files, the program displays a default drawing sheet with a default title block and border. The template IDW file that is selected determines the default drawing sheet, title block, and border. The drawing sheet represents a blank piece of paper on which you place the border, title block, and drawing views. There is no limit to the number of sheets that can exist in the same drawing, but you must have at least one drawing sheet. To create a new sheet, click the New Sheet tool on the Drawing Views Panel Bar as shown in the following image, or right-click in the Browser or on the current sheet in the graphics window and select New Sheet from the menu.

Figure 5-7

A new sheet will appear in the Browser and the new sheet will appear in the graphics window with a default border and title block. You can rename the sheet by slowly double-clicking on its name in the Browser and then entering a new name or right-clicking on the sheet in the Browser and selecting Edit Sheet from the menu. You can then enter a new name in the Edit Sheet dialog box. To create a sheet of a different size, you can double-click on one of the Sheet Formats in Drawing Resources in the Browser, as shown in the following image, or right click in the graphics window and select New Sheet from the menu.

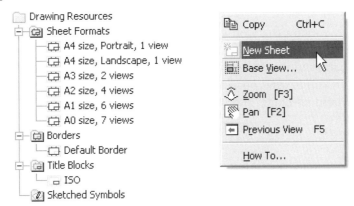

Figure 5-8

If a sheet is selected from the Sheet Formats list, predetermined drawing views will be created. If the necessary sheet size is not on the list, right-click on the sheet name in the Browser and select Edit Sheet from the menu. Then select a size from the list as shown in the following image. To use your own values, select Custom Size from the list and enter values for height and width.

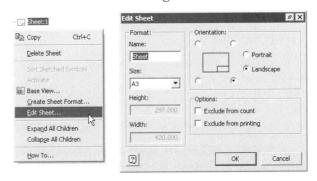

Figure 5-9

NOTE The sheet size is inserted full scale (1=1) and should be plotted at 1=1. The drawing views will be scaled to fit the sheet size.

BORDER CREATION

The four lines and zone labels that surround the edges of the sheet make up the border. To insert the default border in a blank drawing sheet, expand Drawing Resources in the Browser and double-click on Default Border. To insert a customized border with specified margins, expand Drawing Resources in the Browser, right-click on Default Border, and select Insert Drawing Border. The Default Drawing Border Parameters dialog box will appear, similar to the following image. Choose the More button (>>) to access the additional options and settings. Select the options, enter the values that you want, and then click the OK button. The border will appear on the sheet. The sheet margins will be offset from the edges of the sheet. You cannot add a new border to a drawing sheet that already contains a border. To delete an existing border, expand the sheet name in the Browser, right-click on the border's name, and select Delete from the menu.

Figure 5-10

TITLE BLOCKS

To add a title block to the drawing sheet, you can either insert a default title block or construct a customized title block and insert it into the drawing sheet.

INSERTING A DEFAULT TITLE BLOCK

To insert a default title block, follow these steps:

- Make the sheet active, then place the title block by double-clicking on its name in the Browser.

- Insert the title block by expanding Drawing Resources > Title Blocks in the Browser and then either double-clicking on the title block's name, as shown in the following image, or right-clicking on the title block's name and selecting Insert from the menu.

Figure 5-11

EDIT PROPERTY FIELDS DIALOG BOX

The time will come when you will need to fill in title block information. Expanding the default title block in the Browser and double-clicking on the Field Text category will display the Edit Property Fields dialog box. By default, the following information will already be filled in: Sheet Number, Number of Sheets, Author, Creation Date, and Sheet Size. In order to fill in other title block information such as Part Number, Company Name, Checked By, and so on, select Properties–Drawing from the drop-down list and click the iProperties button as shown in the following image.

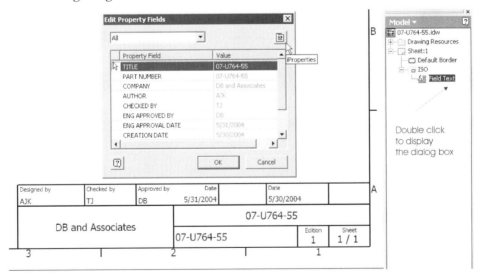

Figure 5-12

The Properties dialog box will appear and you can fill in the information as needed. You can find most title block information under the Summary, Project, and

Status tabs. The following image shows the Properties dialog box and the Status tab with the Checked By and Eng. Approved By categories filled in.

Figure 5-13

CREATING A NEW TITLE BLOCK

This section will introduce you to the methods of creating a customized title block. This book, however, will utilize the default title blocks. If you want to create a customized title block, you can either edit an existing title block or create one from scratch. You also have the ability to bring in AutoCAD data to be used as a title block.

To edit an existing title block, right-click on the title block name in the Browser under Drawing Resources > Title Block and select Edit from the menu. The title block will appear with all of its attributes shown. Edit as needed and when finished, right-click and select Save Title Block. You will have the option to save the modified title block as another name. If you save the title block with another name, the new title block's name will appear in the Browser under Drawing Resources > Title Block. The new title block will not be available to new drawings unless it is saved in a template file.

If you want to create a new title block from scratch, right-click on Title Blocks in the Browser under Drawing Resources, and select Define New Title Block. Next, using the tools from the Sketch toolbar, create the title block. Create the title block using the sketch tools in the same fashion as you would when creating a sketch. To create text that is populated from the file properties, use the Property Field tool. When you are done creating the title block, right-click and select Save

Title Block from the menu. The title block will appear in the Browser under Drawing Resources > Title Block. The new title block will not be available to new drawings unless it is saved in a template file. See Help to learn how to customize a title block's properties.

REORDER AND SORT BY NAME FOR DRAWING RESOURCES

All Drawing Resources headings (Sheet Formats, Borders, Title Blocks, and Sketched Symbols), can have their contents reordered or sorted by name. In the following image, the example on the left illustrates a reordering operation. Pick the item (Allen, Inc.), then press and drag the item to a new location. In the middle example, you can perform a sorting operation on the entire contents of a Drawing Resources folder, in this case Title blocks. Right-click on the folder to display the menu and select Sort by Name to reorganize the folder as shown in the example on the right.

Figure 5-14

STYLES

Autodesk Inventor uses styles to control how objects appear. Styles are saved within an Autodesk Inventor file or to a project library locationand can be saved to a network location where many users can access the same styles. This section introduces you to styles. Autodesk Inventor uses Extensible Markup Language (XML) files for storing style information externally from Autodesk Inventor documents. Autodesk Inventor does not support the editing or use of these XML files with anything other than tools provided inside Autodesk Inventor and the Style Library Manager. Once a style is used in a document, it is stored in the document.

STYLE NAME/VALUE

Autodesk Inventor uses a style's name as the unique identifier of the style. Only one style name for the same style type can exist. For example, in a drawing, only one dimension style with a specific name can exist "Default ANSI". However, a text style could exist in the same file with the name "Default ANSI".

SUB-STYLES

Autodesk Inventor styles reference other styles. When a style references another style it is known as a sub-style. For example, a dimension style can have up to three sub-styles:

- Primary Text Style—Used to format primary and alternate units in a dimension.
- Tolerance Text Style—Used to format the tolerances of the primary and alternate units.
- Leader Style—Used to format Hole Notes and Leader Notes.

CONTROLLING STYLE LIBRARY CHANGES

CAD administrators may want to prevent users from making changes to a style library in order to keep formatting standards under document control. There are two methods to make a style library read-only:

- Set Project file—Set the Use Styles Libraryoption in the Autodesk Inventor project to read-only as shown in the following image. When this is set to Read Only, only an administrator can control the style library. To allow changes to the style library, create a second project file for the administrator with the Use Styles Library option set to Yes. When set to yes, styles can be written back to the style library.

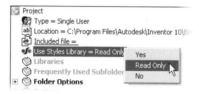

Figure 5-15

- Set Windows file property—Use Windows Explorer to select all of the XML files that make up the style library, right-click, select Properties, and click Read Only. If a single XML file is read-only, Autodesk Inventor treats the entire library as read-only because of sub-styles referencing.

When the Use Styles Library option is set to No, no style library will be used. If you want to use a style with this setting, you must load the style into the document using one of these methods:

- Create a style in the document using the Style Editor tool.
- Load a style into the document using the Style Editor, Import option.
- Start a document from a template file that has the needed styles.

DOCUMENTS, STYLE LIBRARIES, AND TEMPLATES

Autodesk Inventor uses documents, style libraries, and templates to store styles. This section explains how styles are used in each.

Documents

All styles are created and managed while inside a document (part, assembly, drawing, or presentation file). When a style is used in a document, it is copied locally (cached) automatically to that document; this ensures that the document will always have the needed style information. Most of the tools that you use to edit and create styles are available on the Format menu as shown in the following image. If tools are grayed out, the project file has Use Styles Library option set to No or Read Only.

Figure 5-16

Style Libraries

Autodesk Inventor uses a style library as the central storage location of styles for all projects or for a specific project. A style library allows users to access the most current style information and provides a method for sharing styles between documents. Follow these rules when working with styles:

- The local style in a document takes precedence over a style library unless the local style is updated from the style library. This includes editing a style in the style editor. Editing always takes place in a document. A style library can never be edited directly.

- In drawings, all styles are filtered by the active standard. Each standard has a list of available styles. These styles appear at the top of any list of styles in drawing documents. However, you can use other styles in the document. The following image shows the available balloon styles in a drawing file.

- A local (cached) style in a document is always available for use. This allows you to use styles that are not part of the style library in a document, or if you are unable to access the style library.

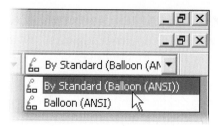

Figure 5-17

Autodesk Inventor includes predefined styles: Balloon, Center Mark, Colors, Datum Target, Dimension, Feature Control Frame, Hatch, Hole Table, ID, Layers, Leader, Lighting, Material, Object Defaults, Parts List, Standard, Surface Texture, Text, Weld Bead and Weld Symbol. All of these (XML) files must be present in the directory that is specified in the Styles Library Folder Options of the current Autodesk Inventor project file. The following image shows the default Style Library location set in the project file.

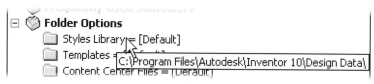

Figure 5-18

Templates

Autodesk Inventor template files specify what styles are to be used by default when a new document is created. If the styles library is set to Yes or Read Only, the style definitions are pulled from the style library when you use the library for the first time in a document. The default styles that are defined in a document can be found by clicking Document Settings on the Tools menu and clicking the Standard tab or by clicking Active Standard on the Format menu. The following is a list of the styles controlled in different document types:

Parts and Weldments

Define lighting and material. Weldments also have a Standard setting on their Active Standard tab. This is not the same as the Standard style used by drawings. This is one of six international standards (ANSI, ISO, DIN, BSI, JIS, or DIN) that define how the weld features are characterized).

Sheet Metal

Define lighting and material (the default material is set by the active Sheet Metal style in the document).

Assemblies and Presentations

Define lighting.

Drawings

Define the active standard.

MANAGING STYLES

Autodesk Inventor automates the task of managing styles between documents and style libraries. When a style from a style library is used, it is copied into the document automatically along with all necessary sub-styles. Styles that are copied automatically into a document are removed automatically from a document whenever they are no longer in use by any object in the document or set as an active standard in a base standard. Automatic cleanup occurs when Autodesk Inventor is sure that no style data will be lost if it removes the style. A style is removed from a document automatically under the following conditions: the style was copied by Autodesk Inventor automatically into the document from a style library to format an object, the style was copied from a style library to perform an edit using the Style Editor.

CREATING A NEW STYLE

To create a new style, follow these steps:

1. Click the Styles Editor tool on the Format menu.

2. Click and expand the style section for which you want to create a new style.

3. Right-click the style on which the new style will be based and select New Style. The following image shows a new style being created from the Default (ANSI) dimension style.

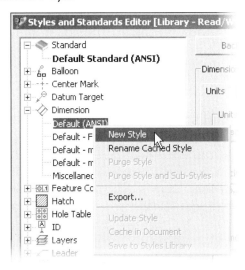

Figure 5-19

4. The New Style Name dialog box will appear as shown in the following image. Enter a style name, and if you do not want the style to be used in the standard, uncheck Add to standard.

5. Make changes to the style and save the changes.

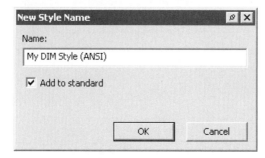

Figure 5-20

If you uncheck the Add to standard option, the style is not added to the standard as an active style. When an object selected in the graphics window has a local style that was not added to the standard, it will appear in the style drop-down list below the dotted line. The following image shows a dimension style that was not added to the standard.

Figure 5-21

EDIT STYLES MANUALLY IN DOCUMENTS

- The most common tasks involving styles are editing an existing style or creating a new style. You edit and create styles in Autodesk Inventor documents. Styles you edit are copied automatically from a style library into the current document for editing. You must save an edited or newly created style to a style library from within a document, and this can only be done if the Use Style Library option in the current project file is set to Yes.

To edit a style, follow these steps:

1. Click the Styles Editor tool on the Format menu.

2. Click and expand the style section for which you want to edit.

3. Click the style you want to edit.
4. Make change to the style area, each tab has different options you can set.
5. Click Save at the top of the dialog box.

The following image shows My DIM Style (ANSI) ready for edits.

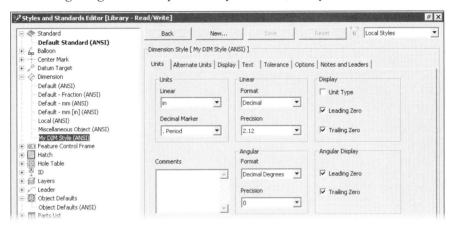

Figure 5-22

SETTING OBJECT DEFAULTS

You control which style is applied to your objects by mapping your objects and styles in the Object Defaults style. After a style has been saved you set specific objects to use that style. Edit styles by clicking Styles Editor on the Format menu and expand the Object Defaults on the left side of the dialog box. Then in the middle of the dialog box click the object type that you want to set the style for, then set the object style. The following image shows the Linear Dimension being changed to My DIM Style. Then all new linear dimensions will have that style.

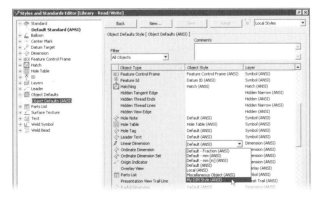

Figure 5-23

OVERRIDING AN OBJECT'S STYLE

After an object has been placed, for example a dimension, its style can be changed. Select the object(s) in the graphics window and then from the style drop-down list click the new style. The following image shows a dimension's style being changed to My DIM Style.

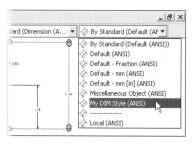

Figure 5-24

 NOTE For more information about styles and standards click Getting Started on the File menu and click White papers in the Autodesk Inventor Launchpad area of the dialog box.

EXERCISE 5-1 Sheets, Borders, and Title Blocks

In this exercise, you set up an A2 (420 x 594mm) drawing sheet with a border and title block. You also create a drafting standard to match ANSI metric standards.

1. Open the file *ESS_E05_01.idw*. This drawing consists of one empty sheet.

2. You will first create a new drafting standard. From the Format menu click Styles Editor as shown in the following image. The Styles and Standards Editor dialog box will display.

Figure 5-25

3. You will now create a new text style. In the left pane expand the Text style.

4. Right-click on the Current-ANSI text style and click New Style from the menu as shown in the following image.

Figure 5-26

5. In the New Style Name dialog box type Border Text, verify that the Add to standard box is checked and then click OK.

6. In the Font list, select Arial (see the following image).

7. In the Size list, select 5.00 mm (see the following image).

8. Click the Save button, then click Done.

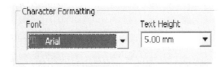

Figure 5-27

9. In the next series of steps, you will resize the drawing sheet and insert a border and title block. First, right-click Sheet:1 in the Browser and select Edit Sheet.

10. Under Size, select A2 then click OK.

11. In the Browser, expand Drawing Resources, and then expand Borders.

12. Right-click Default Border and select Insert Drawing Border. The Default Drawing Border Parameters dialog box is displayed.

13. Click the More (>>) button to expand the dialog box.

14. Under Text Style, select Border Text to use the text style you previously created, then click OK. The text around the border uses the Arial font and has a size of 5.00 mm.

15. Expand Title Blocks in the Browser. Right-click ANSI - Large, and then select Insert. The title block is inserted.

16. You will now change an existing text parameter and create new text parameters for drawing scale and project.

17. In the Browser, right-click ANSI - Large under Title Blocks, and then select Edit. The title block is displayed on a temporary sheet for editing.

18. Zoom in on the title block. Your display should appear similar to the following image.

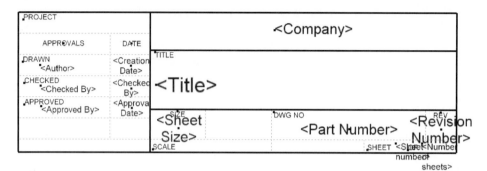

Figure 5-28

19. Select the property field < Checked By > located in the Date column below the property field < Creation Date > as shown in the following image.

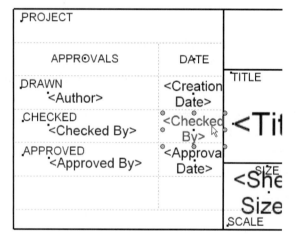

Figure 5-29

20. The next series of steps cover the editing process of a text parameter. Right-click the selected text, and then choose Edit Text. The Format Text dialog box is displayed.

21. Highlight the < Checked By > text in the Format Text dialog box.

22. Select CHECKED DATE from the Property drop-down list.

23. Click the Add Text Parameter button to display the field in the Format Text dialog box. Then click OK.

24. You will now Add a new text parameter. Begin by clicking the Text tool.

25. Select a point below PROJECT and vertically aligned with the center of the APPROVALS text parameters as shown in the following image. Drag to the right and create a rectangle.

Chapter 5 Creating and Editing Drawing Views

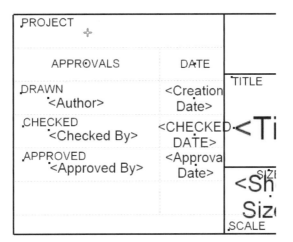

Figure 5-30

26. In the Format Text dialog box, click the Center Justification button.

27. In Type, select Properties-Drawing and from the Property list select PROJECT then click the Add Text Parameter button. Click OK to place this text parameter. If necessary, reposition the text parameter as shown in the following image.

 NOTE A text parameter can be selected and dragged to reposition it in the title block.

Figure 5-31

28. You will now create a prompted entry text parameter. First, Click the Text tool.

29. Select a point to the right of the SCALE text as shown in the following image and then create a rectangle for the text parameter.

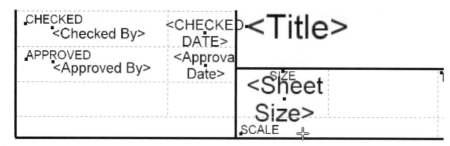

Figure 5-32

30. Under Type, select Prompted Entry.

31. Type SCALE to replace < Enter Prompt for Field > in the title block, and then click OK.

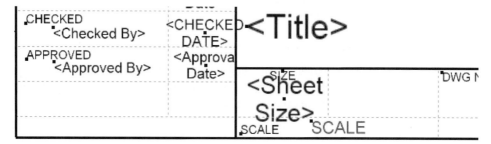

Figure 5-33

32. From the Format menu, select Save Title Block. Click Yes to save your edits to the ANSI-LARGE title block. The temporary sheet is removed.

33. When the Edit Property Fields dialog box appears, enter 1:1 in the Value column for the Scale Property Field and then click OK.

34. Next, you will complete the title block using information derived from the Drawing Properties dialog box. Enter drawing property values for the remaining text parameters.

35. In the Browser, right-click ESS_E05_01.idw, and then select iProperties. The Drawing Properties dialog box is displayed.

36. Type the designer's initials in the Author field, and then click OK. Your display should appear similar to the following image.

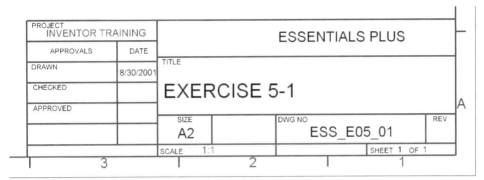

Figure 5-34

37. End of exercise.

TEMPLATES

After you have created a drawing sheet, border, and title block, and have set various styles, you can save the file as a template and use this template as a basis for new drawing files. To save a file as a template, click Save Copy As on the File menu, enter a new template name, and save the file to the appropriate Autodesk Inventor template folder. This new file is now available as a template when creating new drawings.

CREATING DRAWING VIEWS

After you have set the drawing sheet format, border, title block, and styles, you can create drawing views from an existing part, assembly, or presentation file. The file from which you will create the views does not need to be open when a drawing view is created. It is suggested, however, that both the file and the associated drawing file be stored in the same directory and that the directory be referenced in the project file. When creating drawing views, you will find that there are many different types of views you can create. The following sections describe these view types.

Base View This is the first drawing view of an existing part, assembly, or presentation file. It is typically used as a basis to generate the following dependent view types. You can create many base views in a given drawing file.

Projected View This is a dependent orthographic or isometric view that is generated from an existing drawing view.

- Orthographic (Ortho View)–A drawing view that is projected horizontally or vertically from another view.

- Iso View–A drawing view that is projected at a 30° angle from a given view. An isometric view can be projected to any of the four quadrants.

Auxiliary View This is a dependent drawing view that is perpendicular to a selected edge of another view.

Section View This is a dependent drawing view that represents the area defined by a slicing plane or planes through a part or assembly.

Detail View This is a dependent drawing view in which a selected area of an existing view will be generated at a specified scale.

Broken View This is a dependent drawing view that shows a section of the part removed, but the ends remain. Any dimension that spans over the break will reflect the actual object length.

Break Out View This is a drawing view that has a defined area of material removed in order to expose internal parts or features.

Overlay View This drawing view uses positional representations to show an assembly in multiple positions in a single view.

SELECTING DRAWING VIEW COMMANDS

You have numerous methods from which to choose when creating drawing views, as shown in the following image. You can access all drawing view commands via a command button located on the Drawing Views Panel Bar. You could also select Model Views on the Insert menu to access all drawing commands. Right-clicking in the graphics window will also display a menu. In the image on the right (in the following image), only Base View is listed. This is because an existing view was not selected. Had an existing view been selected, the additional drawing view commands would appear.

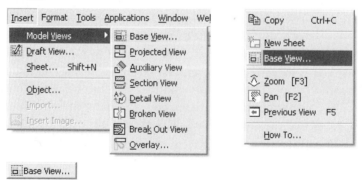

Figure 5-35

USING THE DRAWING VIEW DIALOG BOX

You use the Drawing View dialog box to create the various drawing views just described. Activate the dialog box by clicking the Base View tool on the Drawing Views Panel Bar, as shown in the following image, or by right-clicking in the graphics window and selecting Base View from the menu.

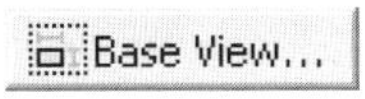

Figure 5-36

The Drawing View dialog box, shown in the following image, has two tabs: Component and Options. The following sections describe these tabs.

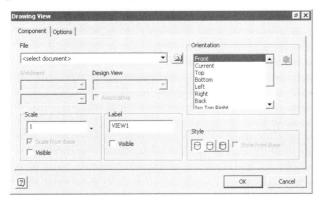

Figure 5-37

The Component Tab

The following categories are available under the Drawing View Component tab.

File   Any open part, assembly, or presentation files will appear in the drop-down list. You can also click the Explore directories icon and navigate to and select a part, assembly, or presentation file.

Design View/ Presentation View   After selecting an assembly file that has design views, the names of the design views will appear in the drop-down list. If you selected a presentation file that has presentation views, the names of the presentation views will appear in the drop-down list.

Orientation   After selecting the file, choose the orientation in which to create the view. After selecting an orientation, a preview image will appear in the graphics window attached to your cursor. If the preview image does not show the view orientation that you want, select a different orientation view by selecting another option. The following view orientations are available:

- Front
- Current (current orientation of part, assembly, or presentation in its file)
- Top
- Bottom
- Left
- Right
- Back

- Iso Top Right
- Iso Top Left
- Iso Bottom Right
- Iso Bottom left
- Saved Camera (if placing a view of a presentation file with saved camera views)

Scale Enter a number for the scale in which to create the view. Note that the drawing sheet will be plotted at full scale (1:1) and the drawing views are scaled as needed to fit the sheet. You can edit the scale of the views after the views have been generated.

Scale from Base Click to set the scale of a dependent view to be the same as that of its base view. To change the scale of a dependent view, clear the option's box and modify the scale value for that view.

Visible Click to display or hide the view scale. Clicking the check box displays the scale, leaving the box unchecked hides the scale.

Weldment You enable this area when you select a document that is a weldment. Weldments have four states–Assembly, Preparations, Welds, and Machining.

Label Use to include and/or change the label for the selected view. When you create a view, a default label is determined by the active drafting standard. To change the label, select the label in the box and enter the new label.

Visible Click to display or hide the view label. Clicking the check box displays the label and leaving the box clear hides the label.

Style Choose how the view will appear. There are three choices–Hidden Line, Hidden Line Removed, and Shaded. The preview image will not update to reflect the style choice. When the view is created, the chosen style will appear. The style can be edited after the view has been generated.

Style from Base Click to set the display style of a dependent view to be the same as that of its base view. To change the display style of a dependent view, clear the check box and modify the style for that view.

The Options Tab

The following image shows the Options tab of the Drawing View dialog box. The following sections describe the tab's options.

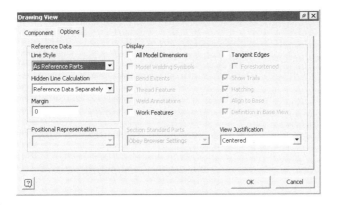

Figure 5-38

Areas to be found in the Options tab include the following:

Reference Data

Line Style The drop-down list includes three options: As Parts, As Reference Parts, and Off. When you select As Parts, reference data will have no special display characteristics. When you select As Reference Parts (for wireframe view types), the reference data and product data will not hide each other. The reference data will appear on top of product data, and the reference data will appear with tangent edges turned off and all other edges as the phantom line type. When you select As Reference Parts (for shaded view types), the reference data will appear as wireframe with tangent edges turned off and all other edges as phantom lines. The reference data will appear on top of the product data that is shaded. When you select Off, reference data will not appear.

Hidden Line Calculation Specifies if hidden lines are calculated for All Bodies or for Reference Data Separately.

Margin To see more reference data, set the value to expand the boundaries by a specified value on all sides.

Positional Representation Use to base a drawing view on a Positional Representation which is defined in an assembly file.

Display

All Model Dimensions Click to see model dimensions in the view (only active upon initial view creation). If the option's box is clear, model dimensions will not be placed upon view creation automatically. When checked, only the dimensions that are parallel to this view, and have not been retrieved in existing views on the sheet, will appear.

Model Welding Symbols This box is active only if you are creating a drawing view of a weldment. Click to retrieve and use welding symbols placed in the model in the drawing.

Bend Extents This box is active only if you are creating a drawing view of a sheet metal part flat pattern. Click to control the visibility of bend extent lines/edges.

Thread Feature This box is active only if you are creating a view of an assembly model. Click to set the visibility of thread features in the view.

Weld Annotations This box is active only if creating a view of a weldment. Click to control the display of weld annotations.

Work Features Click to display work features in the drawing view.

Tangent Edges Click to set the visibility of tangent edges in a selected view. Checking the box displays tangent edges; leaving the option's box clear hides them.

Show Trails Click to control the display of trails in drawing views based on presentation files.

Hatching Click to set the visibility of the hatch lines in the selected section view. Checking the box displays hatch lines; leaving the option's box clear hides them.

Align to Base Click to remove the alignment constraint of a selected view to its base view. When the box is checked, alignment of views exists. Leaving the option's box clear breaks the alignment and labels the selected view and the base view.

Definition in Base View Click to display or hide the projection line for a dependent detail or section view. Checking the box displays the line or circle and text label; leaving the option's box clear hides the line or circle and text label.

View Justification Use to control the position of drawing views when the size or position of the model changes. This area is especially helpful with creating drawing views from assembly models. This area contains two modes: Center and Fixed. The Center mode keeps the model image centered in the drawing view. If, however, the models view increases or decreases in size, the drawing view will shift on the drawing sheet. In some cases, this shift could overlap the drawing border or title block. The Fixed mode keeps the drawing view anchored on the drawing sheet. In the event the model image increases, an edge of the drawing view will remain fixed.

BASE VIEWS

A base view is the first view that you create from the selected part, assembly, or presentation file. When you create a base view, the scale is set in the dialog box. From the base view, you can project other drawing views. There is no limit to the number of base views you can create in a drawing (based on different parts, assemblies, or presentation files). As you create a base view, you can select the orientation of that view from the Orientation list found on the Component tab of the Drawing View dialog box.

To create a base view, follow these steps:

1. Click the Base View tool on the Drawing Views Panel Bar, or right-click in the graphics window and select Base View from the menu as shown in the following image. You can also select Base View found in Model Views on the Insert menu.

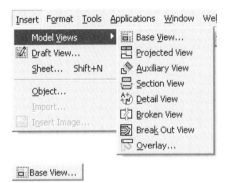

Figure 5-39

2. The Drawing View dialog box will appear. On the Component tab, click the Explore directories icon to navigate to and select the part, assembly, or presentation file from which to create the base drawing view. After making the selection, a preview image will appear attached to your cursor in the graphics window. Do not place the view until the desired view options have been set.

3. Select the type of view to generate from the Orientation list.

4. Select the scale for the view.

5. Select the style for the view.

6. Locate the view by selecting a point in the graphics window.

PROJECTED VIEWS

A projected view can be an orthographic or isometric view that you project from a base view or any other existing view. When you create a projected view, a preview image will appear, showing the orientation of the view you will create as the cursor moves to a location on the drawing. There is no limit to the number of projected views you can create. To create a projected drawing view, follow these steps:

1. Click the Projected View tool on the Drawing Views Panel Bar, as shown in the following image, select Projected View in Model Views on the Insert menu, or right-click inside the bounding area of an existing view box (shown as dashed lines when the cursor moves into the view) and select Create View > Projected from the menu.

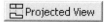

Figure 5-40

2. If you selected the Projected View tool click inside the desired view to start the projection.

3. Move the cursor horizontally, vertically, or at an angle to get a preview image of the view you will generate. Keep moving the cursor until the preview matches the view that you want to create, then press the left mouse button. Continue placing projected views.

4. When finished, right-click and select Create from the menu.

EXERCISE 5-2 Creating a Multiview Drawing

In this exercise, you will create an independent view to serve as the base view, and then you will add projected views to create a multiview orthographic drawing. Finally, you will add an isometric view to the drawing.

1. Open the file *ESS_E05_02.idw*. This drawing file contains a single sheet with a border and title block as shown in the following image.

Figure 5-41

2. Create a base view by clicking on the Base View tool. This will display the Drawing View dialog box.

3. Under the File area of the Component tab, click the Explore directories button and double-click the file *ESS_E05_02.ipt* to use it as the view source.

4. In the Orientation area, verify that Front is selected.

5. In the Scale list, select 1.

6. In Style, click the Hidden Line button.

7. Click the Options tab and ensure that All Model Dimensions is not checked.

8. Position the view preview in the lower left corner of the sheet (in Zone C6) then click to place the view as shown in the following image.

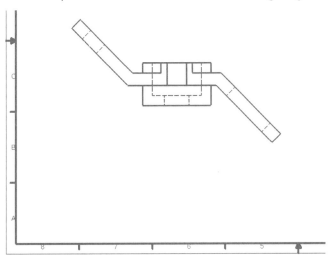

Figure 5-42

9. Click the Projected View tool in the panel bar. This tool will be used to create the top and right side views from the front view.

10. Click the base view and move the cursor vertically to a point above the base view. Click in Zone E6 to place the top view.

11. Move the cursor to the right of the base view. Click in Zone C2 to place the right-side view.

12. Right-click and choose Create from the menu to create the new views as shown in the following image.

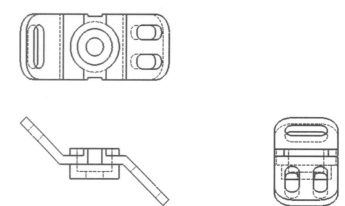

Figure 5-43

13. The views are crowded with a scale of 1:1. In the steps that follow, you will reduce the size of all drawing views by setting the base view scale to 1:2. The dependent views will update automatically. Right-click the base view (front view) and select Edit View.

 NOTE To activate this menu, right-click on the view border or inside the view. Do not right-click on the geometry.

14. When the Drawing View dialog box displays, select 1:2 from the Scale list and click OK.

15. The scale of all views updates as shown in the following image.

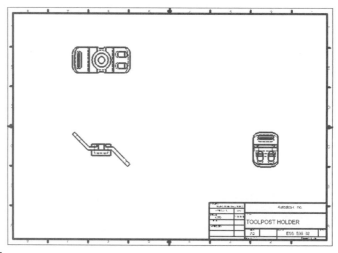

Figure 5-44

16. You will now change the drawing scale in the title block. Begin this process by expanding Sheet:1 in the Browser.

17. While still in the Browser, expand ANSI-Large under Sheet:1.

18. Right-click on the Field Text icon and choose Edit Field Text from the menu.

19. The Edit Property Fields dialog box will display. In the SCALE cell, click on 1:1 and enter a new scale of **1:2**.

20. Click OK. The title block updates to the new scale as shown in the following image.

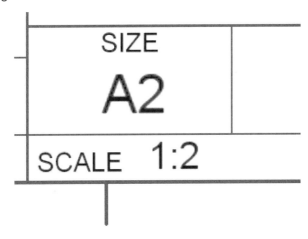

Figure 5-45

21. To complete the multiview layout, an isometric view will be created. Begin by clicking the Projected View tool in the panel bar.

22. Select the base view (front view) and move your cursor to a point above and to the right of the base view.

23. Click in Zone E3 to place the isometric view. Right-click the sheet and choose Create. Your drawing should appear similar to the following image.

Figure 5-46

24. Right-click the isometric view and choose Edit View. When the Drawing View dialog box is displayed, click the Options tab.

25. In the Display area, clear the check box for Tangent Edges, and then click OK.

26. You will complete this exercise by moving drawing views to better locations. Select the right-side view and drag it to Zone C4. Select the isometric view and drag it to Zone E4. Your drawing should appear similar to the following image.

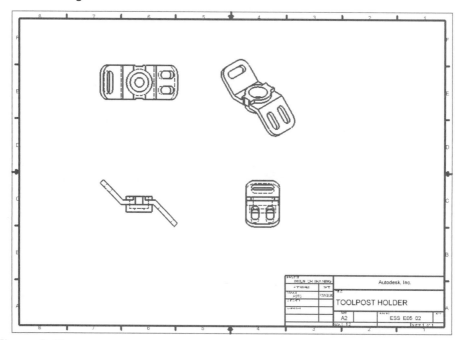

Figure 5-47

27. End of exercise.

AUXILIARY VIEWS

An auxiliary view is a view that is projected perpendicular to a selected edge or line in a base view. It is designed primarily to view the true size and shape of a surface that appears foreshortened in other views.

To create an auxiliary drawing view, follow these steps:

1. Click the Auxiliary View tool on the Drawing Views Panel Bar, as shown in the following image. You can also right-click inside the bounding area of an existing view box (shown as a dotted box when the cursor moves into the view) and then select Create View > Auxiliary from the menu.

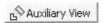

Figure 5-48

2. If you selected the Auxiliary View tool, click inside the view from which the auxiliary view will be projected. The Auxiliary View dialog box will appear, as shown in the following image. Type in a name for Label and a value for Scale, and then select one of the Style options (hidden, hidden line removed, or shaded).

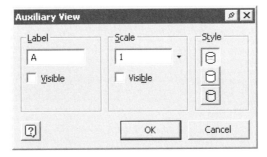

Figure 5-49

3. In the selected drawing view, select an edge or line from which the auxiliary view will be perpendicularly projected, as shown in the following image.

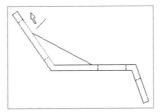

Figure 5-50

4. Move the cursor to position the auxiliary view, as shown in the following image. Notice that the view takes on a shaded appearance as you position it.

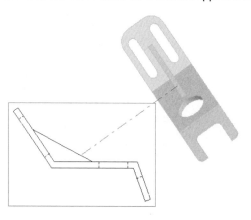

Figure 5-51

5. Click a point on the drawing sheet to create the auxiliary view. The following image illustrates the completed auxiliary view layout.

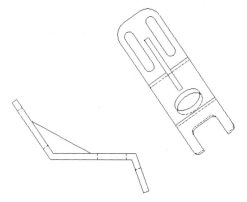

Figure 5-52

SECTION VIEWS

A section view is a view you create by sketching a line or multiple lines that will define the plane(s) that will be cut through a part or assembly. The view will represent the surface of the cut area and any geometry shown behind the cut face from the direction being viewed. When defining a section, you sketch line segments that are horizontal, vertical, or at an angle. You cannot use arcs, splines, or circles to define section lines. When you sketch the section line(s), geometric constraints will automatically be applied between the line being sketched and the geometry in the drawing view. You can also infer points by moving the cursor over (or scrubbing) certain geometry locations (such as centers of arcs, endpoints of lines, and so on) and then moving the cursor away to display a dotted line showing that you are inferring or tracking that point. To place a geometric constraint between the drawing view geometry and the section line, click in the drawing when a green circle appears; the glyph for the constraint will appear. If you do not want the section lines to have constraint(s) applied to them automatically when they are created, hold down the CTRL key when sketching the line(s). Because the area in the section view that is solid material appears with a hatch pattern by default, you may want to set the hatching style before creating a section view.

To create a section drawing view, follow these steps:

1. Click the Section View tool on the Drawing Views Panel Bar, as shown in the following image. You can also right-click inside the bounding area of an existing view box (shown as a dotted box when the cursor moves into the view) and select Create View > Section from the menu or select Section View from the Insert > Drawing Views menu.

Figure 5-53

2. If you selected the Section View tool, click inside the view from which to create the section view.

3. Sketch the line or lines that define where and how you want the view to be cut. In the following image, a vertical line is sketched through the center of the object.

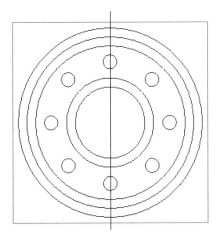

Figure 5-54

4. When you finish sketching the section line(s), right-click and select Continue from the menu.

5. The Section View dialog box will appear as shown in the following image. Fill in the information for how you want the label, scale, and style to appear in the drawing view.

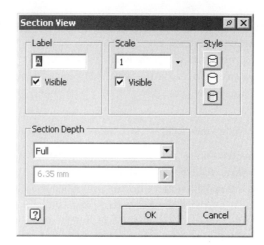

Figure 5-55

6. Move the cursor to position the section view, as shown in the following image, and select a point to place the view.

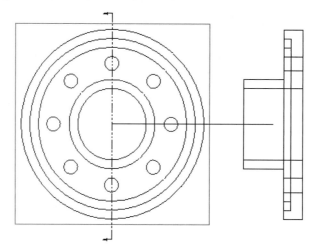

Figure 5-56

7. After you have created the view, you can add or delete constraints and/or dimensions to or from the section line(s) to better define the section cut. To add or delete a constraint or dimension, select the section line, right-click, and select Edit from the menu, as shown in the following image. You can also edit the section line by selecting the section line sketch located in the Browser.

8. Use the General Dimension and Constraint tools in the Drawing Sketch panel to add or delete constraints and dimensions.

9. When you are done editing the section line, right-click and select Finish Sketch or Return on the standard toolbar.

 NOTE If you did not constrain the section lines, you can drag them to show the section in a new location. After you drag the section lines, the dependent section view will be updated to reflect the change.

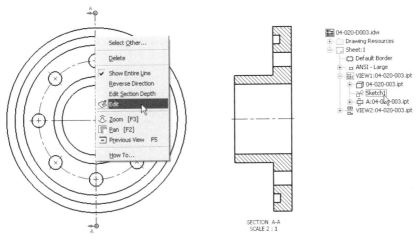

Figure 5-57

10. The following image illustrates the completed section view. Since the cutting plane cuts through the entire view, this is called a full section. Centerlines have also been added to the section view and to the bolt circles in the front view in this image.

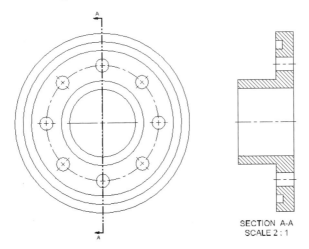

Figure 5-58

Half Sections

You can construct the cutting plane line in various configurations, depending on the desired results. When the cutting plane line cuts halfway through the object (as shown in the following image), this is referred to as a *half* section. The section view displays crosshatching through half of the surfaces.

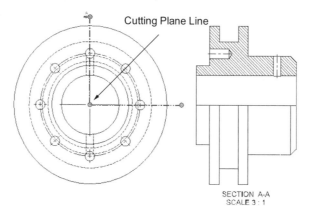

Figure 5-59

Aligned Sections

Aligned sections take into consideration the angular position of features in a drawing. Instead of the cutting plane line being drawn vertically through the object, it is angled or aligned with the same angle as the feature (in this case, the hole), as shown in the following image. This allows aligned sections to produce a clearer drawing. Also notice in the figure the creation of a solid line that runs through the center of the section view. This line is generated by the change in direction of the cutting plane line. If you want to add a centerline to this area, right-click on the solid horizontal line and turn off its visibility.

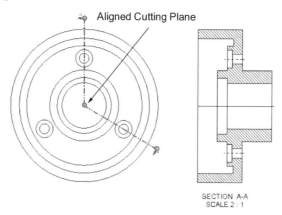

Figure 5-60

Offset Sections

Offset sections take their name from the offsetting of the cutting plane line that passes through certain details in a given view. If the cutting plane passes straight through an object, some features are exposed while others remain hidden. By offsetting the cutting plane line as shown in the following image, you can control the cutting plane line direction and the specific features of a part it passes through.

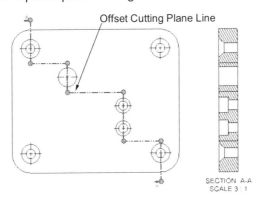

Figure 5-61

Modifying a Hatch Pattern

You can edit the section view hatch pattern by right-clicking on the hatch pattern in the section view and selecting Modify Hatch from the menu. This will launch the Modify Hatch Pattern dialog box shown in the following image. Make the desired changes in the Pattern, Scale, Angle, Shift, Line Weight, Color and Double mode areas. Click the OK button to reflect the changes in the drawing.

Figure 5-62

 NOTE The Shift mode of the Modify Hatch Pattern dialog box shifts the hatch pattern to offset it slightly from the hatch pattern on a different part. This option would be ideal for creating sections that involve assembly models. Enter the distance for the shift.

DETAIL VIEWS

A detail view is a drawing view that isolates an area of an existing drawing view and can reflect a specified scale. You define a detailed area by a circle or rectangle and can place it anywhere on the sheet. To create a detail drawing view, follow these steps.

1. Click the Detail View tool on the Drawing Views Panel Bar, as shown in the following image. You can also right-click inside the bounding area of an existing view box (shown as dashed lines when the cursor moves into the view) and select Create View > Detail from the menu.

Figure 5-63

2. If you selected the Detail View tool, click inside the view from which you will create the detail view.

3. The Detail View dialog box will appear as shown in the following image. Fill in information according to how you want the label, scale, and style to appear in the drawing view. Do not click the OK button at this time, as doing so will generate an error message.

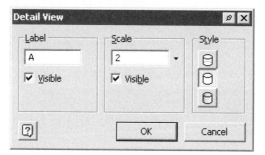

Figure 5-64

4. In the selected view, select a point to use as the center of the circle that will describe the detail area (as shown in the following image).

Chapter 5 Creating and Editing Drawing Views 249

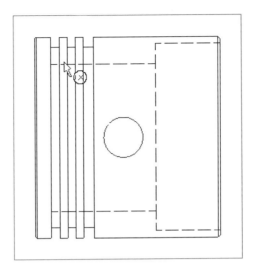

Figure 5-65

 NOTE You can also right-click at this point and select Rectangular Fence to change the bounding area of the detail area definition.

5. Select another point that will define the radius of the detail circle. As you move the cursor, a preview circle of the specified size will appear, as shown in the following image.

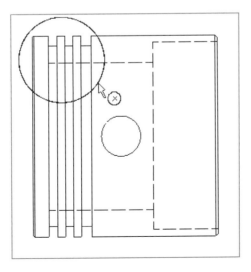

Figure 5-66

6. Select a point on the sheet where you want to place the view. There are no restrictions on where you can place the view. The following image illustrates the completed detail view.

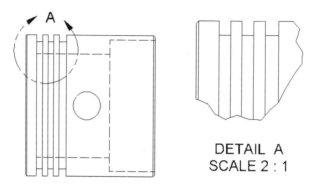

Figure 5-67

BROKEN VIEWS

When creating drawing views of long parts, you may want to remove a section or multiple sections from the middle of the part and show just the ends. This type of view is referred to as a broken view. You may, for example, want to create a drawing view of a 50 x 50 x 6 angle, shown in the following image, which is 1500 mm long and only has the ends chamfered. When you create a drawing view, the detail of the ends is small and difficult to see because the part is so long.

Figure 5-68

In this case, you can create a broken view that removes a 1000 mm section from the middle of the angle and leaves 250 mm on each end. When you place an overall length dimension that spans the break, it appears as 1500 mm, and the dimension line has a break symbol to note that it was derived from a broken view (see the following image).

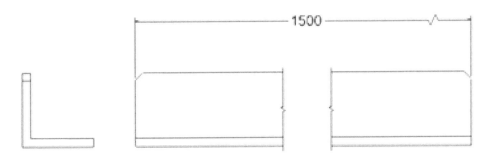

Figure 5-69

You create a broken view by adding breaks to an existing drawing view. The view types that can be changed into broken views are the following: part views, assembly views, projected views, isometric views, auxiliary views, section views, break-out views, and detail views. After creating a broken view, you can move the breaks dynamically to change what you see in the broken view.

To create a broken drawing view, follow these steps:

1. Create a base or projected view (one that will be shown eventually as broken).

2. Click the Broken View tool on the Drawing Views Panel Bar, as shown in the following image. You can also right-click inside the bounding area of an existing view box (shown as a dotted rectangle when the cursor is moved into the view). Select Create View > Broken from the menu.

Figure 5-70

3. If you selected the Broken View tool from the Drawing Views Panel Bar, click inside the view from which to create the broken view.

4. The Broken View dialog box will appear, as shown in the following image. Do not click the OK button at this time, as this will end the operation.

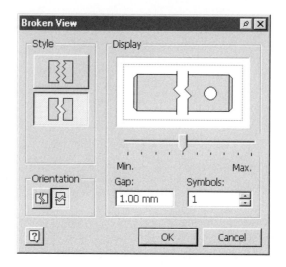

Figure 5-71

5. In the drawing view that will be broken, select a point where the break will begin (see the following image).

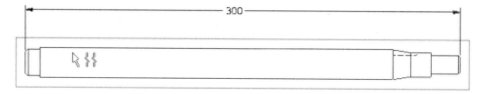

Figure 5-72

6. Then select a second point to locate the second break, as shown in the following image. As you move the cursor, a preview image will appear to show the placement of the second break line.

Figure 5-73

7. The following image illustrates the results of creating a broken view. Notice the dimension line appears with a break symbol to signify the view is broken.

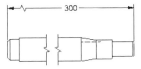

Figure 5-74

8. To edit the properties of the broken view, move the cursor over the break lines and a green circle will appear in the middle of the break. Right-click and select Edit Break from the menu. The same Broken View dialog box will appear. Edit the data as needed and click the OK button to complete the edit.

9. To move the break lines, click on one of the break lines and, with the left mouse button pressed down, drag the break line to a new location, as shown in the following image. The other break line will follow to maintain the gap size.

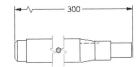

Figure 5-75

BREAK-OUT VIEWS

For cases when you want to expose internal components or features, a method in Autodesk Inventor is available that allows you to cut or peel away a body and expose those internal parts. This method is called a break-out view. It is not unusual for assemblies to have housings or covers that hide internal components. Break-out views make it possible to expose these components. You can create break-out views on assemblies as well as part files.

To use break-out views you will need to be able to define two items: a closed profile boundary over the area to break, and the depth of material to remove or the portion of a component to remove in order to see other components. As with broken views, the break-out view is defined and displayed on the same view.

To create a break-out view, click the Break Out View tool on the Drawing Views Panel Bar, as shown in the following image.

Figure 5-76

After you select the view in which the break out will occur, the Break Out View dialog box will appear, as shown in the following image. The termination options, From Point, To

Sketch, To Hole, and Through Part, are available to give you control over how the break-out section is created. The following sections explain these options.

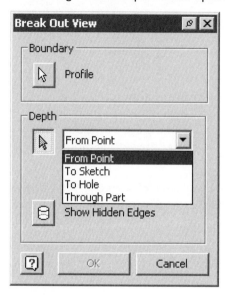

Figure 5-77

All four options are explained as follows:

From Point   Select this to make the break out occur at a specified distance from a point located in a view. The point could be located in the base view or in a projected view.

To Sketch   Select this to use sketched geometry associated with another view to define the depth of the break out.

To Hole   Select this to base the break out on a hole feature whose axis determines the termination depth of the break.

Through Part   Select this to remove drawing content from inside a closed profile through selected components located in the Browser. This termination option is especially useful when you want to reveal internal components or features.

Creating a Break-Out View Using the From Point Option

To create a break-out view using the From Point option, follow these steps:

1. Click on the view in which to create the break-out section.

2. Click on the Sketch button on the standard toolbar and sketch a closed profile over the area you want broken out, as shown in the following image. When done, right-click and select Finish Sketch from the menu.

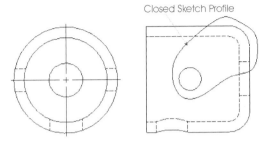

Figure 5-78

3. Click the Break Out View button; this will activate the Break Out View dialog box. Select the view in which the break out will occur and select the defined boundary. Select the sketch you just created as the boundary in the view that will contain the break out.

4. Select the From Point option in the dialog box, click on the Depth arrow, and select a point in the adjacent projected view. This point will be used to calculate the depth of cut based on the distance (see the following image).

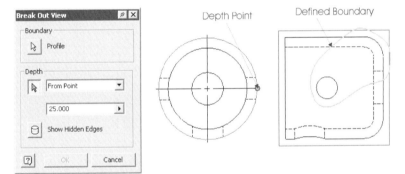

Figure 5-79

5. The following image illustrates the results of using the From Point option for creating a break out. The depth of 25 units from the selected point, cuts the view at the middle of the circular features, thus displaying the wall thickness of the part.

 NOTE You can also select a point at the center of the circle for the depth of the cut. In this case the distance value would change to 0 because the depth of the cut was specified by a point.

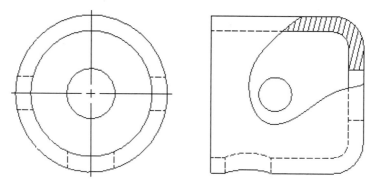

Figure 5-80

Creating a Break-Out View Using the To Sketch Option

To create a break-out view using the To Sketch option, follow these steps:

1. Activate the view in which the break out will occur and sketch a closed profile over the area you want broken out. In the example shown in the following image, two closed profiles will be used to create the break-out view. When finished with this operation, right-click and select Finish Sketch.

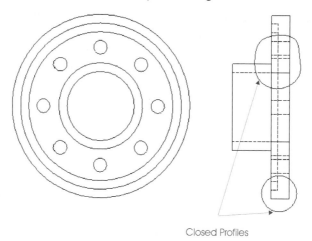

Closed Profiles

Figure 5-81

2. Activate the adjacent projected view and create another sketch. This sketch will be used to determine the depth of the cut for the break out, as shown in the following image. When finished with this operation, right-click and select Finish Sketch.

Chapter 5 Creating and Editing Drawing Views 257

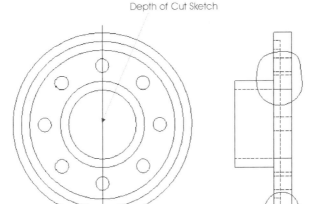

Figure 5-82

3. Click the Break Out View tool and select a view as the base for the break-out view.

4. When the Break Out View dialog box appears, select the initial sketch or sketches to use as the boundary profile. In the Depth area of the dialog box, change the option to Sketch. In the adjacent projected view, select the sketch that will determine the depth of the cut for creating the break-out view (see the following image).

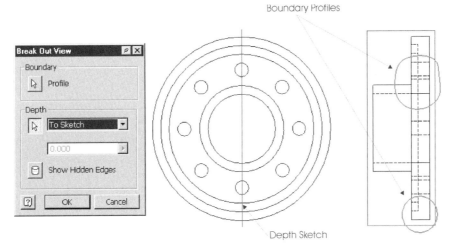

Figure 5-83

5. The following image shows the break-out view created based on the first group of sketched boundaries and the second sketch, which determined the depth of the cut.

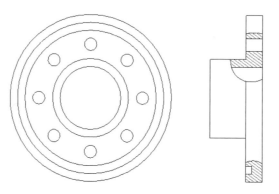

Figure 5-84

Creating a Break-Out View Using the To Hole Option

The To Hole option for creating break-out views is similar to the To Sketch option. Instead of basing the break out on a sketch, it will base it on a hole feature. It is the axis of the hole feature that determines the termination depth.

To create a break-out view using the To Hole option, follow these steps:

1. Click the Break Out View button on the Drawing Views Panel Bar.

2. Activate the view in which the break out will occur and sketch a closed profile around the hole you want broken out.

3. In the graphics area, click to select the view. When the Break Out View dialog box appears, click to select the defined boundary. The boundary profile must be on a sketch associated to the selected view. In the Break Out View dialog box, click the arrow next to the Depth type box and select the To Hole option, as shown in the following image.

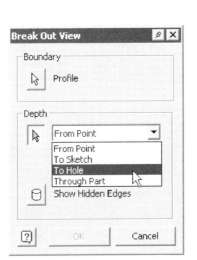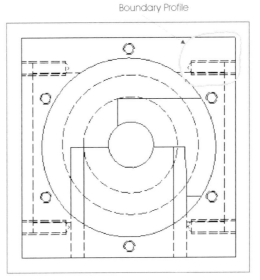

Boundary Profile

Figure 5-85

4. Click the select arrow, then click to select the hole feature in the graphic window, as shown in the following image. The depth is defined by the axis of the hole.

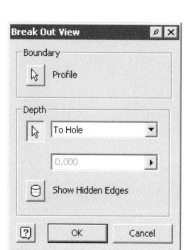
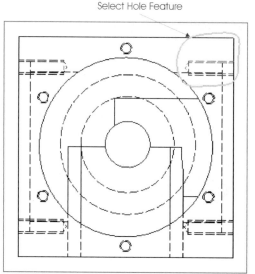

Figure 5-86

 NOTE If the hole feature is hidden, edit the view, and click the Show Hidden Edges button to temporarily display hidden lines.

5. When the view is defined fully, click OK to create the view (as shown in the following image).

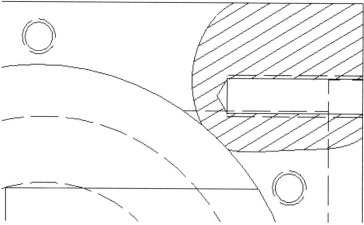

Figure 5-87

Creating a Break Out View Using the Through Part Option

In the following image, a housing hides internal details of an assembly. You can cut away a segment of the housing to expose obscured parts or features using the Through Part option of the Break Out View tool.

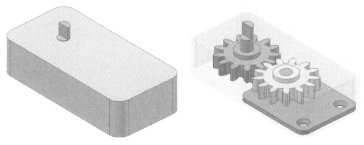

Figure 5-88

To create a break-out view using the Through Part option, follow these steps:

1. Activate the view through which you wish to cut and click on the Sketch button in the standard toolbar.

2. Next, sketch a closed profile (as shown in the following image) as the area through which you wish to cut. When finished, right-click and select Finish Sketch.

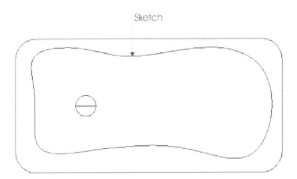

Figure 5-89

3. Click on the Break Out View tool. Selecting the view in which to perform the operation will activate the Break Out View dialog box. The boundary profile consisting of the previous sketch will be selected automatically. In the Depth area of the dialog box, click on the Through Part option (see the following image).

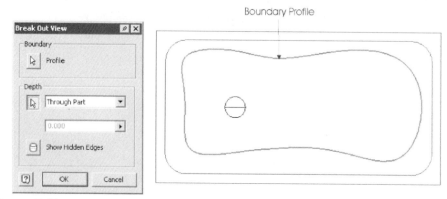

Figure 5-90

4. Move over to the Browser and select the part on which to perform the penetration operation—in this case, the Rotation_Gear Box part, as shown in the following image.

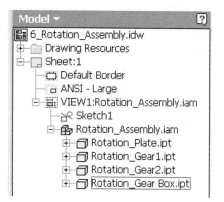

Figure 5-91

5. When finished, click the OK button. The following image illustrates the results of using the Through Part option. The gear cover has been sliced away based on the sketched boundary to expose the internal workings of the gear mechanism.

Figure 5-92

The Through Part option can be very effective when applied to an isometric view. In the example shown in the following image, a spline sketch was created on an isometric view. Notice how the sketch covers the top and sides of the isometric. The image on the right in the figure illustrates the completed break-out view. All sides surrounded by the spline sketch break out to display the internal components.

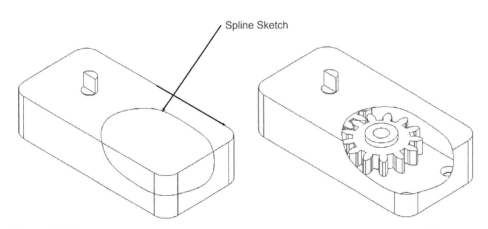

Figure 5-93

SECTION DEPTH

The section depth function allows you to create sliced section views by floating a slicing line a set distance away from the cutting plane line. The following image illustrates a typical full section. In the Section View dialog box, the Section Depth has been set to Full. This creates a section view in addition to displaying all geometry beyond the cutting plane line.

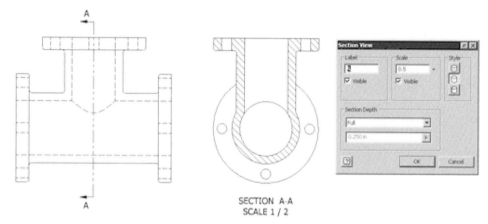

Figure 5-94

In the following example, the Section Depth has been set to Distance. The default distance is 0.25. This creates a slice line a distance of 0.25 units from the cutting plane line. The following image illustrates the results. The section view created does not show any geometry more than 0.25 units beyond the cutting plane.

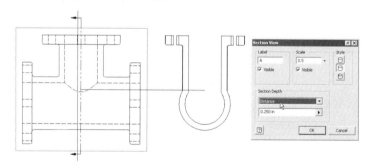

Figure 5-95

The following image illustrates the results of creating the section by Section Depth cut. Increasing the distance of Section Depth will show more geometry past the cutting plane line.

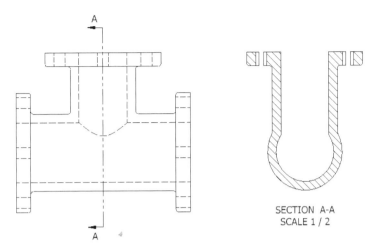

Figure 5-96

EDITING THE SECTION VIEW DEPTH

The following image shows a section view with the Section Depth set to Full. To set a distance for this section view, move your cursor over the view, right-click, and select Edit Section Depth from the menu.

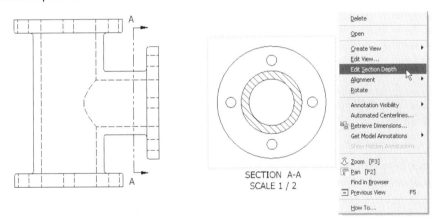

Figure 5-97

This will display the Edit Section Depth dialog box as shown in the following image. Set the Section Depth to Distance and enter a value for the distance (in this example, **0.25**).

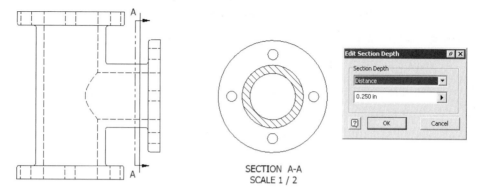

Figure 5-98

The following image illustrates the results. Only the surfaces cut by the cutting plane line appear. All geometry 0.25 units past the cutting plane line is ignored.

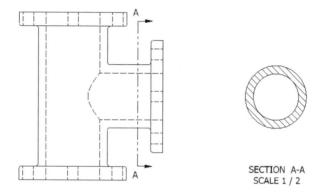

Figure 5-99

CREATING DRAFT VIEWS

Another type of drawing view is a draft view. You do not create a draft view from a 3D part. Draft views contain one or more associated 2D sketches. You can create a draft view on its own sheet, in its own Autodesk Inventor drawing file, or you can use a draft view in an existing sheet with other drawing views to provide detail that is missing in a model. You can use the Drawing Sketch Panel Bar to add geometry, constraints, parametric dimensions, text, and other annotations in the draft view. All these objects will be placed in the draft view and, if the draft view moves, so will all the objects associated with it. When you import an AutoCAD 2D file into an Autodesk Inventor drawing, the geometry is placed in a draft view.

To create a draft view, follow these steps:

1. Click the Draft View tool on the Drawing Views Panel Bar, as shown in the following image.

Chapter 5 Creating and Editing Drawing Views 267

Figure 5-100

2. The Draft View dialog box will appear, as shown in the following image. Fill in information for the Label and Scale of the draft view as you want them to appear in the drawing. Then click the OK button to continue. The Scale setting that you set will be the scale to which you are drafting the objects.

Figure 5-101

3. Using the tools on the Drawing Sketch Panel Bar, create 2D geometry and add constraints, dimensions, and annotations.

4. When you have finished creating the geometry that will make up the draft view, you can either right-click and select Finish Sketch from the menu, as shown in the following image, or click the Return button on the standard toolbar. The draft view will now act like a regular drawing view that you can edit.

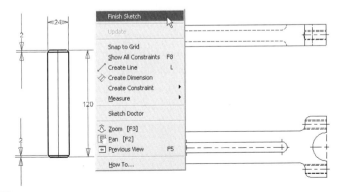

Figure 5-102

CREATING PERSPECTIVE VIEWS

In addition to creating isometric drawing views, you can create perspective drawing views based on part, assembly, presentation, and custom view information. Perspective views provide a more natural or realistic view of an assembly or component.

Perspective views have a very narrow and specific use, they are not treated as normal views because other views are not expected to be projected or developed from them. Therefore, when you select a perspective view, view creation commands that require a base view will not be enabled. You will also be unable to apply dimensions to perspective views.

To create a perspective view, follow these steps:

1. While in drawing mode, click on the Base View button on the Drawing Views Panel Bar.

2. While in the Drawing View dialog box, select the model file to use for the perspective view. Then click on the Custom View button, as shown in the following image (this button is labeled Change view orientation).

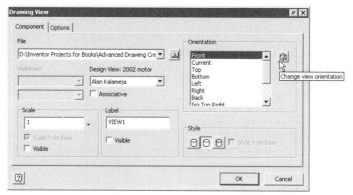

Figure 5-103

3. The Custom View window containing the current model image will appear. Use Pan, Zoom, Rotate, or other viewing tools on the standard toolbar to position the view, as shown in the following image. The figure also illustrates the Incremental View Rotate dialog box, which is used for fine-tuning the position of the view.

Chapter 5 Creating and Editing Drawing Views 269

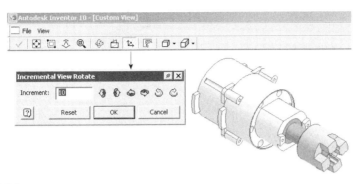

Figure 5-104

4. Use the Perspective Camera tool, as shown in the following image, to switch the view to a perspective.

 NOTE You can resize the Custom View panel by double-clicking on its title bar.

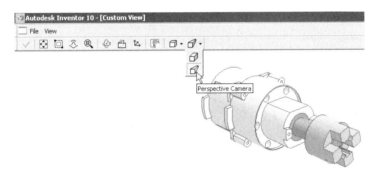

Figure 5-105

5. When you are pleased with the view displayed in the Custom View window, click on the green check mark located on the standard toolbar to accept this view and close the window, as shown in the following image. You are returned to the Drawing View dialog box, where you can make any additional changes, such as scale, before placing the view inside the drawing border. The right side of the following image illustrates the created perspective drawing view.

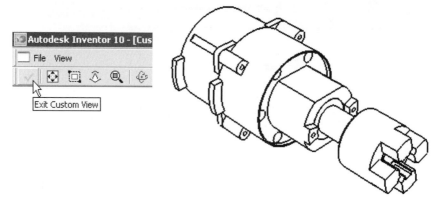

Figure 5-106

 NOTE If the model file is open and the current display is set to Camera Perspective, you can set up the perspective view in the drawing by selecting Current from the Orientation list of the Drawing View dialog box, and then placing the view in the drawing.

EXERCISE 5-3 Complex Drawing View Techniques

In this exercise, you create a variety of complex drawing views from a model of a cover.

1. Open the file *ESS_E05_03.idw*. Use the Zoom tool to zoom in on the base view as shown in the following image.

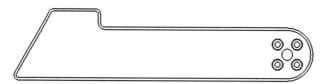

Figure 5-107

2. You will now create a broken view. First, click the Broken View tool.
3. In the graphics window, select the lower left view to break.
4. Select the start and end points for the material to be removed as shown in the following image.

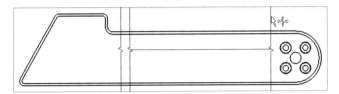

Figure 5-108

5. The results are displayed as shown in the following image. The base and projected views are broken.

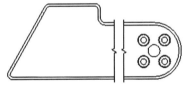

Figure 5-109

6. Next, you will create a section view. First, click the Section View tool.

7. In the graphics window, select the lower left broken view.

8. To define the first point of the section line, hover the cursor over the center hole, then click a point directly above the hole as shown in the following image.

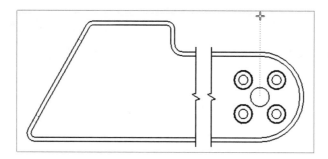

Figure 5-110

9. Complete the three-segment section line as shown in the following image, then right-click and select Continue.

10. Place the section view to the right of the broken view as shown in the following image.

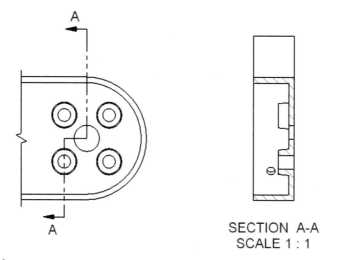

Figure 5-111

11. Next, you will create an auxiliary view. First, click the Auxiliary view tool.

12. In the graphics window, select the broken view.

13. Select the outside angled line, as shown below, to define the projection direction.

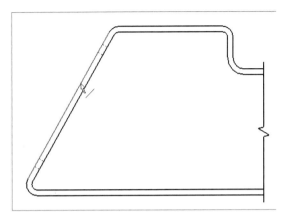

Figure 5-112

14. Place the view to the left of the broken view as shown in the following image.

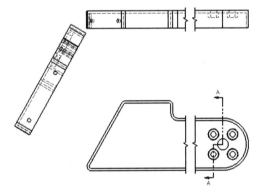

Figure 5-113

15. Next, you break the alignment of the auxiliary view to the parent view and reposition the auxiliary view. In the graphics window, double-click near the auxiliary view geometry to edit the auxiliary view.

16. In the Drawing View dialog box, click the Options tab, and clear the Align to Base option; then select OK.

17. In the graphics window, select and drag the auxiliary view to the right side of the drawing sheet as shown in the following image.

18. Adjust the length of the Auxiliary View cutting plane line by clicking and dragging on the green handles.

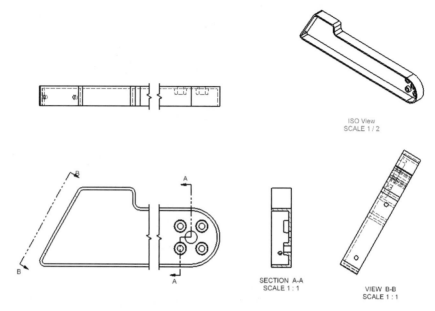

Figure 5-114

19. Finally, you will create a detail view. First, click the Detail View tool.
20. In the graphics window, select the lower left broken view.
21. Select a point near the left corner of the view to define the center of the circular boundary of the detail view.
22. Drag the circular boundary to include the entire lower left corner of the view geometry as shown in the following image, then click when finished.

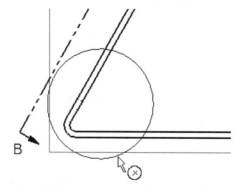

Figure 5-115

23. Click to position the detail view below the broken view as shown in the following image.

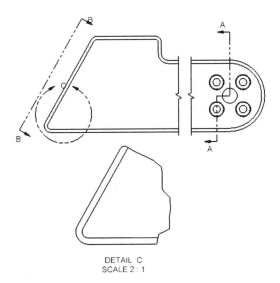

Figure 5-116

24. To finish this exercise, reposition the drawing views as required. The finished drawing is displayed as shown in the following image.

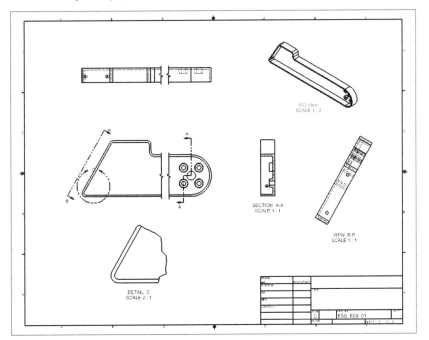

Figure 5-117

25. End of exercise.

EDITING DRAWING VIEWS

After creating the drawing views, you may need to move, edit properties of, or delete a drawing view. The following sections discuss these options.

MOVING DRAWING VIEWS

To move a drawing view, first click inside the view until a bounding box consisting of red dotted lines appears, as shown in the following image. Then move your cursor to the red dotted boundary, press and hold down the left mouse button, and move the view to its new location. Release the mouse button when you are finished. As you move the view, a rectangle will appear that represents the bounding box of the drawing view. If you move a base view, any projected (children or dependent) views will also move with it as required to maintain view alignment. If you move an orthographic or auxiliary view, you will only be able move it along the axis in which it was projected from the part edge or face. You can move detail and isometric views anywhere in the drawing sheet.

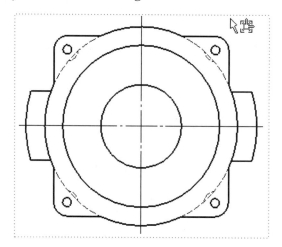

Figure 5-118

 NOTE You can also move a view by picking the view boundary and dragging it directly without even activating the view with a pick first.

EDITING DRAWING VIEW PROPERTIES

After creating a drawing view, you may need to change the label, scale, style, or hatching visibility; break the alignment constraint to its base view; or control the visibility of the view projection lines for a section or auxiliary view. To edit a drawing view, follow one of these steps:

- Double-click in the bounding area of the view.

- Double-click on the icon of the view you want to edit in the Browser.

Right-click in the drawing view's bounding area or on its name, and select Edit View from the menu, as shown in the following image.

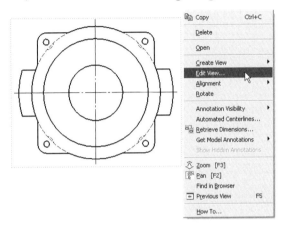

Figure 5-119

- Right-click on the drawing view's name in the Browser and click Edit View from the menu.

When you perform the operations listed above, the Drawing View dialog box will appear as shown in the following image. This is the same dialog box used to create drawing views. Depending on the view that you selected, certain options may be grayed-out from the dialog box. Make the necessary changes and then click the OK button to complete the edit. When you change the scale in the base view, all of the dependent views will be scaled as well.

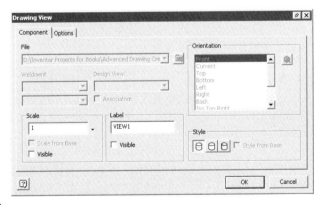

Figure 5-120

See the Using the Drawing View Dialog Box section in this chapter for detailed descriptions of Component and Options tab elements.

DISPLAYING TANGENT EDGES

The following image illustrates an isometric drawing view where the faces of the object change direction. The directional changes are filleted instead of forming sharp corners. However, when you project the view, the true projection leaves out any traces of the fillets. While this is correct, it gives the appearance that something is missing from the view. On the Options tab of the Drawing View dialog box is an option to turn on tangent edges. This feature is illustrated by the middle graphic in the following image. Lines are formed where the fillets begin and end. This gives a better idea of the form and purpose of the object. With Tangent Edges turned on, you can also have the edges foreshortened. Notice the tangent edges in the last image in the figure do not touch the edges of the object. Rather a small gap is present to help further define the design intent of the view.

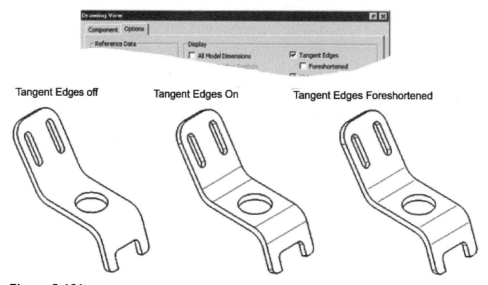

Figure 5-121

DELETING DRAWING VIEWS

To delete a drawing view, either right-click in the bounding area of the drawing view or on its name in the Browser and select Delete from the menu. You can also click in the bounding area of the drawing view and press the DELETE key on the keyboard. A dialog box will appear asking to confirm the deletion of the view. If the selected view has a view that is dependent on it, you will be asked if the dependent views should also be deleted. By default, the dependant view will be deleted. To exclude a dependant view from the delete operation, expand the dialog box and click on the Delete cell to change the selection to No.

DRAWING VIEW ALIGNMENTS

There may be times when you will need to reestablish the alignment of views. In the following image, the section view of the flanged pipe tee was generated from the top view as identified by the cutting plane line. The side view is not aligned with any view and appears to float on the drawing sheet. This view needs to be in alignment with the front view in section. To perform this operation, right-click on the bounding box of the side view and select Alignment from the menu. Select Horizontal because the side view is along the right of the front view. This describes a horizontal direction.

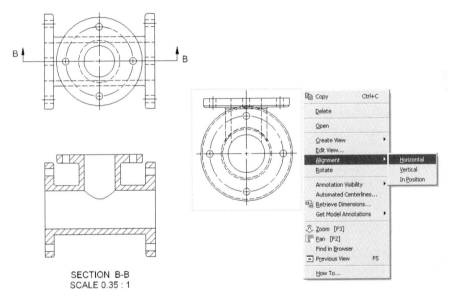

Figure 5-122

Next you pick a base view to use as reference for the horizontal alignment. As shown in the following image, select the section view as the base.

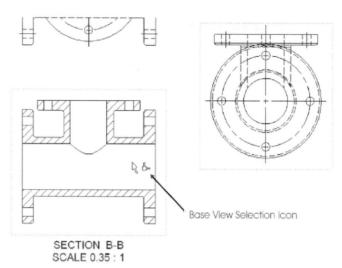

Figure 5-123

The following image illustrates the results. You can now move the side view horizontally and maintain the view alignment.

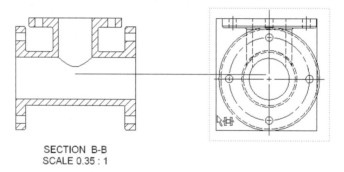

Figure 5-124

EXERCISE 5-4 Editing Drawing Views

In this exercise, you delete the base view while retaining its dependent views. The base view is not required to document the part, but its dependent views are. Next, you align the section view with the right-side view to maintain the proper orthographic relationship between the views.

You also modify the hatch pattern of the section to better represent the material.

> 1. Open the drawing file *ESS_E05_04.idw*. This drawing contains three orthographic views, an isometric view, and a section view as shown in the following image.

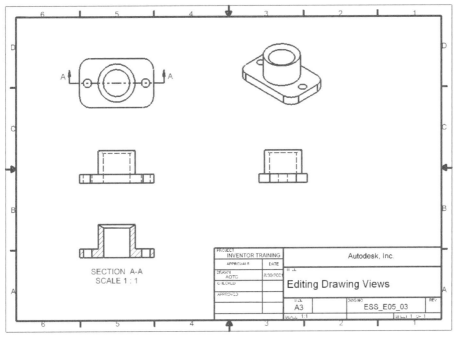

Figure 5-125

2. Right-click on the base view as shown in the following image and choose Delete from the menu.

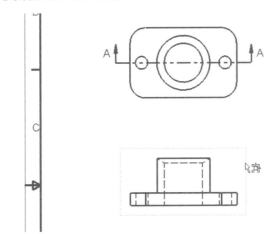

Figure 5-126

3. In the Delete View dialog, notice that the projected views are highlighted on the drawing sheet. Select the More (>>) button in the dialog box.

4. Click on the word Yes in the Delete column for each dependent view to toggle each to No. This will delete the base view but leave the dependent views present on the drawing sheet.

5. Click OK to delete the base view and retain the two dependent views. Your drawing should appear similar to the following image.

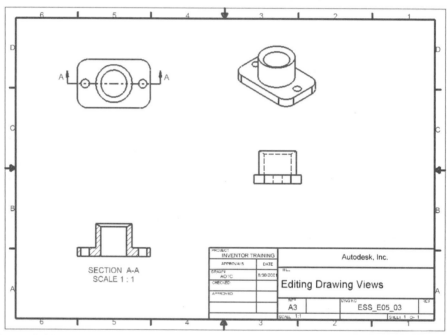

Figure 5-127

With the front view deleted, the section view needs to be identified as the new base view and aligned with the right-side view.

6. Right-click the right-side view, and choose Alignment > Horizontal from the menu.

7. Select the section view as the base view.

8. Select the section view and drag the view vertically to the location previously occupied by the front view. Notice how the right-side view remains aligned to the section view.

9. Right-click the isometric view and select Alignment > In Position.

10. Select the section view as the base view. Move the section view and notice that the isometric view now moves with the section view. Your drawing should appear similar to the following image.

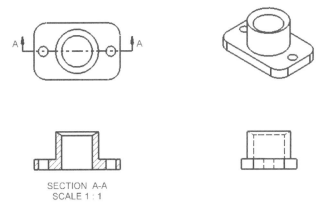

Figure 5-128

You now edit the section view hatch pattern to represent the material as bronze using the ANSI 33 hatch pattern.

11. Right-click the hatch pattern in the section view and choose Modify Hatch from the menu.

12. The Modify Hatch Pattern dialog box is displayed. Select ANSI 33 from the Pattern list and click OK. The hatch pattern changes as shown in the following image.

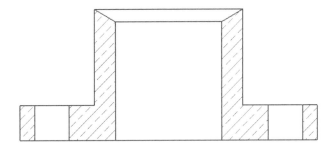

Figure 5-129

13. End of exercise.

DIMENSIONS

Once you have created the drawing view(s), you may need to change the value of a model dimension. If the dimensions did not appear when you created the view, you may want to get the model dimensions so they appear in a view, hide certain dimensions, add drawing (general) dimensions, or move dimensions to a new location. The following sections describe these operations.

RETRIEVING MODEL DIMENSIONS

Model dimensions may not appear automatically once you have created a drawing view. You can use the Retrieve Dimensions tool to select valid model dimensions for display in a drawing view. Only those dimensions that you placed in the model on a plane parallel to the view will appear.

To activate this command, use one of the following three methods:

- Right-click in the bounding area of a drawing view and select Retrieve Dimensions.
- Right-click on the name of the view in the Browser and select Retrieve Dimension from the menu.
- Click on the Retrieve Dimension button located at the bottom of the Drawing Annotations Panel Bar.

The following image shows all three methods.

The appropriate dimensions for the view that were used to create the part will appear.

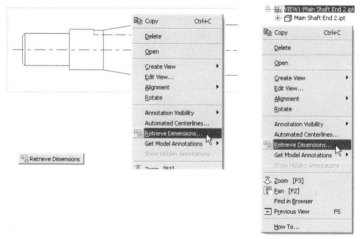

Figure 5-130

Selecting Retrieve Dimensions will display the Retrieve Dimensions dialog box shown in the following image. Through this dialog box you first select a drawing view in which to retrieve the model dimensions. The Select Source area of the dialog box allows you to retrieve model dimensions based on an entire part or based upon part features. The Select Parts option allows you to retrieve all model dimensions based on a part. When you use either of these methods, only valid model dimensions appear in preview mode. Once you have previewed the model dimensions, click the Select Dimensions button to pick the dimensions you want to retrieve.

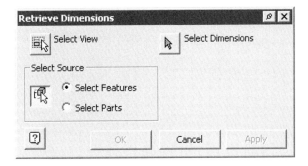

Figure 5-131

To retrieve model dimensions, follow these steps:

1. Click Retrieve Dimensions on the Drawing Annotations Panel Bar.

2. Select the drawing view in which to retrieve model dimensions.

3. To retrieve model dimensions based on one or more features, click the Select Features radio button. Then select the desired features.

4. If you want to retrieve model dimensions based on the entire part in a drawing view, click the Select Parts radio button. Then select the desired part or parts. You can select multiple objects by holding down the SHIFT or CTRL keys. You can also select objects with a selection window or crossing selection box.

5. When the model dimensions appear in the drawing view in preview mode, click the Select Dimensions button and select the desired dimensions. The selected dimensions will be highlighted.

6. Click the Apply button to retrieve the model dimensions and leave the dialog box active for further retrieval operations.

7. Click the OK button to retrieve the selected dimensions and dismiss the Retrieve Dimensions dialog box.

The following image shows the effects of retrieving model dimensions using the Select Features mode. In this example, the large rectangle of the engine block was selected. This resulted in the horizontal and vertical model dimensions being retrieved.

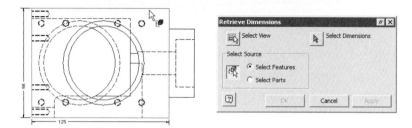

Figure 5-132

The following image shows the effects of retrieving model dimensions using the Select Parts mode. In this example, the entire engine block was selected. This resulted in all model dimensions parallel to this drawing view being retrieved.

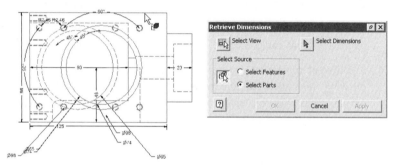

Figure 5-133

You can also retrieve model dimensions automatically during the drawing view creation process. To perform this task, follow these steps:

1. Launch the Options dialog box by clicking Tools > Application Options.

2. Click on the Drawing tab and locate the option Retrieve all model dimensions on view placement, as shown in the following image.

3. Click in the option's box to automatically retrieve model dimensions in all placed views. (Clear the box if you want to create drawing views without retrieving model dimensions automatically.)

4. When you want to retrieve model dimensions in a base drawing view, click the All Model Dimensions option on the Options tab of the Drawing View dialog box.

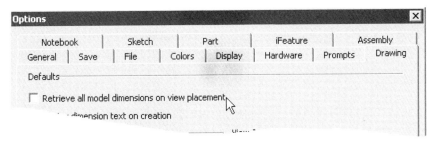

Figure 5-134

DIMENSION VISIBILITY

To hide all of the model or drawing dimensions for a drawing view, right-click in the bounding area of a drawing view and select Annotation Visibility from the menu, as shown in the following image. The appearance of a check means the specific feature is turned on. Clicking the check adjacent to Model Dimensions will remove the check and turn off the model dimensions in a drawing. Use the same operation to turn off drawing dimensions.

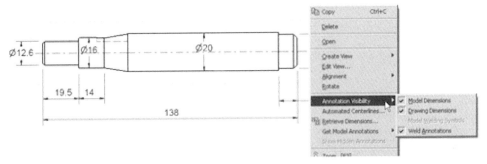

Figure 5-135

CHANGING MODEL DIMENSION VALUES

While working in a drawing view, you may find it necessary to change the model (parametric) dimensions of a part. You can open the part file, change a dimension's value and save the part, and the change will then be reflected in the drawing views. You can also change a dimension's value in the drawing view by right-clicking on the dimension and selecting Edit Model Dimension from the menu, as shown in the following image.

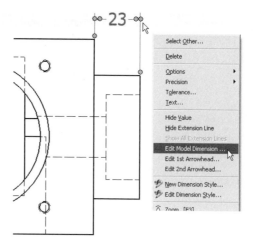

Figure 5-136

The Edit Dimension text box will appear. Enter a new value and click the check mark, as shown in the following image, or press ENTER. The associated part will be updated and saved and the associated drawing view(s) will be updated automatically to reflect the new value.

Figure 5-137

 NOTE It is possible to edit model dimensions found in drawing views. This feature is made possible when you install Autodesk Inventor and enable the option to edit model dimensions from the drawing. It is considered, however, poor practice to make major dimension changes from the drawing. It is highly recommended that you make all dimension changes through the part model.

GENERAL DIMENSIONS

After laying out the drawing views, you may find that a dimension other than an existing model dimension or an additional dimension is required to better define the part. You can go back and add a parametric dimension to the sketch if the part was underconstrained, add a driven dimension if the sketch was fully constrained, or add a drawing (general) dimension to the drawing view. A general dimension is not a parametric dimension, however, it is associative to the geometry to which it is referenced. The general dimension reflects the size of the geometry being dimensioned. After you create a general dimension, and the value of the geometry that you dimensioned changes, the general dimension will be updated to reflect the change. You add a general dimension with the General Dimension tool on the

Drawing Annotation Panel Bar, as shown in the following image. Create a general dimension in the same way you would create a parametric dimension.

Figure 5-138

ADDING GENERAL DIMENSIONS TO A DRAWING VIEW

The following image illustrates a simple object with various different dimension types. General dimensions can take the form of linear dimensions illustrated at A and B, diameter dimensions at C, radius dimensions at D, and angular dimensions at E. When using the General Dimension tool, Inventor automatically chooses the dimension type depending on the object you chose. In this example, the counterbored hole is dimensioned using a special dimension tool called a hole note. This tool will be described later in this chapter.

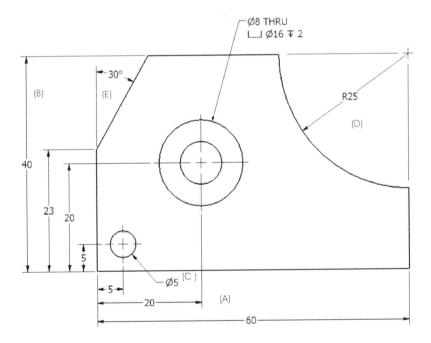

Figure 5-139

GENERAL DIMENSIONS–LINEAR

The following image illustrates one method to create a linear general dimension. Select the horizontal line; this will activate the linear general dimension mode. Then locate the dimension by moving your cursor as shown in the lower image. As you move your cursor, the dimension text will become centered and the dimension line will snap to a predefined distance away from the object. At this point, the dimension and extension lines will take on a highlighted appearance. Click again to place the dimension; the highlighting will be removed and the dimension will be placed.

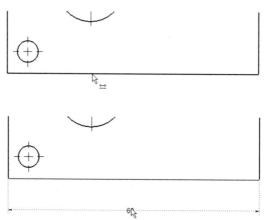

Figure 5-140

Another method to create linear general dimensions is to select the endpoint of the line found at the corner of the object at A followed by the green grip of the center mark as shown in the following image. Locate the dimension as in the previous example.

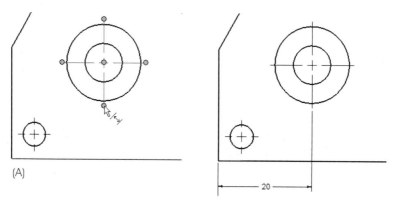

Figure 5-141

GENERAL DIMENSIONS–RADIUS

To create a radius general dimension, click the arc as shown in the following image. Position the leader in a convenient location.

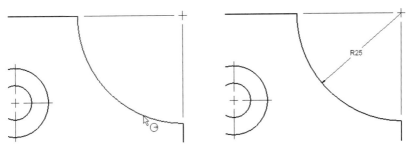

Figure 5-142

GENERAL DIMENSIONS–DIAMETER

To create a diameter general dimension, click the circle as shown in the following image. Position the leader in a convenient location.

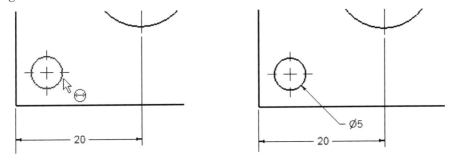

Figure 5-143

GENERAL DIMENSIONS–ANGULAR

To create an angular general dimension, click the line near its midpoint at A and at B, as shown in the following image. Inventor will interpret this combination of two choices as an angle automatically. Pick a point to place the dimension in a convenient location as shown on the right side of the image.

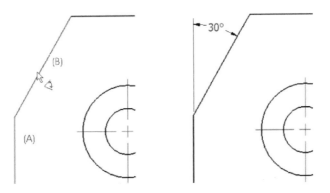

Figure 5-144

MOVING AND CENTERING DIMENSION TEXT

To move a dimension by either lengthening or shortening the extension lines or moving the text, position the cursor over the dimension until the dimension becomes highlighted, as shown in the following image. An icon consisting of four diagonal arrows will appear attached to your cursor. Notice also the appearance of a centerline. This represents the center of the dimension text. You can drag the dimension text up, down, right, or left in order to better position it.

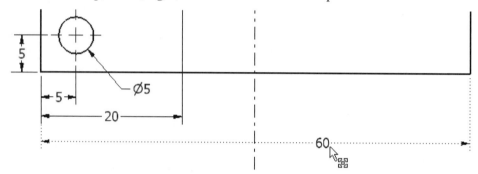

Figure 5-145

If you want to re-center the dimension text, slide the text towards the centerline. The dimension will snap to the centerline. In the following image, the centerline changes to a dotted line to signify the dimension text is centered. This centerline action will work on any type of linear dimension text including horizontal, vertical, and aligned. This feature is not active when repositioning the text of radius, diameter, or hole note dimensions.

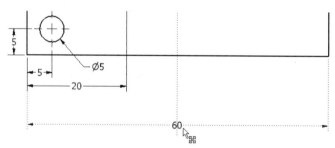

Figure 5-146

EDITING GENERAL DIMENSION EXTENSION LINES

At times, you may need to edit the origin of an extension line in a general dimension in order to provide a more correct dimension. In the following image, a general dimension mistakenly was added from the bottom of the plate to the beginning of the chamfered edge. The dimension needed to extend to the top of the plate. Rather than erase and then recreate the dimension, click on the dimension to display the green grips. Move your cursor over the green grip located at the origin of the extension line. Notice the four arrows that signify move mode, as shown in the following image. Click on this green grip and dragging the extension line to the intersection to the top of the plate will recalculate the general dimension.

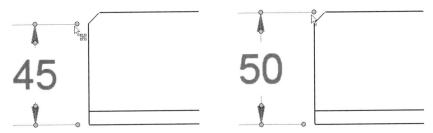

Figure 5-147

EDITING MODEL EXTENSION LINES

You can also edit the extension lines of model dimensions in a drawing; however, there are limitations to their editing. You can only relocate the extension line endpoints in such a way that will not violate or change their dimension value.

EDITING THE ARROW TERMINATOR

You can easily change arrowheads to a different terminator type by moving your cursor over the arrow as shown on the left side of the following image. Notice a glyph consisting of three different terminators. Double-clicking on the arrow will display the Change Arrowhead toolbar as shown in the middle figure of the image. Clicking in the arrow edit box will display all arrows supplied with Inventor. Select the desired arrow type from this list to change the terminator.

 NOTE You can also right-click on an arrowhead and select the Edit 1st Arrowhead and/or 2nd Arrowhead commands from the menu as shown on the right side of the following image.

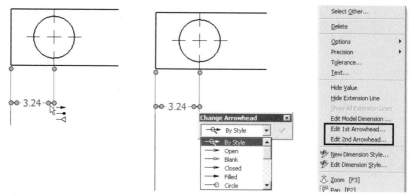

Figure 5-148

SELECTING DRAWING OBJECTS

Managing drawing objects can be a tedious task, especially when the drawings are complex. At times, you need to move a number of views to a new location; or you may need to change a number of dimensions to a different dimension style. Rather than selecting the objects or views individually, you may use a window selection mode or crossing window selection mode to assist you in the selection of multiple objects.

In the example shown in the following image, you want to select all drawing objects that make up the shaft without selecting any dimensions. Clicking and dragging the cursor from the left at A to the right at B will form a solid box representing a window. This window will select all drawing objects that are completely enclosed within it. The selected geometry will become highlighted to distinguish it from unselected geometry.

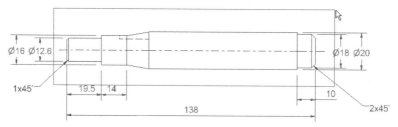

Figure 5-149

The following image illustrates the results of selecting drawing objects with a window.

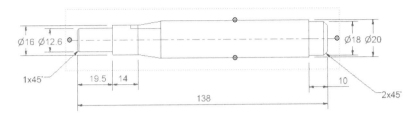

Figure 5-150

The example illustrated in the following image shows how you can select objects and dimensions by a crossing window. If you click and drag the cursor from the right at A to the left at B, a dashed box will form. This dashed window will select all drawing geometry that is completely surrounded by and intersected by it.

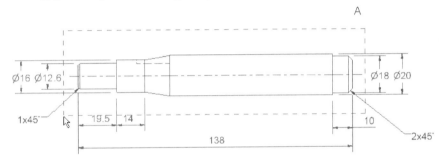

Figure 5-151

The following image illustrates the results of using a crossing window, with all drawing view lines and dimensions selected.

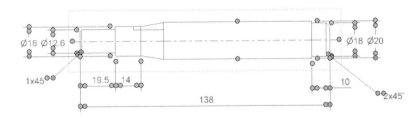

Figure 5-152

 NOTE To remove a selected object from a selection set, deselect it by holding down the SHIFT or CTRL keys as you reselect the object. You can also create multiple window and crossing window selections by pressing and holding down the SHIFT or CTRL keys as you construct the selection box.

ANNOTATIONS

To complete an engineering drawing, you must add annotations such as centerlines, surface texture symbols, welding symbols, geometric tolerance symbols, text, bill of materials, and balloons. Before adding annotations to a drawing, make the desired drawing style active and add annotations to the drawing as needed.

CENTER MARKS AND CENTERLINES

When you need to annotate the centers of holes, circular edges, or the middle (center axis) of two lines, there are four methods you can use to construct the needed centerlines. Use the Center Mark, Center Line Bisector, Center Line, and Centered Pattern tools on the Drawing Annotation Panel Bar as shown in the following image. The centerlines are associated to the geometry that you select when you create them. If the geometry changes or moves, the centerlines will update automatically to reflect the change. This section outlines the steps for creating the different types of centerlines.

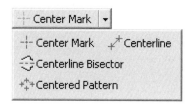

Figure 5-153

Center Marks

To add a center mark, follow the steps and refer to the following image:

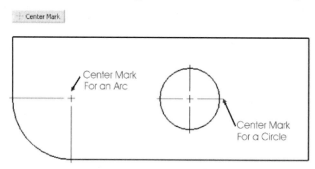

Figure 5-154

1. Click the Center Mark tool on the Drawing Annotation Panel Bar.

2. In the graphics window, select the circle and/or arc geometry in which you want to place a center mark.

3. Continue placing center marks by selecting geometry.

4. To complete the operation, right-click and select Done from the menu.

 NOTE If the features form a circular pattern, the center mark for the pattern is automatically placed when you have selected all of the members.

Centerline Bisector

To add a centerline bisector, follow the steps and refer to the following image:

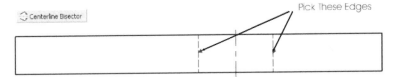

Figure 5-155

1. Click the Centerline Bisector tool on the Drawing Annotation Panel Bar.

2. In the graphics window, select two lines between which you want to place the centerline bisector.

3. Right-click and select Create from the menu to place the centerline bisector.

4. Continue placing centerline bisectors by selecting geometry.

5. To complete the operation, right-click and select Done from the menu.

Centerline

To add a centerline, follow these steps and refer to the following image:

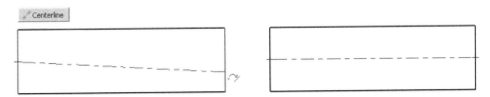

Figure 5-156

1. Click the Centerline tool on the Drawing Annotation Panel Bar.

2. In the graphics window, select a piece of geometry for the start of the centerline.

3. Click a second piece of geometry for the ending location.

4. Right-click and select Create from the menu to create the centerline. The centerline will be attached to the midpoints of the selected geometry.

5. Continue placing centerlines by selecting geometry.

6. To complete the operation, right-click and select Done from the menu.

Centered Pattern

To add a centered pattern, follow the steps and refer to the following image:

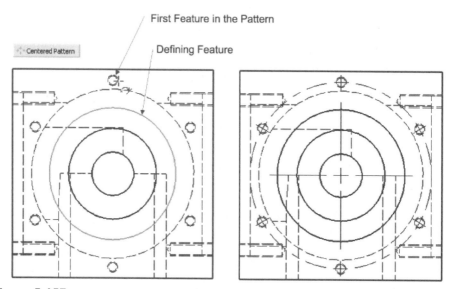

Figure 5-157

1. Click the Centered Pattern tool on the Drawing Annotation Panel Bar.

2. In the graphics window, select the defining feature for the pattern to place its center mark.

3. Click the first feature of the pattern.

4. Continue selecting features in a clockwise direction until all of the features are added to the selection set.

5. Right-click and select Create from the menu to create the centered pattern.

6. To complete the operation, right-click and select Done from the menu.

AUTOMATED CENTERLINES FROM MODELS

The ability to create centerlines automatically can eliminate a considerable amount of work. You can control which features automatically get centerlines and marks and in which views these occur.

You can apply automated centerlines and marks as a drawing template default by clicking Active Standard on the Format menu as shown in the following image. This will activate the Document Settings dialog box for the active document. Click the Automated Centerline Settings button on the Drawing tab, as shown in the figure. You can also set automated centerlines for an individual drawing view by right-clicking on the view boundary to display the menu and selecting Automated Centerlines.

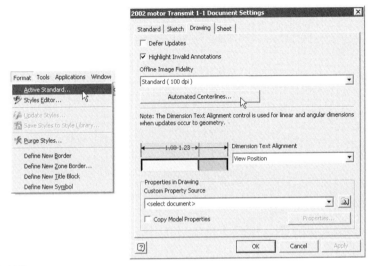

Figure 5-158

Clicking the Automated Centerline Settings button will launch the Centerline Settings dialog box, as shown in the following image. Use this dialog box to set the type of feature(s) to which you will apply automated centerlines (such as holes, fillets, cylindrical features, etc.) and to choose the projection type (plan or profile). The following sections will discuss these areas in greater detail.

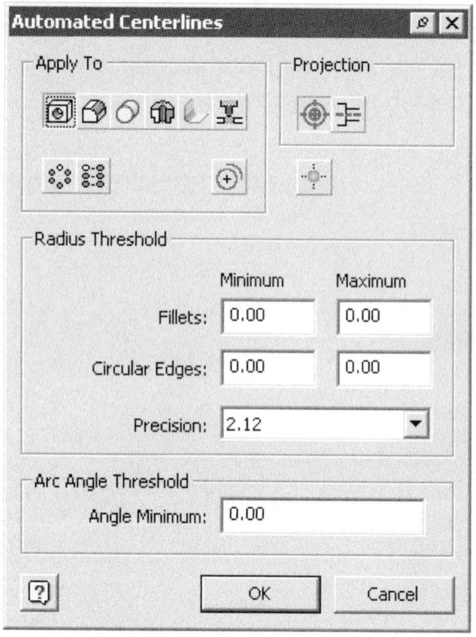

Figure 5-159

Apply To

This area controls the feature type to which you want to apply automated centerlines. Feature types include holes, fillets, cylindrical features, revolved features, circular patterns, rectangular patterns, sheet metal bends, punches, and circular sketched geometry. Click on the appropriate button to depress it, and automated centerlines will be applied to all features of that type in the drawing. You can click on multiple buttons to apply automated centerlines to multiple features. To disable centerlines in a feature, click on the feature button a second time.

Projection

Click on the projection buttons apply automated centerlines to plan (axis normal) and/or profile (axis parallel) views.

Threshold

Thresholds are minimum and maximum value settings and are provided for fillet features, arcs, and circles. Any object residing within a range should get the appropriate center mark. The values are based upon the model values, not the drawing values. This allows you to know what will or will not receive a centerline regardless of the view scale in the document. For example, if you set a minimum value of 0.50 for the fillet feature, a fillet that has a radius of 0.495 will not receive a center mark. A zero value on both threshold settings (min/max) denotes no restriction. This means that center marks will be placed on all fillets regardless of size.

TEXT AND LEADERS

To add text to the drawing, click either the Text or Leader Text tool on the Drawing Annotation Panel Bar as shown in the following image.

Figure 5-160

The Text tool will add text while the Leader Text tool will add a leader and text. Select the desired text tool and define the leader points and/or text location. Once you have chosen the location in the graphics window, the Format text dialog box will appear. When placing text through the Format Text dialog box, as shown in the following image, select the orientation and text style as needed and type in the text. Click the OK button to place the text in the drawing. To edit the text or text leader position, move your cursor over it, click one of the green circles that appear, and drag to the desired location. To edit the text or text leader content, right-click on the text and select Edit Leader Text or Edit Text from the menu.

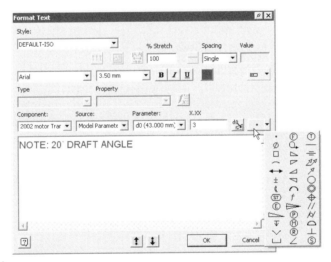

Figure 5-161

TEXT POSITIONING

At times, you may need a way to align general note text, field text, and sketch text objects with each other to concatenate multiple text box objects.

To begin the process of positioning text, first select multiple text objects by pressing the SHIFT or CTRL keys as you select the text or using a window selection method. Next, right-click to display the menu containing the Align option (as shown in the following image). Selecting Align from this menu will display the Align Text dialog box also shown in the following image.

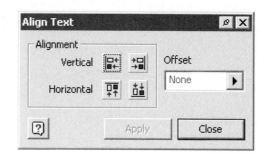

Figure 5-162

Alignment

Two buttons for vertical and horizontal alignment are available to assist in the positioning of text. The Vertical alignment buttons consist of left- and right-justified controls. The Horizontal alignment buttons consist of top- and bottom-justified controls.

Offset

This edit box allows you to apply a line-spacing value in the form of an offset to the horizontal and vertical text positions.

The first selected text object becomes the anchor point for the alignment of the other text objects. In an example of positioning text vertically, the top text object in the following image was selected first, followed by the others (after the first text string is selected, the order of selecting the other text strings is not important).

FIRST TEXT SELECTED

TEXT OBJECT
TEXT OBJECT

Figure 5-163

Clicking on the Vertical Align Left button displays the results shown in the following image. In this image, the vertical base point of the first selected text object becomes the new base point for all other selected text objects.

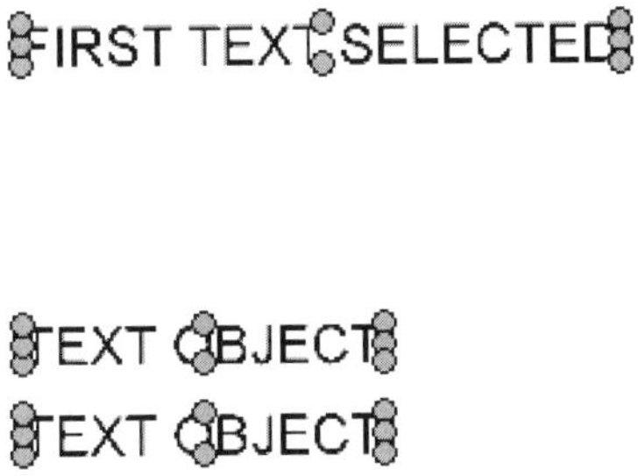

FIRST TEXT SELECTED

TEXT OBJECT
TEXT OBJECT

Figure 5-164

If you enter an offset value such as 0.30 and again click on the Vertical Align Left button, all selected text objects share the vertical spacing between each other, as illustrated in the following image.

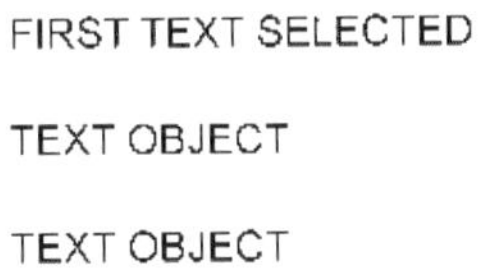

FIRST TEXT SELECTED

TEXT OBJECT

TEXT OBJECT

Figure 5-165

Horizontal positioning of text is similar to vertical, in that the horizontal base point for the first selected text object becomes the horizontal base point for all other selected text objects. In the following image, the top text object was selected first, followed by the others.

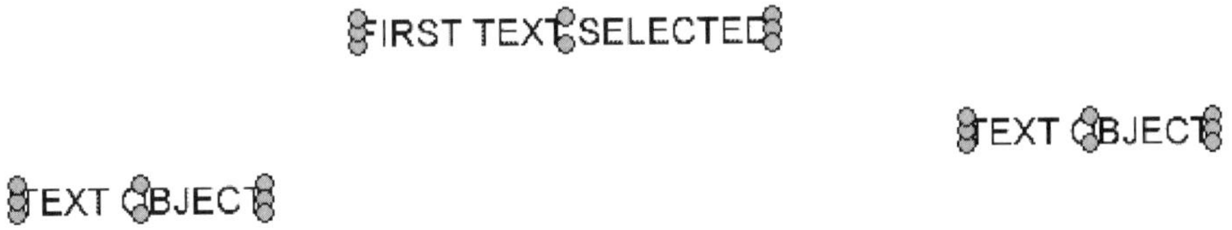

FIRST TEXT SELECTED

TEXT OBJECT

TEXT OBJECT

Figure 5-166

Clicking the Horizontal Top Align button displays the results in the following image. A Bottom Align button is also available if you are dealing with lowercase letters.

TEXT OBJECT   FIRST TEXT SELECTED   TEXT OBJECT

Figure 5-167

ADDITIONAL ANNOTATION TOOLS

To add more detail annotations to your drawing, you can add surface texture symbols, welding symbols, feature control frames, feature identifier symbols, datum identifier symbols, and datum targets by clicking the corresponding tool on the Drawing Annotation Panel Bar, as shown in the following image.

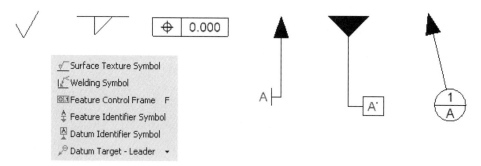

Figure 5-168

Follow these general steps for placing these symbols:

1. Click the tool on the Drawing Annotation Panel Bar.
2. Select a point at which the leader will start.
3. Continue selecting points to position the extension lines.
4. Right-click and select Continue from the menu.
5. Fill in the information as needed in the dialog box.
6. When done, click the OK button in the dialog box.
7. To complete the operation, right-click and select Done from the menu.
8. To edit a symbol, position the cursor over it. When the green circles appear, right-click and select the corresponding Edit option from the menu.

HOLE AND THREAD NOTES

Another annotation you can add is a hole or thread note. Before you can place a hole or thread note in a drawing, a hole or thread feature must exist. You cannot annotate extruded circles using the Hole or Thread Note tool. If the hole or thread feature changes, the note will be updated automatically to reflect the change. The following image illustrates examples of counterbore, countersink, and threaded/tapped hole notes.

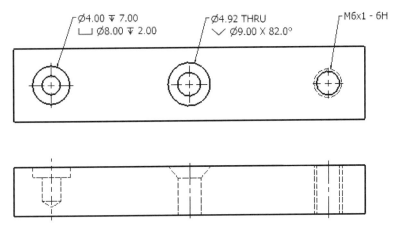

Figure 5-169

To create a hole or thread note, follow these steps:

1. Click the Hole/Thread Notes tool on the Drawing Annotation Panel Bar as shown in the following image.

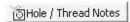

Figure 5-170

2. Select the hole or thread feature to annotate.

3. Select a second point to locate the leader and the note.

4. To complete the operation, right-click and select Done from the menu.

5. To edit a note, position the cursor over it. When the green circles appear, right-click and select the corresponding Edit option from the menu.

HOLE NOTE STYLES

Hole note styles enable you to define global formatting of a hole note, as well as edit existing hole notes and apply specific formatting on an as-needed basis. This will allow you the flexibility to conform to a standard or create your own standard for hole notes.

To control a hole note as part of an existing dimension style, follow these steps:

1. In the Styles and Standards Editor, select an existing dimension style and click on the Notes and Leaders tab as shown in the following image.

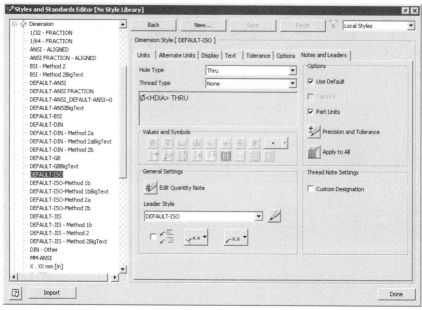

Figure 5-171

2. Either modify the hole note as part of an existing dimension style or click the new button to create a new dimension style.

3. Select the hole type to format from the Hole Type drop-down list.

4. Change the Value or Symbol to reflect the desired hole format.

5. Click the Save button to save any changes to the new or existing dimension style.

EDITING HOLE NOTES

1. A hole note edit is invoked by right-clicking on an existing hole note to display the menu shown in the following image. Selecting Edit Hole Note from this menu will display the Edit Hole Note dialog box also illustrated in the following image. You can perform edits on a single hole note.

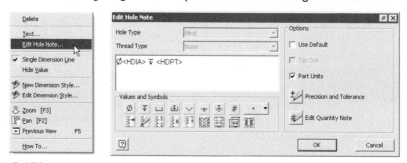

Figure 5-172

EXERCISE 5-5 Creating Text and Dimension Styles

In this exercise, you create new text and dimension styles.

1. Open the file *ESS_E05_05.idw*. The drawing consists of three orthographic views in addition to an isometric view.

2. Begin the process of creating a new text style. From the Format menu click Styles Editor as shown in the following image.

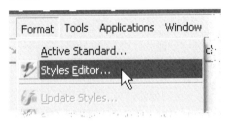

Figure 5-173

3. When the Styles and Standards Editor dialog box appears, expand the Text style in the left pane. Then, right-click on Current-ANSI and click New Style as shown in the following image.

Figure 5-174

4. In the New Style Name dialog box type **My ANSI Text** and then click OK.

5. In the Font list, select Arial and in the Size list, select 3.50 mm as shown in the following image.

6. When you have finished modifying this new text style, click the Save button.

Figure 5-175

7. Begin the process of creating a new dimension style. In the left pane expand the Dimension style. Then right-click on the Metric style and click New Style from the menu as shown in the following image.

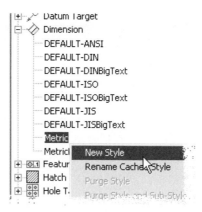

Figure 5-176

8. In the New Style Name dialog box type **My Metric** and then click OK.

9. Uncheck the Leading Zero for the Display and Angular Display as shown in the following image

Figure 5-177

10. Next, click the Display tab. Type **3.00 mm** for Extension(A) and Origin Offset(B) as shown in the following image.

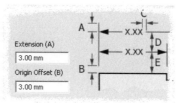

Figure 5-178

11. Next, click the Text tab. In the Primary Text Style list, select My ANSI Text as shown in the following image.

12. When finished, click the Save button followed by the Done Button.

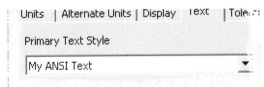

Figure 5-179

13. You will now change a number of dimensions from one dimension style to another. Zoom in on the top view and examine a few dimensions. Notice the text style and leading zeros.

14. Select all the dimensions in the top view. To accomplish this task, hold down the Ctrl key while adding dimensions to the selection set (do not select any of the edges).

15. From the dimension style drop-down list, click My Metric as shown in the following image.

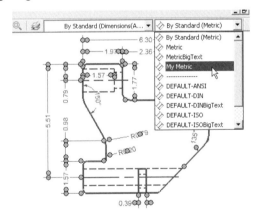

Figure 5-180

16. Zoom in and examine the updated dimensions. Your display should appear similar to the following image.

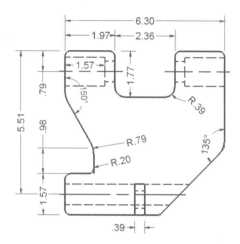

Figure 5-181

 NOTE The new dimension style only pertains to this drawing. To add a new style to the style library, the Autodesk Inventor Project file Use Styles Library option must be set to Yes. Then from the Format menu click Save Styles to Style Library.

17. End of exercise.

EXERCISE 5-6 Adding Dimensions and Annotations

In this exercise, you add dimensions and annotations to a drawing of a clamp that is used to hold a work piece in position during machining operations.

Both model dimensions and drawing dimensions are used to document feature size.

1. Open the drawing *ESS_E05_06.idw*. The drawing file contains a single sheet with 4 drawing views.

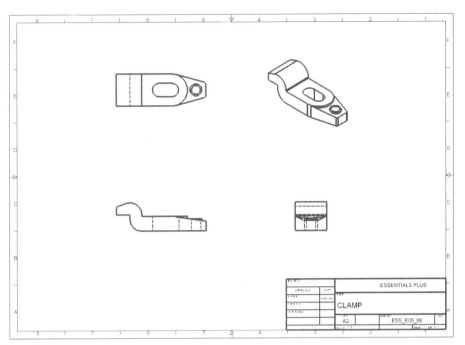

Figure 5-182

2. In this step you will retrieve model dimensions for use in the drawing view. Begin by zooming in on the front view.

3. Right-click the front view and choose Retrieve Dimensions from the menu. This will launch the Retrieve Dimensions dialog box.

4. In this dialog box, click the Select Dimensions button. The model dimensions will appear on the view.

5. Construct a selection box around all dimensions. This will select the dimensions to retrieve. When finished, click the OK button. The model dimensions that are planar to the view are displayed as shown in the following image.

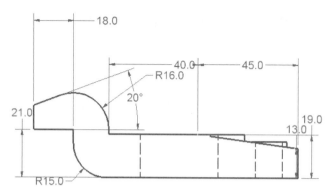

Figure 5-183

6. You will now delete certain model dimensions that do not accurately describe the feature being dimensioned. First, right-click the 45.0 horizontal dimension and choose Delete from the menu. Then right-click the 40.0 horizontal dimension and choose Delete from the menu. Your display should appear similar to the following image.

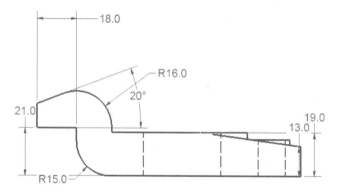

Figure 5-184

7. Reposition the model dimensions to better locations by dragging the dimensions until they appear as shown in the following image.

 NOTE To reposition dimension text, click a dimension text object and drag it into position. The dimension will be highlighted when it is a preset distance from the model.

 NOTE Radial dimensions can be repositioned by selecting the handle at the annotation end of the leader.

To drag dimensions make sure there is no command active.

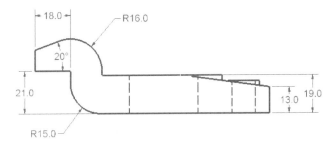

Figure 5-185

8. Centerlines and center marks need to be added to aid in the placement of drawing dimensions. To display the center mark annotation tools, click the title area of the panel bar and select Drawing Annotation Panel.

9. Click the Center Line Bisector tool in the panel bar and select the two hidden lines that represent the drilled hole through the boss. Your display should be similar to the following image.

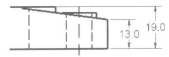

Figure 5-186

10. Pan to display the top view. Then click the Center Mark tool in the panel bar.

11. Click the outer circle of the boss and the two arcs of the slot to place center marks on these features as shown in the following image.

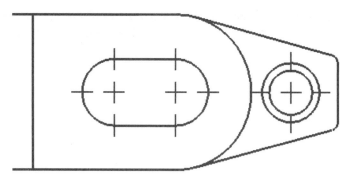

Figure 5-187

12. You will now add a number of general drawing dimensions complete the documentation of the model. First, pan back to display the front view.

13. Click the General Dimension tool in the panel bar.

14. Click the right endpoint of the bottom edge, and then click the right endpoint of the top of the boss.

15. Move the cursor to the right and place the 16.0 dimension between the 13.0 and 19.0 vertical dimensions as shown in the following image.

 NOTE You may have to reposition the 19.0 dimension.

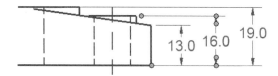

Figure 5-188

16. Next, you will manually add dimensions to the top view. Pan to display the top view.

17. Click the General Dimension tool in the panel bar.

18. Click the endpoints of the centerlines in the slot and drag above the model to place the 17 dimension.

19. Click the arc on the slot and place the radial dimension as shown.

20. Place the 16 diameter dimension on the boss.

21. Click the top horizontal line then click the sloped line and place the 15 degree angular dimension. Your display should appear similar to the following image.

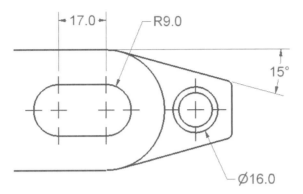

Figure 5-189

22. Add the 13.0, 45.0, and 40.0 horizontal dimensions as shown in the following image.

 TIP To align a dimension when dragging it, move the cursor over an existing dimension and acquire an alignment point. Move the cursor back to the dimension being placed. The dotted line indicates an alignment inference. Click to place the dimension.

23. Add the 21.0 radial dimension. When finished, right-click and choose Done.

24. Drag the 16.0 diameter dimension so the leader does not cross the extension lines. Your display should appear similar to the following image.

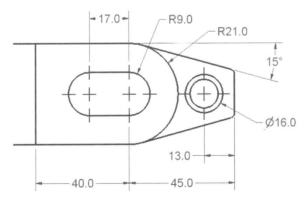

Figure 5-190

25. You need to add a note to the end of the angle dimension. Right-click the 15° dimension and choose Text.

26. At the insertion point, press the space bar then type **TYP** as shown in the following image on the left and click OK. Your display should appear similar to the following image on the right

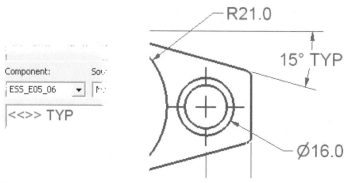

Figure 5-191

27. Additional notes need to be added to other dimensions. Right-click the 16.0 diameter dimension and choose Text.

28. At the insertion point, press the space bar, and then enter **BOSS** and press ENTER.

29. Select the diameter symbol from the symbol list in the dialog box as shown in the following image on the left and type **12.0 THRU**. When finished, click OK. Your display should appear similar to the following image on the right.

Figure 5-192

30. In the following steps, you add a general note and use leader text. First, click the Text tool in the panel bar.

31. Click a point below and to the right of the top view.

32. In the text entry area, enter **TOLERANCE FOR** and press ENTER.

33. On the next line, enter **ALL DIMENSIONS** then press the space bar and select ± from the symbol list.

34. Type **0.5** then click OK. Right-click and choose Done. Your display should appear similar to the following image.

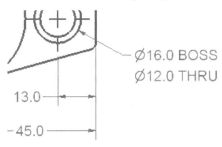

Figure 5-193

35. You will now add a note attached to a leader. First, click the Leader Text tool in the panel bar.

36. Select the bottom arc on the right end to define the leader start point.

37. Click a point below and to the right to define the end of the leader, right-click, and select Continue.

38. Enter **ROUNDS R2** then click OK. Right-click and select Done. Your display should appear similar to the following image.

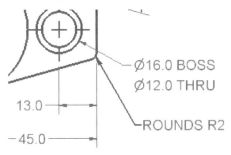

Figure 5-194

39. You will now finish documenting this drawing by using drawing properties to complete the title block information.

40. From the File menu, select iProperties. This will display the Drawing Properties dialog box.

41. On the Summary tab, enter your name in the Author field.

42. Click on the Status tab and select the current date from the Checked Date list.

43. Enter your initials in the Checked By field and click OK to update the title block. Zoom in to the title block to examine the changes.

44. End of exercise.

REVIEW SEQUENCE FOR CREATING DRAWING VIEWS

The following is a review of the steps needed to create a new sheet with drawing views:

1. Create a new sheet or select an existing drawing sheet from the Browser.
2. Add a border.
3. Add a title block.
4. Create drawing views based on a part, assembly, or presentation file.
5. Get model dimensions if not done in conjunction with Step 4.
6. Edit drawing views, if necessary.
7. Add additional dimensions and annotations.

CHAPTER SUMMARY

To	Do This	Tool
Set drafting styles	Click Format > Styles Editor.	
Set sheet size	Right-click the sheet and select Edit Sheet.	
Insert a border	Double-click the border in the Browser.	
Insert a title block	Double-click the title block in the Browser.	
Edit a title block	Right-click the title block in the Browser's Drawing Resources section and select Edit.	
Create an independent view	Click the Base View tool on the Drawing Views Panel Bar or on the Drawing Views toolbar.	▣
Create a projected view	Click the Projected Views tool on the Drawing Views Panel Bar or on the Drawing Views toolbar. Select	▦

To	Do This	Tool
	the base view, right-click and select Create View > Projected.	
Create an auxiliary view	Click the Auxiliary View tool on the Drawing Views Panel Bar.	
Create a section view	Click the Section View tool on the Drawing Views Panel Bar.	
Create a detail view	Click the Detail View tool on the Drawing Views Panel Bar.	
Create a broken view	Click the Broken View tool on the Drawing Views Panel Bar.	
Create a break out view	Click the Break Out View tool on the Drawing Views Panel Bar.	
Create an overlay view	Click the Overlay View tool on the Drawing Views Panel Bar.	
Creating a draft view	Click the Draft View tool on the Drawing Views Panel Bar.	
Retrieve a model dimension	Right-click a view and select Retrieve Model Dimensions.	
Place a drawing dimension	Click the General Dimension tool on the Drawing Annotation Panel Bar or on the Drawing Annotation toolbar.	
Place center marks and centerlines	Click one of the four tools on the Drawing Annotation Panel Bar or on the Drawing Annotation toolbar.	
Place a text note	Click the Text tool on the Drawing Annotation Panel Bar or on the Drawing Annotation toolbar.	
Place a leader	Click the Leader Text tool on the Drawing Annotation Panel Bar or on the Drawing Annotation toolbar.	
Enter drawing properties	Click File > iProperties.	
Print a drawing	Click File > Print.	

Applying Your Skills

SKILL EXERCISE 5-1

In this exercise, you create a working drawing of a part.

1. Begin by starting a new drawing based on the *metric ANSI (mm).idw* template.
2. Use Edit Sheet to select an A2 size sheet.

3. Use the Base View and Projected View tools to create the required views of ESS_E05_07.ipt.

4. Use a scale of 2 for all views.

5. Add a center mark and centerline bisector to the drawing views.

6. Use the Retrieve Dimensions tool to display the model dimensions.

7. Delete dimensions as required then use the General Dimension tool to add the dimensions.

8. Use the Leader Text tool to place the thread note; M8x1.25 – 6H.

9. Continue on with the Leader Text tool by placing the depth symbol followed by 20.00 TYP.

Your display should appear similar to the following image.

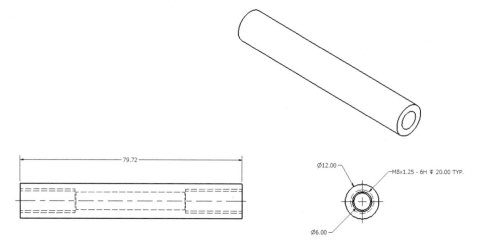

Figure 5-195

SKILL EXERCISE 5-2

In this exercise, you create a drawing for a drain plate cover.

1. Begin by starting a new drawing using a *Metric ANSI (mm).idw* template.

2. Create three views of the part *ESS_E05_08.ipt* on an A2 sheet.

3. Use a scale of 2 for all views.

4. Add center marks to the views.

5. Create a new dimension style. Set all text to align horizontally. Set all trailing zeros to display. Set all text height to 3.50 mm.

6. Add dimensions and annotations. Your display should appear similar to the following image.

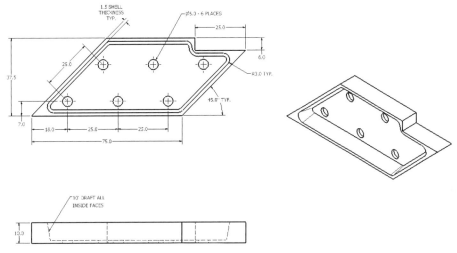

Figure 5-196

CHECKING YOUR SKILLS

Use these questions to test your knowledge of the material covered in this chapter.

1. **True__ False__** A drawing can have an unlimited number of sheets.

2. **True__ False__** A drawing's sheet size normally is scaled to fit the size of the drawing views.

3. **True__ False__** There can only be one base view per sheet.

4. **True__ False__** An isometric view can only be projected from a base view.

5. **True__ False__** Drawing dimensions can drive dimensional changes parametrically back to the part.

6. Explain how to shade an isometric drawing view.

7. **True__ False__** When creating a hole note using the Hole/Thread Notes tool, circles that are extruded to create a hole can be annotated.

CHAPTER 6

Creating and Documenting Assemblies

In the first three chapters you learned how to create a component in its own file. In this chapter you will learn how to place individual component files into an assembly file. This process is referred to as bottom-up assembly modeling. You will also learn to create components in the contents of the assembly file, which is referred to as top-down assembly modeling. After creating components, you will learn how to constrain the components to one another using assembly constraints, edit the assembly constraints, check for interference, and create presentation files that show how the components are assembled or disassembled.

CHAPTER OBJECTIVES

After completing this chapter, you will be able to

- Understand the assembly options
- Create bottom-up assemblies
- Create top-down assemblies
- Create subassemblies
- Constrain components together using assembly constraints
- Edit assembly constraints
- Create adaptive parts
- Pattern components in an assembly
- Check parts in an assembly for interference
- Drive constraints
- Create design view representations
- Create flexible assemblies
- Create positional representations
- Create a presentation file
- Create individual and automatic balloons
- Create a parts list of an assembly

CREATING ASSEMBLIES

As you learned already, component files have the IPT extension, and they can only have one component each. In this chapter you will learn how to create assembly files (IAM file extension). An assembly file holds the information needed to assemble the components together. All of the components in an assembly are *referenced in*, meaning that each component exists in its own component IPT file, and its definition is linked into the assembly. You can edit the components while in the assembly, or you can open the component file and edit it. When you have made changes to a component and saved the component, the changes will be reflected in the assembly after you open or update it. There are three methods for creating assemblies: bottom-up, top-down, and middle-out (a combination of both top-down and bottom-up techniques). *Bottom-up* refers to an assembly in which all of the components were created in individual component files and are referenced into the assembly. A *top-down* approach refers to an assembly in which the components are created from within the context of the assembly. In other words, the user creates each component from within the top-level assembly. Each component in the assembly is saved to its own component (IPT) file. The following sections describe the bottom-up and top-down assembly techniques.

To create a new assembly file, click the New icon in What To Do and then click the *Standard.iam* icon on the Default template tab, as shown in the following image. There are two other methods for creating a new assembly: you can click New on the File menu and then click the *Standard.iam* icon, or you can click the down arrow of the New icon on the left side of the standard toolbar and select Assembly. After issuing the new assembly operation, Autodesk Inventor's tools will change to reflect the new assembly environment. The assembly tools will appear on the Panel Bar. These tools will be covered throughout this chapter.

 NOTE There is no correct or incorrect way to create an assembly. You will determine which method works best for the assembly that you are creating based on experience. You can create an assembly using the bottom-up or top-down assembly techniques, or a combination of both. Whether you place or create the components in the assembly, all of the components will be saved to their own individual IPT files, and the assembly will be saved as an IAM file.

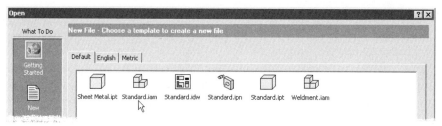

Figure 6-1

ASSEMBLY OPTIONS

Before creating an assembly, let's examine the assembly option settings. On the Tools menu, click Application Options and the Options dialog box will appear. Click on the Assembly tab, and your screen should look similar to the following image. The following sections describe the various assembly settings. These settings are global and will affect how new components are created, referenced, analyzed or placed in the assembly.

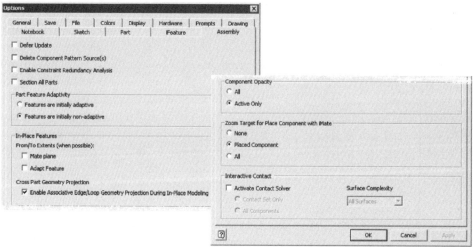

Figure 6-2

Defer Update   Click this option so that when you change a component that will affect the assembly, you will have to click the Update button manually to get the assembly up-to-date. If you leave the option's box clear, the assembly will update automatically to reflect the change to a component.

Delete Component Pattern Source(s)   Click to delete the source component of a component pattern when you delete the pattern. If you leave the option's box clear, the source component will not be deleted when you delete the pattern.

Enable Constraint Redundancy Analysis   Click to perform a secondary analysis of all assembled component constraints. You are notified when redundant constraints exist. The component Degrees of Freedom are also updated but not displayed.

Section All Parts   Click to section standard parts placed in an assembly from the included library when you create a section drawing view through them. To prevent the sectioning of standard parts, leave this box unchecked.

 NOTE Part Feature Adaptivity, In-Place Features, and Cross Part Geometry Projection will be covered later in this chapter under the heading of Adaptivity.

Component Opacity

In this section you determine if all components or only the active component will be opaque when an assembly cross-section is displayed.

All   Click to make all components in the assembly opaque when in shaded mode.

Active Only   Click to make only the active component in the assembly opaque when in shaded mode; the other assembly components will be dimmed.

Zoom Target for Place Component with iMate

Set the default zoom behavior for the graphics window when placing components with iMates.

None   Click to perform no zooming. This leaves the graphics display as is.

Placed Component   Click to zoom in on the placed part so that it fills the graphics window.

All   Click to zoom in on the assembly so that all elements in the model fill the graphics window.

Interactive Contact

Activate Contact Solver   Activates the Contact Solver for those components added to a contact set.

ASSEMBLY TOOLS

Once active in the assembly environment, the Panel Bar changes to reflect the tools illustrated in the following image. Note that this figure shows the Panel Bar with Expert mode turned on. A brief description of each tool follows.

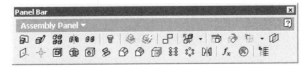

Figure 6-3

Assembly Tools

Button	Tool	Function
	Place Component	Places individual components or subassemblies into the current assembly file.
	Create Component	Allows you to create a new part file in the context of an assembly.
	Pattern Component	Creates multiple copies of parts in an assembly as a pattern. Typical patterns include rectangular and circular.
	Mirror Component	Creates mirrored components or subassemblies in an assembly file. This tool is ideal when you need left-hand, right-hand versions of parts or assemblies.
	Copy Component	Creates copies of components or subassemblies in an assembly file.
	Bolted Connection	Simplifies the placement of nuts, bolts and washers in an assembly.
	Content Center	Allows part and feature content to be managed, customized, and published.
	Refresh Standard Components	Used to update an old version of a part in an assembly with a new version saved in a library.
	Constraint	Determines how components in an assembly fit together. This tool activates a dialog box that displays three tabs used for most assembly situations: Assembly, Motion, and Transitional.
	Replace Component	Replaces one assembly component or subassembly with another. A second button is available for replacing all occurrences of a component or subassembly.
	Move Component	Moves an individual assembly component to any location on the display screen.
	Rotate Component	Rotates an individual component on the display screen.
	Quarter Section View	Creates an assembly section view in order to view part of an assembly obscured by other components. Four buttons are available: Quarter, Half, and Three-Quarter. The fourth button removes any applied assembly section view. These tools are covered in Chapter 10.
	Work Plane	Creates a work plane on selected geometry in an assembly file. This work plane does not reside in an individual component file.
	Work Axis	Creates a work axis on selected geometry in an assembly file. This work axis does not reside in an individual component file.
	Work Point	Creates a work point on selected geometry in an assembly file. This work plane does not reside in an individual component file.
	Extrude	Creates an extruded cut feature through one or more parts in an assembly. This feature does not reside in an individual component file.

Assembly Tools (continued)

Button	Tool	Function
	Revolve	Creates a revolved cut feature through one or more parts in an assembly. This feature does not reside in an individual component file.
	Hole	Creates a hole feature through one or more parts in an assembly. This feature does not reside in an individual component file.
	Sweep	Creates a sweeping cut feature through one or more parts in an assembly. This feature does not reside in an individual component file.
	Fillet	Creates a fillet round feature in an assembly. This feature does not reside in an individual component file.
	Chamfer	Creates a chamfer feature in an assembly. This feature does not reside in an individual component file.
	Move Face	Moves one or more faces on a base solid or a feature by a specified distance and direction or by a planar move to specific coordinates.
	Rectangular Pattern	Produces a rectangular pattern of an assembly feature. This operation does not affect the individual component files.
	Circular Pattern	Produces a circular pattern of an assembly feature. This operation does not affect the individual component files.
	Mirror Feature	Creates a mirrored assembly feature. This operation does not affect the individual component file.
f_x	Parameters	Displays the Parameters dialog box, which you use to define and list parameters used in an assembly. You can create user-defined parameters, rename existing parameters, add equations, and link parameters to a Microsoft Excel file. The techniques for using this tool are covered in Chapter 7.
	Create iMate	Creates a predefined constraint or group of iMate constraints (Composite iMates) on a component to specify how parts will connect when they are referenced into an assembly.
	Bill of Materials	Used for editing iProperties and BOM properties.

THE ASSEMBLY BROWSER

The Assembly Browser, as shown the following image, displays the hierarchy of all component and subassembly occurrences and assembly constraints in the assembly. Each occurrence of a component is represented by a unique name. In the Browser you can select a component for editing, move components between assembly levels, reorder assembly components, control component status, rename components, edit assembly constraints, and manage design views and representations.

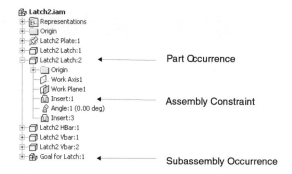

Figure 6-4

BOTTOM-UP APPROACH

The *bottom-up assembly approach* uses existing files that are referenced to an assembly file. To create a bottom-up assembly, create the components in their individual files. If you place an assembly file into another assembly, it will be brought in as a *subassembly*. Any drawing views created for these files will not be brought into the assembly file. You will want to make sure to include the path(s) for the file location(s) of the placed components in the project file; otherwise, Inventor may not be able to locate the referenced component when you reopen the assembly.

To insert a component into the current assembly, click the Place Component tool on the Assembly Panel Bar as shown in the following image, press the hot key P, or right-click in the graphics window and select Place Component from the menu.

Figure 6-5

The Open dialog box will appear, as shown in the following image. If the component you are placing is the first in the assembly, an instance will be placed into the assembly automatically. If the component is not the first, you pick a point in the assembly where you want to place the component. If you need multiple occurrences of the component in the assembly, continue selecting placement points. When done, press the ESC key or right-click and select Done from the menu.

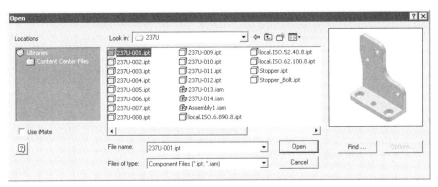

Figure 6-6

TOP-DOWN APPROACH

You can create new components while in an assembly. This method is referred to as the *top-down assembly approach*. To create a new component in the current assembly, click the Create Component tool on the Assembly Panel Bar as shown in the following image, press the hot key N, or right-click and select Create Component from the menu.

Figure 6-7

The Create In-Place Component dialog box will appear, as shown in the following image. Enter a new file name, a file type, a location where you will save it, and the template file on which to base it. Also determine if the component will be constrained to a face on another component in the assembly. If you click the Constrain sketch plane to selected face or plane box option, a flush constraint will be applied between the selected face and the new component's initial sketch plane. If you leave the options box clear, no constraint will be applied. You will still select a face, however, on which to start sketching the new component.

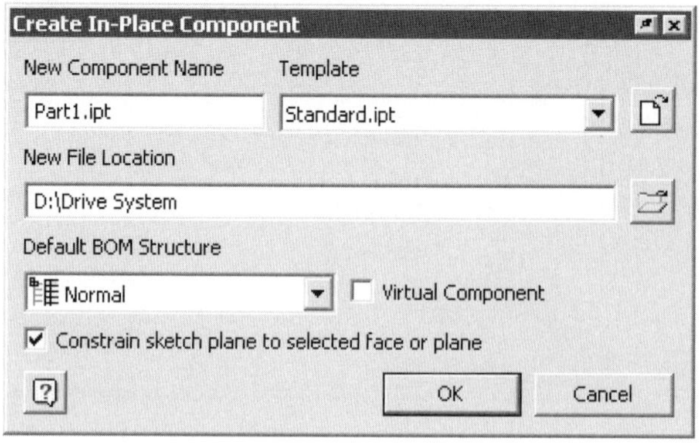

Figure 6-8

OCCURRENCES

An *occurrence*, or instance, is a copy of an existing component and has the same name as the original component with an added number sequenced after a colon. If the original component is named Bracket, for example, the occurrence in the assembly will be Bracket:1 and a subsequent occurrence will be Bracket:2 (see the following image). If the original component or an occurrence of the component changes, all of the components will reflect the change.

To create an occurrence, you place the component. If the component already exists in the assembly, you can click the component's icon in the Browser and drag an additional occurrence into the assembly. You can also use the *copy-and-paste* method to place additional components by right-clicking on the component name in the Browser or on the component itself in the graphics window and then right-clicking and selecting Copy from the menu. You can then right-click and select Paste from the menu. You can also use the Windows shortcuts CTRL-C and CTRL-V keys to copy and paste the selected component. If you want an occurrence of the original component to have no relationship with its source component, you will need to make the original component active, click the Save Copy As tool on the File menu, and enter a new name. The new component will have no relationship to the original. You can place it in the assembly using the Place Component command.

```
Motor.iam
  Representations
  Origin
  Motor Housing:1
  Motor Bushing:1
  Motor Shaft:1
  Motor Gear:1
  Bracket:1
  Bracket:2
```

Figure 6-9

MULTIPLE DOCUMENTS AND DRAG & DROP COMPONENTS

In Inventor you can open as many files as needed. This is referred to as a *multiple document environment.* You can split the screen as needed to show all of the files that are open by using options under the Window menu. You can switch between the open files to model, edit, or interrogate the files as needed. With multiple documents open, you can drag a component from one file into an assembly. To do so, both the assembly and the component files must be open and the screen split so both files are visible. With the source document active, select the component's name in the Browser with the left mouse button, drag it into the target assembly file, and release the mouse button.

 NOTE You can press the **CTRL** and **TAB** keys to switch between various files that are currently open.

ACTIVE COMPONENT

To edit a component while in an assembly, activate the component. Only one component in the assembly can be active at a time. To make a component active, you double-click on the component in the graphics window. You can double-click on the file name or an icon in the Browser, or you can right-click on the component name in the Browser or graphics window and select Edit from the menu. Once the component is active, the other component names in the Browser will be grayed out, as shown in the following image. If Component Opacity–in Application Options on the Assembly tab–is set to Active Only, the other components in the assembly will be dimmed in the graphics window.

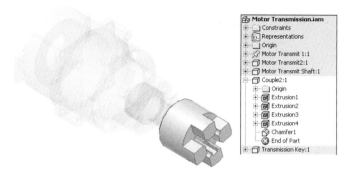

Figure 6-10

You can then edit the component as you learned to in previous chapters and save the changes by using the Save tool. Only the active component will be saved. To make the assembly active, click the Return button on the command bar (as shown in the following image), double-click on the assembly name in the Browser, or right-click in the graphics window and select Finish Edit from the menu.

Figure 6-11

OPEN AND EDIT

Another method for editing a component in the assembly is to open the component in another window. This can be done by clicking the Open tool on the standard toolbar, clicking Open on the File menu, or right-clicking on the component's name in the Browser or the component in the graphics window and then selecting Open from the menu (see the following image). The component will appear in a new window. Edit the component as needed, save the changes, activate the assembly file, and the changes will appear in the assembly.

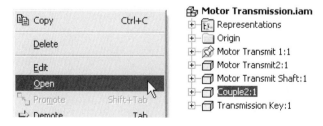

Figure 6-12

GROUNDED COMPONENTS

When assembling components together, you may want to have a component or multiple components that are grounded (stationary), meaning that they will not move. When applying assembly constraints, the unconstrained components will be moved to the grounded component(s). By default, the first component placed in an assembly is grounded. There is no limit to how many components can be grounded. It is recommended that at least one component in the assembly be grounded; otherwise, the assembly will be able to move. A grounded component is represented with a pushpin superimposed on its icon in the Browser. To ground or *unground* a component, right-click on the component's name in the Browser and select or deselect Grounded from the menu, as shown in the following image.

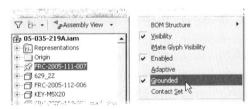

Figure 6-13

INSERTING MULTIPLE PARTS IN A SINGLE OPERATION

Once you have placed the first component into an assembly file, it is possible to place multiple components in a single operation. To perform this task, split your display screen into two separate windows where the Inventor Assembly file and Windows Explorer are both visible as shown in the following image. In Windows Explorer, press and hold down the SHIFT or CTRL key to select multiple component files as shown on the right side of the following image.

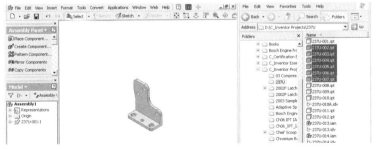

Figure 6-14

Then drag and drop the selected components from Windows Explorer into the Inventor window. The results are illustrated in the following image.

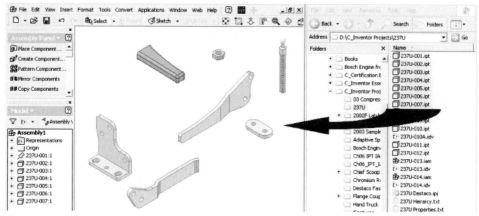

Figure 6-15

SUBASSEMBLIES

While working, you may want to group components together to define a subassembly. You can use two methods to create a subassembly. You can place an existing assembly into another assembly or create a new assembly from within the assembly. To create a subassembly from *within* the assembly, issue the Create Component tool and change the file type to Assembly, as shown in the following image. Any subsequent component that you place or create will be a component of the active subassembly.

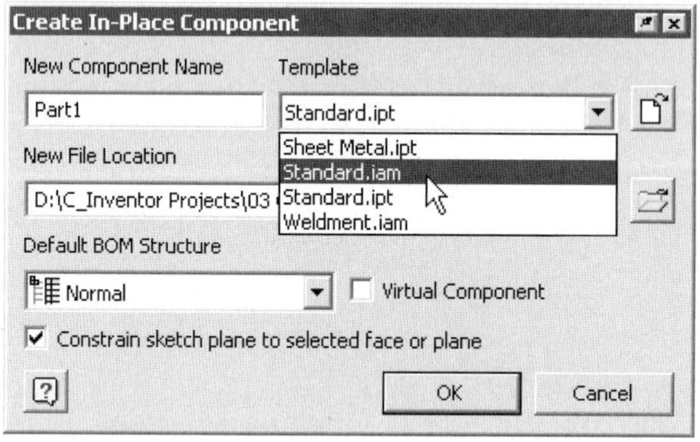

Figure 6-16

To make a subassembly active, double-click on the subassembly's name in the Browser. To make the top-level assembly active, click the Return tool on the command bar or double-click on the top-level assembly's name in the Browser. When working in the top level of the assembly, any subassembly will act as a single component when selected. You can promote subassembly components into an assembly or demote them into another subassembly, and you can demote assembly components into a subassembly. Do this by clicking a component's name in the Browser and selecting either Promote or Demote from the menu shown in the following image.

Figure 6-17

RESTRUCTURING COMPONENTS

After components are in an assembly, you can restructure or move them from the main assembly to another subassembly, from subassembly to subassembly, or from subassembly to the main assembly. To restructure a component, click the component to be restructured in the Browser, press and hold the left mouse button, and move the component up or down in the Browser. A line appears to show where it will be placed. Keep moving the cursor until the line is positioned in the correct place, then release the left mouse button. The following image shows how the Browser looks when a component is being restructured in a subassembly. In this example, the component 680bar:5 is being restructured into the Robot Base Sub:2 assembly. When restructuring components into or from a subassembly, a dialog box may appear stating that assembly constraints may be deleted. After you complete the restructuring, examine the constraints of the restructured component and assembly to verify that they are still valid.

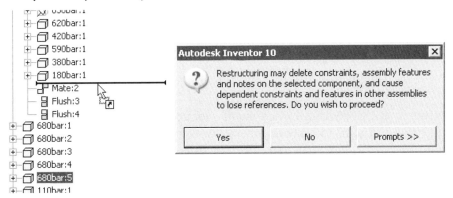

Figure 6-18

REORDERING COMPONENTS

As you assemble components, the Browser keeps track of the order that you used to insert or instance components into an assembly. Due to design intent, you may want to perform a reordering operation, which will move a component adjacent to another in the Browser. To reorder a component, click the component to be reordered in the Browser, press and hold the left mouse button, and drag the component up or down in the Browser. A line appears to show where it will be placed. Keep adjusting the position of the line until you are satisfied with the components new Browser position, then release the mouse button. In the following image, the component 237U-005:1 is being reordered between components 237U-004:1 and 237U-006:1.

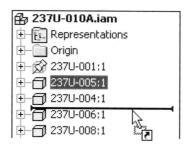

Figure 6-19

ASSEMBLY CONSTRAINTS

In previous sections you learned how to create assembly files, but the components did not have any relationship to one another except for the relationship that was defined when you created a component from the context of an assembly and in reference to a face on another component. If you placed a bolt in a hole, for example, and the hole moved, the bolt would not move to the new hole position. Use assembly constraints to create relationships between components. With the correct constraint(s) applied, if the hole moves, the bolt will move to the new hole location.

In a previous chapter you learned about geometric constraints. When you apply geometric constraints to sketches, they reduce the number of dimensions or constraints required to constrain a profile fully. When you apply assembly constraints, they reduce the degrees of freedom (DOF) that allow the components to move freely in space.

There are six degrees of freedom: three translational and three rotational. Translational means that a component can move along an axis: X, Y, or Z. Rotational means that a component can rotate about an axis: X, Y, or Z. As you apply assembly constraints, the number of DOF decreases.

Autodesk Inventor does not require components to be fully constrained. By default, the first component created or added to the assembly will be grounded and will have zero DOF. More than one component can be grounded, as discussed earlier in this chapter. Other components will move in relation to the grounded component(s). To see a graphical display of the DOF remaining on all of the components in an assembly, select Degrees of Freedom on the View menu, as shown in the following image.

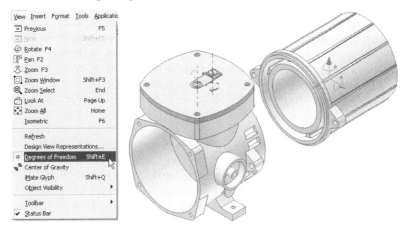

Figure 6-20

An icon will appear in the center of the component that shows the DOF remaining on the component. The line and arrows represent translational freedom, and the arc and arrows represent rotational freedom. To turn off the DOF icons, again click Degrees of Freedom on the View menu.

 TIP You can turn on the DOF symbols for a single (or a few) component(s) by right-clicking on the component's name in the Browser, selecting Properties from the menu, and then clicking Degrees of Freedom on the Occurrence tab. Following the same steps will toggle the symbols off if they are turned on.

When placing or creating components in an assembly, it is recommended to have them in the order in which they will be assembled. The order will be important when placing assembly constraints and creating presentation views. When constraining components to one another, you will need to understand the terminology. The following is a list of terminology used with assembly constraints:

Line This can be the centerline of an arc, a circular edge or cylindrical face, a selected edge, a work axis, or a sketched line.

Normal This is a vector that is perpendicular to a planar face.

Plane This can be defined by the selection of a plane or face: two noncollinear but coplanar lines or axis, three points, or one line or axis and a point that does not lie on the line or axis. When you use edges and points to select a plane, this creates a workplane, and it is referred to as a construction plane.

Point   This can be an endpoint or midpoint of a line, the center or end of an arc or circular edge, or a vertex created by the intersection of an axis and plane or face.

Offset   This is the distance between two selected lines, planes, points, or any combination of the three.

TYPES OF CONSTRAINTS

Autodesk Inventor has four types of assembly constraints (mate, angle, tangent, and insert), two types of motion constraints (rotation and rotation-translation), and a transitional constraint. You can access the constraints through the Constraint tool found on the Assembly Panel Bar, as shown in the following image, by right-clicking and selecting Constraint from the menu, or by using the hot key c.

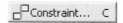

Figure 6-21

The Place Constraint dialog box will appear as shown in the following image. The dialog box is divided into four areas (the following sections describe these areas). Depending upon the constraint type, the option titles may change.

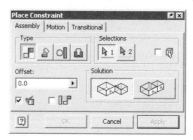

Figure 6-22

The Assembly Tab

Type   Select the type of assembly constraint to apply: mate, angle, tangent, or insert.

Selections   Click the button with the number 1 and select a component's edge, face, point, etc., on which to base the constraint type. Then click the button with the number 2 and select a component's edge, face, point, etc., on which to base the constraint type. By default, the second arrow will become active after you have selected the first input. You can edit an edge, face, point, etc., of an assembly constraint that has already been applied by clicking the number button that corresponds to the constraint and then selecting a new edge, face, point, etc. While working on complex assemblies, you can click the box on the right side of Selections (called Pick part first). If the box has a check, you will then select the component before selecting a component's edge, face, point, etc.

Offset/Angle   Enter or select a value for the offset or angle from the drop-down list.

Solution   Select how the constraint will be applied; the normals will be pointing in either the same or opposite directions.

Show Preview   Click and, when constraints are applied to two components, you will see the underconstrained components previewed in their constrained positions. If you leave the option's box clear, you will not see the components assembled until you click the Apply button.

Predict Offset and Orientation   Click to display the existing offset distance between two components. This allows you to either accept this offset distance or enter a new offset distance in the edit box.

The Motion Tab

Type   Select the type of assembly constraint to apply: rotation or rotation-translation.

Selections   Click the button with the number 1 and select a component's face or axis on which to base the constraint type. You will see a glyph in the graphics window previewing the direction of rotation motion. Then click the button with the number 2 and select the component's axis or face on which to base the constraint type. A second glyph appears showing the direction of rotation.

Ratio   Enter or select a value for the ratio from the drop-down list.

Solution   Select how the constraint type will be applied. The components will rotate in either the same or opposite directions as previewed by the graphics window glyphs.

The Transitional Tab

A transitional constraint will maintain contact between the two selected faces. You can use a transitional constraint between a cylindrical face and a set of tangent faces on another part.

Type   Select transitional as the type of assembly constraint to apply.

Selections   Click the First Selection button and select the first face on the part that will be moving. Click the Second Selection button and then select a face around which the first part will be moving. If there are tangent faces, they will become chained as part of the selected face automatically.

ASSEMBLY CONSTRAINT TYPES

This section explains each of the assembly constraint types.

Mate

There are three types of mate constraints: plane, line, and point.

Mate Plane

The mate plane constraint assembles two components so that the surface normals on the selected planes will be coplanar to and oppose one another. In the following image, the mate condition is being applied (it is selected in the Solution area of the Place Constraint dialog box).

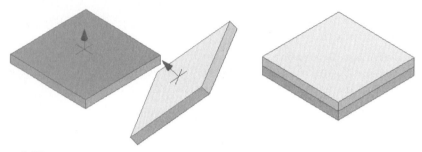

Figure 6-23

Mate Line

The mate line constraint assembles the edges of lines in the following image to be collinear.

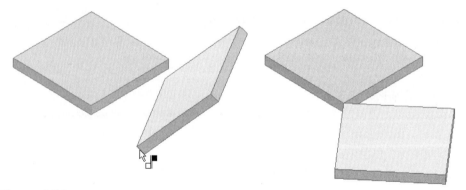

Figure 6-24

You can also use the mate line constraint effectively to assemble the center axis of a cylinder with a matching hole feature or axis, as in the following image.

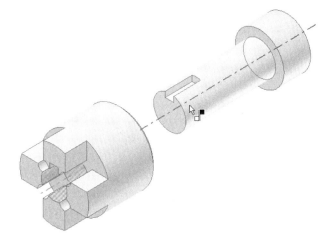

Figure 6-25

Mate Point

The mate point constraint assembles two points to be coincident (center of arcs and circular edges, endpoints, and midpoints), as shown in the following image.

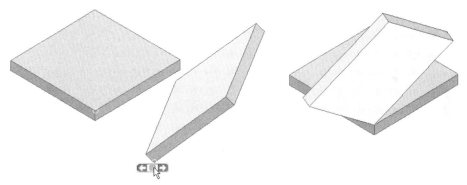

Figure 6-26

Mate Flush Solution

The mate flush solution constraint aligns two components so that the selected planes face the same direction or have their surface normals pointing in the same direction, as shown in the following image. Faces are the only geometry that you can select for this constraint.

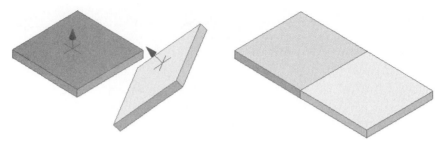

Figure 6-27

Angle

The angle constraint specifies the degrees between selected planes or faces or axes. The following image illustrates the angle constraint with two planes selected and a 30° angle applied.

Two solutions are available when placing an angle constraint: Directed Angle and Undirected Angle. You can experiment with both of these solutions, especially when driving the angle constraint and observing the behavior of the assembly.

Figure 6-28

Tangent

The tangent constraint defines a tangent relationship between planes, cylinders, spheres, cones, and ruled splines. At least one of the faces you select needs to be a curve, and you can apply the tangency to the inside or outside of the curve.

The following image illustrates the tangent constraint applied to one outside curved face and a selected planar face, and the outer and inner solutions applied.

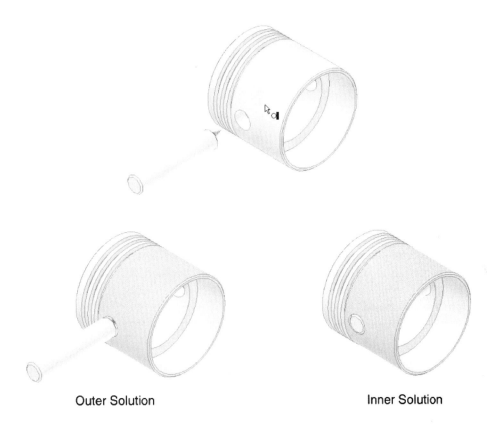

Outer Solution   Inner Solution

Figure 6-29

Insert

The insert constraint takes away five DOF with one constraint, but it only works with components that have circular edges. Select the circular edges of two different components. The centerlines of the components will be aligned, and a mate constraint will be applied to the planes defined by the circular edges. Circular edges define a centerline/axis and a plane. The following image illustrates the insert constraint with two circular edges selected and the opposed solution applied.

Figure 6-30

MOTION CONSTRAINT TYPES

There are two types of motion constraints: rotation and rotation-translation, as shown in the following image. Motion constraints provide the ability to simulate the motion relationships of gears, pulleys, rack and pinions, and other devices. By applying motion constraints between two or more components, you can drive one component and cause the others to move accordingly.

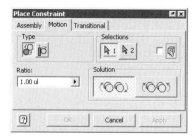

Figure 6-31

Both types of motion constraints are secondary constraints, which means they define motion but do not maintain positional relationships between components. You are to constrain your components fully before you apply motion constraints. You can then suppress constraints that restrict the motion of the components you want to animate.

Rotation

The rotation constraint defines a component that will rotate in relation to another component by specifying a ratio for the rotation between the two components. Use this constraint for showing the relationship between gears and pulleys. Selecting the tops of the gear faces displays the rotation glyph, as shown in the following image. You may also have to change the solution type from Forward to Backward, depending on the desired results.

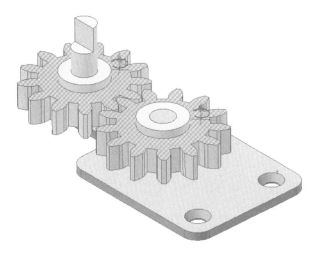

Figure 6-32

Rotation-Translation

The rotation-translation constraint defines the rotation relative to translation between components. This type of constraint is well suited for showing the relationship between rack and pinion gear assemblies. In the image of the rack and pinion assembly in the following image, the top face of the pinion and one of the front faces of the rack are selected. You supply a distance the rack will travel based on the pitch diameter of the pinion gear, and then you can drive the constraints and test the travel distance of the mechanism.

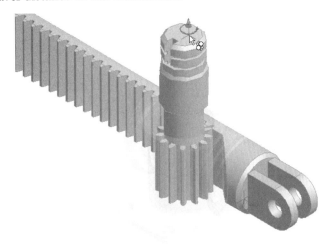

Figure 6-33

TRANSITIONAL CONSTRAINT

The transitional constraint specifies the intended relationship between (typically) a cylindrical part face and a contiguous set of faces on another part, such as a cam follower in a cam slot. The transitional constraint maintains contact between the faces as you slide the component along open DOF. You access this constraint type through the Transitional tab of the Place Constraints dialog box, as shown in the following image.

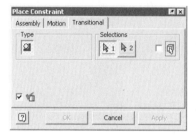

Figure 6-34

Select the moving face first, followed by the transitional face, as in the image of the cam and follower in the following image.

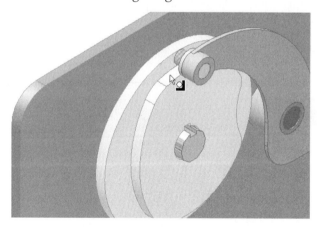

Figure 6-35

APPLYING ASSEMBLY CONSTRAINTS

After selecting the type of assembly constraint that you want to apply, the Selections button with the number 1 will become active. You can click the button if it does not. Position the cursor over the face, edge, point, etc., to apply the first assembly constraint.

You may need to cycle through the selection set using the Select Other tool, shown in the following image, until the correct location is highlighted. Cycle by clicking on the left or right arrows in the Select Other tool until you see the desired constraint condition. You then press the left mouse button (or the green rectangle in the Select Other tool) to place the constraint.

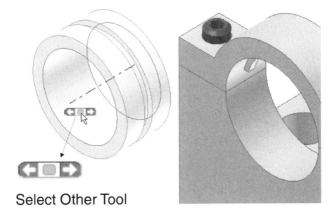

Select Other Tool

Figure 6-36

The next step is to position the cursor over the face, edge, point, etc., to apply the second assembly constraint. Again, you may need to cycle through the selection set until the correct location is highlighted. If the Show Preview option is selected in the dialog box, the components will move to show how the assembly constraint will affect the components, and a snapping sound will be heard when you preview the constraint. To change either selection, click on the button with the number 1 or 2 and then select the new input. Enter a value as needed for the offset or angle and select the correct Solution option until the desired outcome appears. Click the Apply button to complete the operation and leave the Constraint dialog box active to define subsequent constraint relationships.

 NOTE If your mouse is equipped with a rotating middle wheel, you can roll the wheel when the Select Other tool is active to cycle through the selection set of faces, edges, or points in a more efficient manner.

ALT-Drag Constraining

Another method for applying an assembly constraint is to hold down the ALT key while dragging a part edge or face to another part edge or face; no dialog box will appear. The key to dragging and applying a constraint is to select the correct area on the part. Selecting an edge will create a different type of constraint than selecting a face will create. If you select a circular edge, for example, an insert constraint will be applied. To apply a constraint while dragging a part, you cannot have another tool active.

To apply an assembly constraint, follow these steps:

1. Hold down the ALT key, then select the face, edge, and so on, on the part that will be constrained.

2. Select a planar face, linear edge, or axis to place a mate or flush constraint.

Select a cylindrical face to place a tangent constraint.

Select a circular edge to place an insert constraint.

3. Drag the part into position. As you drag the part over features on other parts, you will preview the constraint type. If the face you need to constrain to is behind another face, pause until the Select Other tool appears. Cycle through the possible selection options, and then click the center dot to accept the selection.

To change the constraint type previewed while you drag the part, release the ALT key and enter one of the following shortcut keys.

M or 1 Use to change to a mate constraint. Press the space bar to flip to a flush solution.

A or 2 Use to change to an angle constraint. Press the space bar to flip the angle direction on the selected component.

T or 3 Use to change to a tangent constraint. Press the space bar to flip between an inside and outside tangent solution.

I or 4 Use to change to an insert constraint. Press the space bar to flip the insert direction.

R or 5 Use to change to a rotation motion constraint. Press the space bar to flip the rotation direction.

S or 6 Use to change to a rotation-translation constraint. Press the space bar to flip the translation direction.

X or 8 Use to change to a transitional constraint.

 NOTE A work plane can also be used as a plane with assembly constraints; a work axis can be used to define a line.

MOVING AND ROTATING COMPONENTS

Use the Move Component tool, shown in the following image, to drag individual components in any direction in the viewing plane.

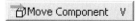

Figure 6-37

To perform a move operation on a component, activate the Move Component command and click and hold the left mouse button on the component to drag it to a new location. Drop the component at its new location by releasing the button.

Moved components will follow these guidelines:

- An unconstrained component remains in the new location when moved until you constrain it to another component.
- A partially constrained component adjusts its location to comply with a constraint that you have already applied.
- You can force a grounded component to move. After the move, the grounded component will remain grounded in its new location. Components constrained to the grounded component will adjust and update to its new location.

Use the Rotate Component tool found on the Assembly Panel Bar, as shown in the following image, to rotate an individual component. This is very useful when constraining faces that are hidden from your view.

Figure 6-38

Follow these steps for rotating a component in an assembly:

1. Activate the Rotate Component command and select the component to rotate. Notice the appearance of the 3D rotate symbol on the selected component in the following image.

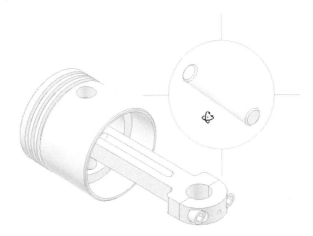

Figure 6-39

2. Drag your cursor until you see the desired view of the component.
- For free rotation, click inside the Dynamic Rotate tool and drag in the desired location.
- To rotate about the horizontal axis, click the top or bottom handle of the Dynamic Rotate tool and drag your cursor vertically.

- To rotate about the vertical axis, click the left or right handle of the Dynamic Rotate tool and drag your cursor horizontally.
- To rotate planar to the screen, hover over the rim until the symbol changes to a circle, click the rim, and drag in a circular direction.
- To change the center of rotation, click inside or outside the rim to set the new center.

Release the mouse button to drop the component in the new, rotated position.

NOTE If you click the Update button after moving or rotating components in an assembly, any components constrained to a grounded component will snap to their constrained positions in the new location. A fully constrained component can be moved or rotated temporarily. Once the assembly constraints are updated, the component will resolve the constraints and return to a fully constrained location/orientation.

EDITING ASSEMBLY CONSTRAINTS

After you have placed an assembly constraint, you may want to edit, suppress, or delete it to reposition the components. There are two methods for editing assembly constraints.

Both methods are executed through the Browser. In the Browser, activate the assembly or subassembly that contains the component you want to edit. Expand the component name and you will see the assembly constraints. You can then double-click on the constraint name, and an Edit Dimension dialog box will appear, allowing you to edit the constraint offset value. You can also right-click on the assembly constraint's name in the Browser and select Edit, Suppress, or Delete from the menu, as shown in the following image.

NOTE When editing offset dimension values, click once on the constraint with the offset value with the left mouse button and change the offset value from the edit box that will appear at the bottom of the Assembly Browser.

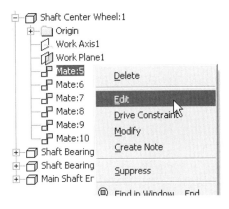

Figure 6-40

If you select Edit, the Edit Constraint dialog box will appear. If you select Suppress, the assembly constraint will not be applied. Select Drive Constraint to drive a constraint through a sequence of steps, simulating mechanical motion. Select Delete and the assembly constraint will be deleted from the component. If you try to place or edit an assembly constraint and it cannot be applied, an alert box will appear that explains the problem. You will have to either select new options for the operation, or suppress or delete another assembly constraint that is conflicting with it.

If an assembly constraint is conflicting with another, a small yellow icon with an exclamation point will appear in the Browser, similar to the following image. To edit a conflicting constraint in the Browser, either double-click on its name or right-click on its name and select Recover from the menu, as shown in the figure. The Design Doctor will appear and walk you through the steps to fix the problem.

Figure 6-41

Another problematic case occurs when you add too many constraints and Autodesk Inventor fails to generate error messages. As you view a list of constraints in the Browser, you may notice a few constraints that are preceded by a small circle with the letter "i" inside, as shown in the following image. The appearance of this icon means you have applied an unnecessary or redundant constraint. While your assembly will not suffer from the presence of these constraints, it is considered good practice to clean up all redundant constraints by deleting them from the Browser.

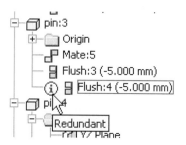

Figure 6-42

OTHER CONSTRAINT TOOLS

Additional tools are available to manipulate, navigate and edit assembly constraints. These tools include Find Other Half, Constraint Tool Tip, and Constraint Offset Value Modification. The following sections describe these tools.

FIND OTHER HALF

You can use the Find Other Half tool to find the matching part that participates in a constraint placed in an assembly. As you add parts over time, you may wish to highlight an assembly constraint and find the part(s) to which it is constrained. In the following image, a mate constraint has been highlighted. Half of this constraint has been applied to a part called Engine Block. To view the part sharing a common constraint, right-click on the constraint in the Browser and select Other Half from the menu. The Browser will expand and highlight the second half of the constraint. In the case of this example, the other half of the mate constraint was made to a part called Cylinder.

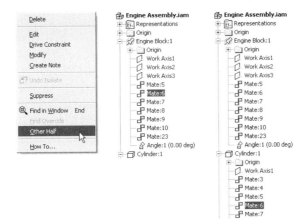

Figure 6-43

CONSTRAINT TOOL TIP

In order to display all property information for a specific constraint, move your cursor over the constraint icon and a tool tip will appear, as shown in the following image.

Figure 6-44

The following information is displayed in the tool tip:

- Constraint name and parameter name (for offset and angle parameters)
- Constrained components (the two part names from the Assembly Browser)
- Constraint solution and type
- Constraint offset or angle value

 NOTE Although the constraint name is highlighted in the above image, you must hover your cursor over the constraint icon to view the tool tip information.

CONSTRAINT OFFSET VALUE MODIFICATION

When editing a constraint offset value, use the standard value edit control. This process is similar to editing work plane offsets and sketch dimensions, and it will allow you to measure while editing constraint offset values. Right-click on the constraint to edit in the Browser and select Modify from the menu, as shown in the following image. The Edit Dimension dialog box will appear, allowing you to edit the offset value.

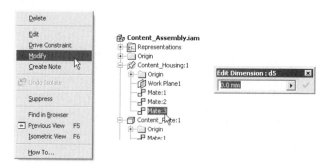

Figure 6-45

 NOTE When modifying an offset value, the offset edit box is also present at the bottom of the Assembly Browser enabling you to make changes to the constraint there.

EXERCISE 6-1 Assembling Parts

In this exercise, you assemble a lift mechanism.

1. Open the assembly file ESS_E06_01.iam as shown in the following image.

Figure 6-46

2. Begin assembling the connector and sleeve. Right-click in the graphics window and select Isometric View then zoom in on the small connector and

sleeve. Drag the connector so the small end is near the sleeve as shown in the following image.

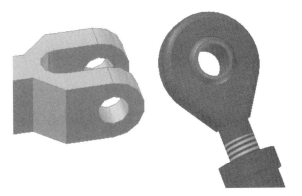

Figure 6-47

3. Next, add a Mate between centerlines of both parts. Click the Place Constraint tool in the panel bar. Mate is the default constraint.

4. Move the cursor over the hole in the arm on the sleeve then click when the center line displays as shown in the following image.

 NOTE If the green dot displays, move the cursor until the center line displays or use Select Other to cycle through the available choices.

5. Move the cursor over the hole in the link then click when the center line displays as shown in the following image.

6. Click Apply to accept this constraint.

Figure 6-48

7. Now add a Mate between the faces of both parts. Click the small flat face on the ball in the link end as shown in the following image.
8. Click the inner face of the slot on the sleeve as shown in the following image.
9. Click Apply to accept this constraint.

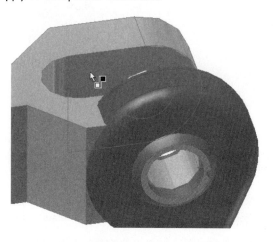

Figure 6-49

10. Click Cancel to close the Place Constraint dialog box.
11. Click on the small link arm and drag it to view the effect of the two constraints as shown in the following image.

Figure 6-50

12. Place the Crank in the assembly by clicking the Place Component tool in the panel bar and opening the file *ESS_E06_01-Crank.ipt*.
13. Move the component near the spyder arm, click to place the part, then right-click and select Done.

Chapter 6 Creating and Documenting Assemblies 357

Figure 6-51

14. Begin the process of constraining the Crank arm by clicking the Place Constraint tool in the panel bar.

15. Move the cursor over the hole in the arm on the crank then click when the center line displays.

16. Click Apply to accept this constraint.

17. Click Cancel to close the Place Constraint dialog box.

18. Drag the crank away from the spyder arm to make it easier to apply the next constraint.

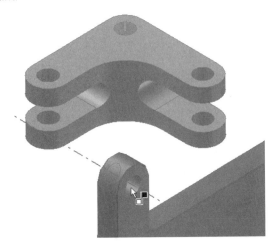

Figure 6-52

19. Next constrain the faces of the crank arm and the spyder.

20. Click the Place Constraint tool in the panel bar.
21. Click the inner face of slot on the crank as shown on the left side of the following image.
22. Rotate the model then click the face of the spyder arm as shown on the right side of the following image or choose the face using the Select Other tool.
23. Click Apply to accept this constraint. Then click Cancel to close the Place Constraint dialog box.

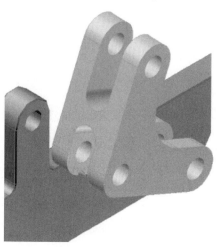 

Figure 6-53

24. You will now assemble the crank and link. First, return to an isometric view.
25. Drag the end of the small link arm close to the crank as shown in the following image.
26. Zoom in to the crank and link.
27. Then place a mate constraint between the centerlines of the two holes as shown in the following image. Then click Cancel in the Place Constraint dialog box.

Figure 6-54

28. You will now place a Claw in the assembly. First click the Place Component tool in the panel bar and open the file ESS_E06_01-Claw.ipt.

29. Move the component near the end of the spyder arm, click to place the part, then right-click and select Done. Your display should appear similar to the following image.

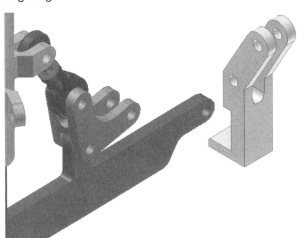

Figure 6-55

30. Constrain the claw to the spyder by first placing a mate constraint between the centerlines of the holes as shown on the left side of the following image.

31. Then place another mate constraint between the two faces as shown on the right side of the following image.

NOTE When the centerline mate constraint is applied, the claw will be oriented incorrectly. The orientation will correct when the face to face mate is applied.

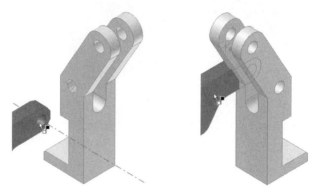

Figure 6-56

32. Now assemble the link rod to the crank and claw. First drag the claw and link rod into the positions shown.

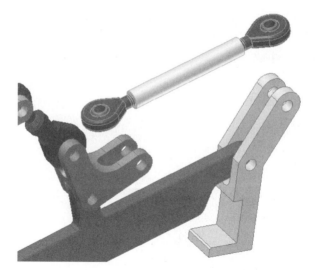

Figure 6-57

33. Assemble the link arm to the claw and crank using mate constraints between centerlines at each end and a mate between faces at one end. Close the Place Constraint dialog box then drag the claw to view the affect of the assembly constraints.

Figure 6-58

34. You will now add bolts and nuts to the spyder assembly. Begin by clicking the Place Component tool in the panel bar and opening the file *ESS_E06_01-Bolt.ipt*.

35. Place 6 bolts near their final position in the assembly as shown in the following image then right-click and select Done.

36. Then click the Place Component tool in the panel bar and open the file *ESS_E06_01-Nut.ipt*.

37. Place 6 nuts in the assembly as shown in the following image then right-click and select Done.

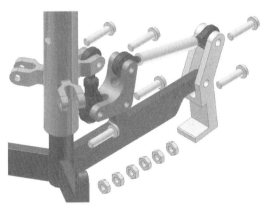

Figure 6-59

38. Insert the bolts into the assembly by clicking the Place Constraint tool in the panel bar.

39. In the Type area, click Insert to select an Insert constraint.

40. Insert each bolt by selecting the top of the bolt's shank then the edge of the hole as shown in the following image.

41. Close the Place Constraint dialog box.

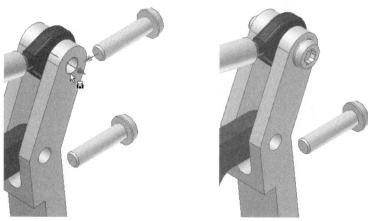

Figure 6-60

42. Now add the nuts to the end of the bolts.

43. Rotate the model so you can see the back of the claw. Then drag 2 nuts near their final location as shown in the following image.

44. Click the Place Constraint tool in the panel bar and in the Type area, click Insert to select an Insert constraint.

45. Insert each nut by selecting the edge of the hole in the nut and the edge of the hole on the claw as shown in the following image.

 NOTE You must select an edge on the face of the nut. Do not select the inner chamfered edge.

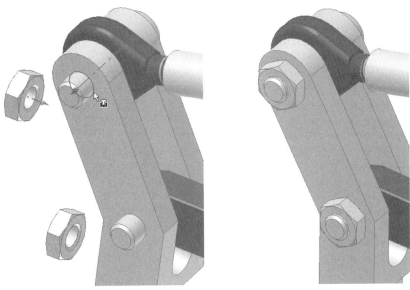

Figure 6-61

46. Assemble the remaining bolts and nuts using the same procedure. The completed assembly is displayed in the following image.

Figure 6-62

47. End of exercise.

DESIGNING PARTS IN-PLACE

Most components created in the assembly environment are created in relation to existing components in the assembly. When creating an in-place component, you can sketch on the face of an existing assembly component or a work plane. You can also click the graphics window background to define the current view orientation as the XY plane. If the YZ or XZ plane is the default sketch plane, you must reorient the view to see the sketch geometry. Click Application Options on the Tools menu, and then click on the Part tab to set the default sketch plane.

When you create a new component, you can select an option in the Create In-Place Component dialog box to constrain the sketch plane to the selected face or work plane automatically. After you specify the location for the sketch, the new part immediately becomes active, and the Browser, Panel Bar, and toolbars switch to the part environment (see the following image).

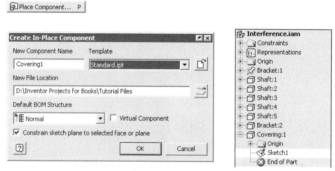

Figure 6-63

Notice also that the 2D Sketch Panel shown in the following image is available to create sketch geometry in the first sketch of your new part.

Figure 6-64

After you create the base feature of your new part, you can define additional sketches based on the active part or other parts in the assembly. When defining a new sketch, you can click a planar face of the active part or another part to define the sketch plane on that face. You can also click a planar face and drag the sketch away from the face to create the sketch plane automatically on the resulting offset work plane. When you create a sketch plane based on a face of another component, Autodesk Inventor automatically generates an adaptive work plane and places the active sketch plane on it. The adaptive work plane moves as necessary to reflect any changes in the component on which it is based. When the work plane adapts, your sketch moves with it. Features based on the sketch then adapt to match its new position.

After you finish creating a new part, you can return to assembly mode by double-clicking the assembly name in the Browser. In assembly mode, assembly constraints become visible in the Browser. If you selected the Constrain Sketch Plane to Selected Face option when you created your new part, a flush constraint will appear in the Assembly Browser. As with all constraints, you can delete or edit this constraint at any time. No flush constraint is generated if you create a sketch by clicking in the graphics window or if the box is clear when selecting an existing part face.

EXERCISE 6-2 Designing Parts in the Assembly Context

In this exercise, you create a lid for a container based on the geometry of the container. This cross-part sketch geometry is adaptive and automatically updates to reflect design changes in the container.

1. You will first start with an existing assembly, then create a new part based on cross-part sketch geometry. Begin by opening the file *ESS_E06_02.iam* as shown in the following image.

Figure 6-65

2. Begin the process of creating a new component in the context of an assembly by clicking the Create Component tool.

3. Under New File Name, enter *ESS_E06_02-LID*

4. Click the Template Browse button and in the Metric tab, click *Standard (mm).ipt* then click OK. There should be a checkmark beside Constrain Sketch Plane to Selected Face or Plane at the bottom of the Create In-Place Component dialog box.

5. Click the OK button to exit the dialog box, then select the top face of the container as shown in the following image. This face becomes the new sketch plane for the new part.

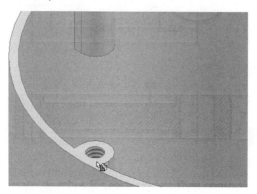

Figure 6-66

6. You will now project all geometry contained in this face. First, click the Project Geometry tool.

7. Move the cursor onto the face of the base part until the profile of the entire face is highlighted as shown in the following image.

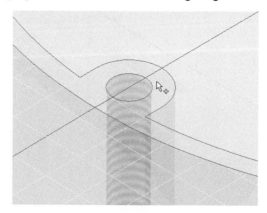

Figure 6-67

8. Click to project the edges. Your display should appear similar to the following image.

Chapter 6 Creating and Documenting Assemblies 367

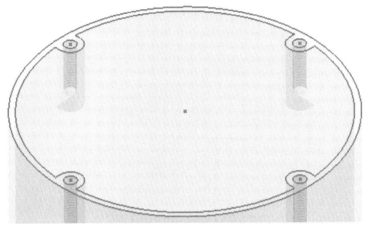

Figure 6-68

9. You will now create a series of clearance holes. First, zoom into a tapped hole.

10. Click the Center point circle tool and click the projected circles center point.

11. Move the cursor then click to create a circle larger than the projected circle as shown in the following image.

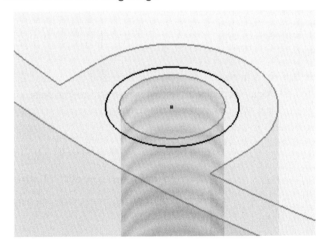

Figure 6-69

12. Create three additional circles centered on the remaining projected holes as shown in the following image.

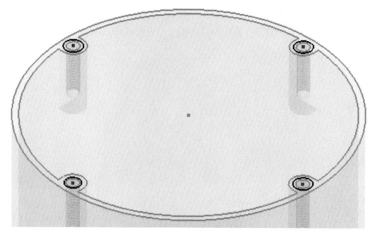

Figure 6-70

13. All circle diameters need to be made equal. Perform this operation by clicking the Equal constraint tool.

14. Click one circle, and then click one of the other circles. Then click the first circle, and click a third circle. Finally, click the first circle, and then click the fourth circle. This will make all circles equal to each other in diameter.

15. Now add a dimension to one of the circles to fully constrain the sketch. First, click the General Dimension tool.

16. Click the edge of one of the circles and click outside the circle to place the dimension.

17. Enter **2.5 mm** in the dimension edit box, and then press Enter to apply the dimension. Your display should appear similar to the following image.

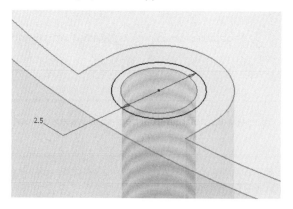

Figure 6-71

18. Next, click the Return button to exit Sketch mode. Then click the Extrude tool and select the profile as shown in the following image.

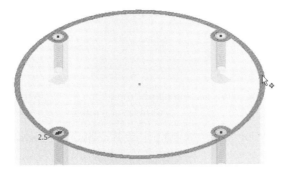

Figure 6-72

19. In the Extrude dialog box, enter a distance of **5** then click OK. Your display should appear similar to the following image.

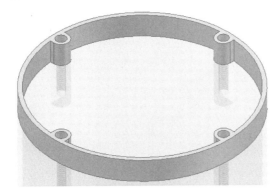

Figure 6-73

20. You will now use the Extrude tool to create the top of the lid. Click the Sketch tool and select the top face of the lid as shown in the following image.

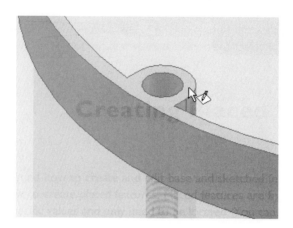

Figure 6-74

21. Press E to start the Extrude tool and select the two profiles as shown in the following images.

Figure 6-75

22. While inside the Extrude dialog box, enter a distance of **3** then click OK to create the top of the lid as shown in the following image.

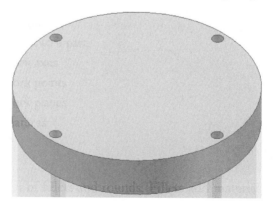

Figure 6-76

23. Click the Return tool to activate the assembly. Your display should appear similar to the following image.

Figure 6-77

24. You will now modify the container base and observe how this affects the lid. First, view the model as Hidden Edge Display as shown on the left side of the following image.

25. In the Browser, right-click *ESS_E06_02-Base:1*, and then select Edit. Your display should appear similar to the right side of the following image.

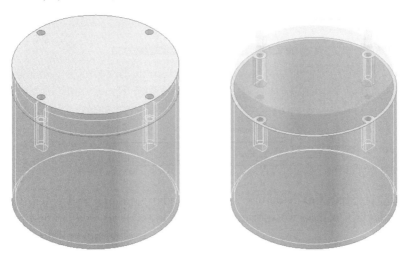

Figure 6-78

26. In the Browser, right-click Extrusion1, and then select Edit Sketch.

27. Double-click the 50 dimension, and then change the value to 60.

28. Click the Return tool. Switch back to Shaded Display as shown on the left in the following image.

29. Click the Return tool again to return to the assembly context. Notice that the lid adapts to the modified dimensions of the base as shown on the right in the following image.

Figure 6-79

30. End of exercise.

ASSEMBLY BROWSER TOOLS

Additional tools are available through the Assembly Browser as a means of better controlling and managing data in an assembly file. These tools include In-Place Activation, Visibility Control, Assembly Reorder, Restructuring an Assembly, Demoting and Promoting assembly components, and the use of Browser filters.

IN-PLACE ACTIVATION

The level of the assembly that is currently active determines whether you can edit components or features. You can take some actions only in the active assembly and its first-level children, while other operations are valid at all levels of the active assembly.

Double-click any subassembly or component occurrence in the Browser to activate it, or right-click the occurrence in the Browser and select Edit. All components not associated with the active component appear shaded in the Browser, as shown in the following image.

NOTE Double-clicking directly on a component in the graphics window will also activate the part for editing.

Chief Plate.ipt

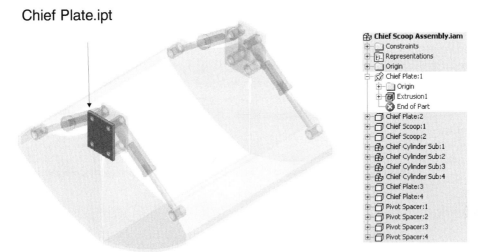

Figure 6-80

If you are working with a shaded display, the active component appears shaded in the graphics window and all other components appear translucent. If you are working with a wireframe display, the active component appears in a contrasting color.

You can perform the following actions on the first-level children of the active assembly:

- Deleting a component
- Displaying the degrees of freedom of a component
- Designating a component as adaptive
- Designating a component as grounded
- Editing or deleting the assembly constraints between first-level components

You can edit the features of an activated part in the assembly environment. The Panel Bar and toolbars change to reflect the part environment when a part is activated.

 NOTE Double-click a parent or top-level assembly in the Browser to reactivate it.

VISIBILITY CONTROL

Controlling the visibility of components is critical to managing large assemblies. You may need some components only for context, or the part you need may be obscured by other components. Assembly files open and update faster when the visibility of nonessential components is turned off.

You can change the visibility of any component in the active assembly, even if the component is nested many layers deep in the assembly hierarchy. To change the visibility of a component, expand the Browser until the component occurrence is visible, right-click the occurrence, and select Visibility, as shown in the following image.

 NOTE You can also right-click on a component in the graphics window and select Visibility.

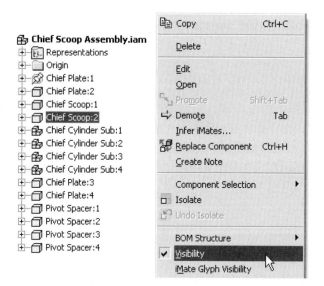

Figure 6-81

ADAPTIVITY

Adaptivity is the functionality in Autodesk Inventor that allows the size of a part to be determined by setting up a relationship between the part and another part in the assembly. Adaptivity allows underconstrained sketches–features that have undefined angles or extents, hole features, and subassemblies (which contain parts that have adaptive sketches or features)–to adapt to changes. The adaptivity relationship is defined by applying assembly constraints between an adaptive sketch or feature and another part. If a sketch is fully constrained, it cannot be made adaptive. However, the extruded length or revolved angle of the part can be. A part can only be adaptive in one assembly at a time. In an assembly that has multiple placements of the same part, only one occurrence can be adaptive. The other occurrences will reflect the size of the adaptive part.

An example of adaptivity would be to determine the diameter of a pin from the size of a hole. In the same example, you could determine the diameter of the hole from the size of the pin. You can turn adaptivity on and off as needed. Once a part's size is determined through adaptivity, and adaptivity is no longer useful, you may want to turn its adaptivity *off*. If you want to create adaptive features, there are options in Autodesk Inventor that will speed up the process of creating them. On the Tools menu, click Application Options. On the Assembly tab, there are three areas that relate to adaptivity, as shown in the following image.

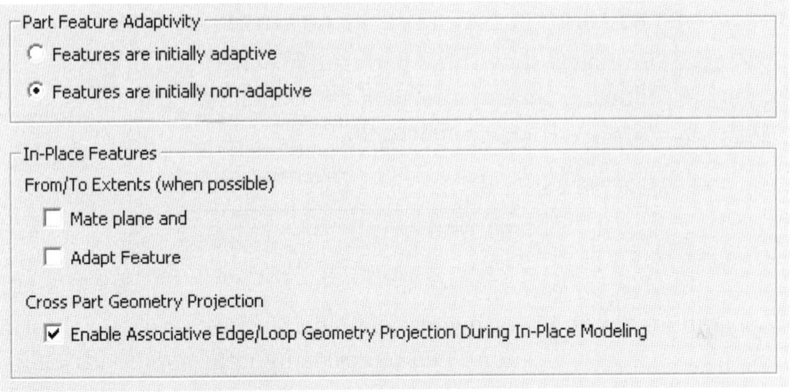

Figure 6-82

ASSEMBLY TAB OPTIONS

The following sections describe the adaptivity options found on the Assembly tab.

Part Feature Adaptivity

In this section you will indicate if new features will be adaptive or non-adaptive when they are created. You can also change the adaptivity of a feature *after* it has been created. Adaptivity means that a feature will be able to change its size according to the relationship it has with another component.

Features are initially adaptive Click to make features adaptive when they are created. This is useful when you are creating many adaptive features.

Features are initially non-adaptive Click to make features non-adaptive when they are created.

In-Place Features From/To Extents (when possible)

Here you will determine if a feature will be adaptive when the To or From/To option is selected for the extrusion extent. If both options are selected, Autodesk Inventor will try to make the feature adaptive. If it cannot, it will terminate at the selected face.

Mate plane and   Click when you create a new component to have a mate constraint applied to the plane on which it was constructed. It will not be adaptive.

Adapt Feature   Click when you want to create a new component to have it adapt to the plane on which it was constructed.

Cross Part Geometry Projection

Enable Associative Edge/Loop Geometry Projection During In-Place Modeling   Click when geometry is projected from another part onto the active sketch to make the projected geometry associative (has sketch associativity) and to update it when changes are made to the parent part. You can use projected geometry to create a sketched feature.

UNDERCONSTRAINED ADAPTIVE FEATURES

The next series of images show how parts adapt when you apply assembly constraints. In the following image, a rectangular sketch for a small plate is not dimensioned (unconstrained) along its length.

Figure 6-83

The extruded feature is then defined as *adaptive* as shown in the following image by right-clicking on the extrusion in the Browser to display the appropriate menu. Selecting Adaptive from the menu applies the adaptive property to the sketch and the extrusion.

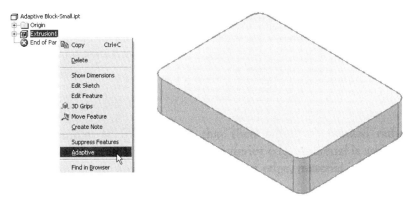

Figure 6-84

Once you have placed the parts in an assembly, right-clicking on the small box in the Browser activates a menu. Selecting Adaptive from the menu adds an icon consisting of two arcs with arrows next to the small box part, as shown in the following image. The presence of this icon in the Browser means the small box part is now adaptive. As flush constraints are placed along the edge faces of the plates, the small plate will adapt its length to meet the length of the large plate.

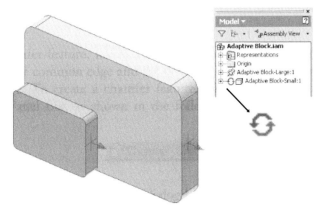

Figure 6-85

The results are shown in the following image.

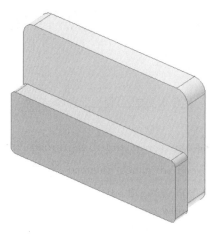

Figure 6-86

When you place or create a part containing adaptive features in an assembly, it is not initially adaptive. To specify the part as adaptive in the assembly context, right-click the component in the Assembly Browser or graphics window and select Adaptive from the menu. You can also use the Assembly Browser to specify a feature of the part as adaptive, and the part will become adaptive automatically.

When you constrain an adaptive part to fixed features on other components, underconstrained features on the adaptive part resize when the assembly is updated.

Only one occurrence of a part can define its adaptive features. If you use multiple placements of the same part in an assembly, all occurrences are defined by the one adaptive occurrence–including placements in other assemblies. If you want to adapt the same part to different assemblies, save the part file with a unique name using Save Copy As before defining any occurrences as adaptive.

ADAPTIVE FEATURE PROPERTIES

You can resize a part to meet assembly constraints if one or more features of the part are defined as adaptive and the related part geometry is underconstrained (The part itself must also be adaptive). You can also specify extruded or revolved features, hole features, and work planes as adaptive.

You can make a feature adaptive in one of two ways. Right-click the feature in the Browser and choose one of the following methods:

- Select Adaptive from the menu to make all available parameters of the feature adaptive.
- Select Properties from the menu and select the parameters or sketch for the feature to be adaptive in the Feature Properties dialog box, as shown in the following image.

Chapter 6 Creating and Documenting Assemblies 379

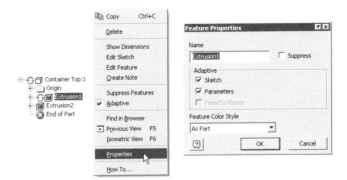

Figure 6-87

 NOTE You can set the default status of newly created features from the Assembly tab of the Options dialog box. Select the Features are Initially Adaptive option to give features adaptive status at the time of their creation.

The following sections describe the available adaptive parameters for each feature type.

Extruded Features

Selecting an extrusion feature from the Browser activates the Feature Properties dialog box, as shown in Figure the following image.

Figure 6-88

This dialog box controls the suppression, feature color styles, and adaptive status of sketched features. Click the appropriate check box to indicate the adaptive status of underconstrained geometry. Clear the check box to remove adaptive status. A brief description of each option in this dialog box follows.

Suppress

Suppresses the feature in the Browser and the graphics window.

Adaptive   Options in this area set the adaptive status of sketched features.

Sketch   You can intentionally leave out specific geometric constraints or dimensions on a sketch and then make it adaptive. An underconstrained length of a line in a sketch, for example, allows the length of the face it defiles to adapt to meet assembly constraints. Removal of parallel or perpendicular constraints may also allow an angle between faces to adapt to meet assembly constraints.

Parameters   This makes the extrusion distance, originally defined as a numeric value, adaptive when selected.

From/To Planes   You can use an adaptive work plane as the termination of an extrusion. If you place an assembly constraint between the adaptive plane and fixed geometry, then the termination face of the extrusion extends and/or tilts to satisfy the applied constraint.

Depending on the termination specified for the extrusion, either the Parameters option or the From/To Planes option is available. Options that do not apply to the feature appear shaded in the dialog box.

Revolved Features

Selecting a revolved feature from the Browser activates the Feature Properties dialog box, as shown in the following image.

Figure 6-89

As with an extruded feature, this dialog box controls the suppression, feature color styles, and adaptive status of sketched features. Click the appropriate check box to indicate the adaptive status of underconstrained geometry. Clear the check box to remove adaptive status. A brief description of each option of this dialog box follows.

Adaptive   Options in this area set the adaptive status of sketched features.

Sketch   You can intentionally leave out specific geometric constraints or dimensions on a sketch and then make it adaptive. An underconstrained length between the revolution centerline and a parallel sketch line, for example, would allow the radius of a feature to adapt, given a suitable assembly constraint.

Parameters   This makes the angle of revolution, originally defined as a numeric value, adaptive when selected.

Hole Features

Selecting a hole feature from the Browser activates the Feature Properties dialog box, as shown in Figure the following image.

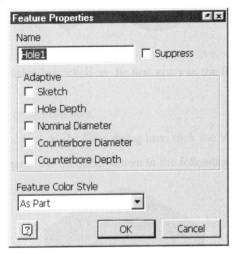

Figure 6-90

Use this dialog box to set adaptive status to parameters of hole features. A brief description of each option of this dialog box follows:

Adaptive   Options in this area set the adaptive status of sketched features.

Sketch   This makes the position of the sketch point defining the hole center adaptive when selected. This point must be underconstrained, meaning one or more located dimensions are not specified.

Hole Depth   This applies to blind termination holes. Specify a Flat Drill Point type in the Holes dialog box so that the bottom of the hole can adapt with a mate or flush constraint.

Nominal Diameter   This makes the diameter of the hole adaptive when selected.

Counterbore Diameter   This applies to counterbored holes only. The diameter of the hole counterbore becomes adaptive when selected.

Counterbore Depth   This applies to counterbored holes only. The counterbore depth becomes adaptive when selected.

Work Planes

The Properties option is not available for work planes, but you can specify the work plane itself as adaptive or not adaptive. The offset value for a work plane or the angle between a work plane and a planer face or another work plane can be adaptive.

NOTE To fix a feature at its current size and shape, right-click the feature in the Browser and then clear the Adaptive check mark.

ADAPTIVE SUBASSEMBLIES

In Autodesk Inventor you can use adaptive subassemblies in your models to control assembly constraints for moving parts inside any subassembly nesting level. When you specify a subassembly occurrence to be adaptive, parts inside the subassembly can adjust their size or position to fit changing conditions automatically (and independently) in a higher level of the assembly.

Subassemblies, when merged into assembly files, typically are defined as rigid bodies. Drag constraints are used to work on underconstrained subassembly components. A typical example of this concept in action is an air cylinder. All parts of the assembly are fully dimensioned; however, the rod can translate along the axis of the cylinder.

In the following image, the air cylinders are constrained to the industrial shovel. Unfortunately, the air cylinder motion is restricted due to the rigid nature of the subassemblies.

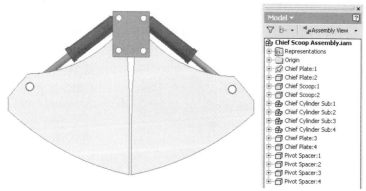

Figure 6-91

In the following image, one of the air cylinders is toggled to adaptive. Notice the appearance of the adaptive icon next to the subassembly in the Browser. Multiple occurrences of the same subassembly are now controlled by the initial subassembly made adaptive.

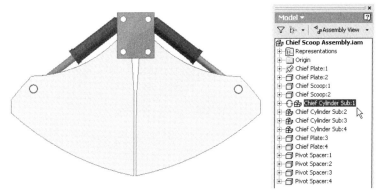

Figure 6-92

With the air cylinder toggled to adaptive, the underconstrained rod of the air cylinder subassembly can now move along the cylinder axis and affect the other shovel components of the assembly, as shown in the following image.

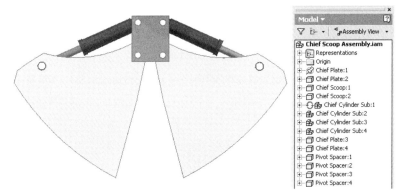

Figure 6-93

 NOTE To turn off the adaptivity for sketches, features, and subassemblies, right-click on the sketch, feature, or subassembly and deselect Adaptivity on the menu or in the Feature Properties dialog box.

ADAPTING THE SKETCH OR FEATURE

After making a sketch or feature adaptive, you must make the part itself adaptive at the assembly level. To make a part adaptive, make the assembly (where the part exists) active. Right-click on the part's name and select Adaptive from the menu. Apply assembly constraints that will define the relationship for the adaptive sketch or feature. As you are applying the assembly constraints, degrees of freedom are being removed.

NOTE You cannot make parts that are imported from an SAT or STEP file format adaptive because they are static and do not have underconstrained sketches and features. However, you can make assemblies created from these parts adaptive.

In assemblies that have multiple adaptive parts, two updates may be required to solve correctly.

For revolved features, use only one tangency constraint.

Avoid offsets when applying constraints between two points, two lines, or a point and a line. Incorrect results may occur.

EXERCISE 6-3 Creating Adaptive Parts

In this exercise, you create a link arm that adapts to fit between existing components in an assembly.

1. This exercise begins by first creating a link arm. The length of the link and the diameters of the circles are not dimensioned in the sketch so they can adapt to fit components in the assembly.

2. Click the New tool and select the Metric tab, then double-click *Standard(mm).ipt*. A new part is created.

3. Click the Line tool and click in the graphics screen to start the line.

4. Move the cursor to the right, and then click when a horizontal symbol is shown.

5. Click and drag off the point to create a 180 degree arc.

6. Move the cursor to the left, and then click to create a line the same length, and parallel to the first line.

7. Drag off the point to create an arc and close the profile. Your link arm should appear similar to the following image.

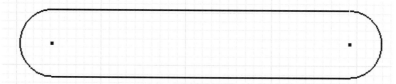

Figure 6-94

8. Continue with the creation of the link arm by clicking the down arrow beside the Constraint tool and then click the Tangent constraint tool.

9. Click the last arc and first line.

10. Next, click the Center Point Circle tool and on the left side of the sketch select the center point, then create a circle. On the right side of the sketch select the center point, then create a circle. Your sketch should appear similar to the following image.

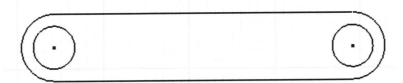

Figure 6-95

11. You will now dimension the arc. Click the General Dimension tool and add a 6 mm dimension to the left arc as shown in the following image. This is the only dimension needed for this new part file. The length of the link arm and the diameter of the holes will be determined when the arm is assembled.

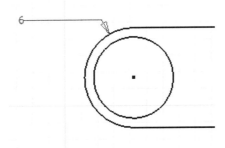

Figure 6-96

12. With the sketch completed but underconstrained, continue by right clicking in the graphics window and selecting Isometric View.

13. Create an extruded feature by first pressing E to start the Extrude tool.

14. Select the profile to extrude. The inside of the holes should not be selected.

15. Enter a value of **3** for the depth of the extrusion and click the OK button. The extrusion is created as shown in the following image and an entry is added to the Browser.

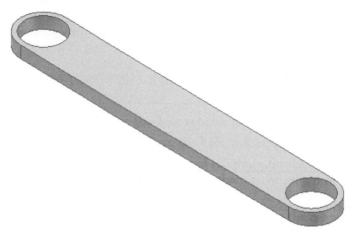

Figure 6-97

16. Change the color of the new part by selecting Metal-Titanium from the Color list on the Standard toolbar. Your part should appear similar to the following image.

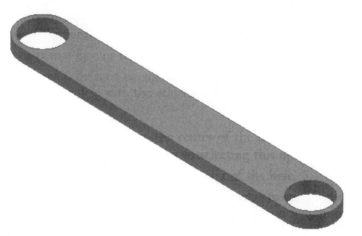

Figure 6-98

17. Save this new part file as ESS_E06_03-Link.ipt.

 NOTE It is acceptable if your part looks different when compared to the one in the previous illustration.

18. Close the part file and continue on with the assembly phase of this exercise.
19. Open the assembly file ESS_E06_03.iam as shown in the following image.

Figure 6-99

20. Place the link arm in the assembly by clicking the Place Component tool and opening the file *ESS_E06_03-Link.ipt*.

21. Click to place an occurrence of the link arm in the assembly as shown in the following image. Then right-click in the graphics window and select Done.

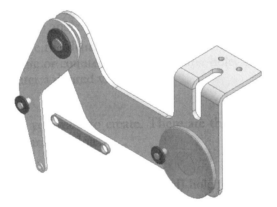

Figure 6-100

22. You will now make the link arm adaptive inside of the assembly. First click the Zoom Window tool then define a window around the link arm and lower-left pin as shown in the following image.

23. In the Browser, right-click *ESS_E06_03-Link.ipt* and select Edit from the menu

24. Right-click Extrusion1 then select Adaptive from the menu. The adaptive symbol is added to the extrusion and the part in the Browser.

25. Click the Return tool.

Figure 6-101

26. Now use assembly constraints to constrain the link arm to the main assembly. First click the Place Constraint tool.

27. Select the inside cylindrical surface of the hole in the link arm as shown in the following image.

 NOTE Make sure you do not select the centerline by mistake.

28. Then select the outside surface of the pin and click Apply.

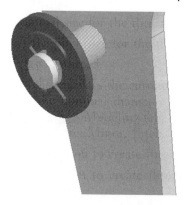

Figure 6-102

29. The link arm assembles to the pin and the diameter of the hole adapts to fit the pin as shown in the following image.

Chapter 6 Creating and Documenting Assemblies   389

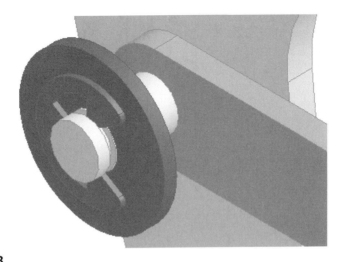

Figure 6-103

30. Continue assembling the link arm by adding a mate constraint between the face of the link arm and the main assembly arm. Do this by selecting the front face of the main assembly arm and the front face of the link arm with the mate constraint command active as shown on the left side of the following image.

31. Click Apply to accept the constraint, then click Cancel to exit the Constraints dialog box. Your display should appear similar to the right side of the following image.

Figure 6-104

32. Click the Zoom All tool on the Standard toolbar to view the entire assembly. Then click the Zoom Window tool on the Standard toolbar then define a window around the right end of the link arm and the pin on the eccentric as shown in the following image.

Figure 6-105

33. Constrain this end of the link arm to the pin on the eccentric. Complete this task by clicking the Place Constraint tool in the panel bar or from the Assembly toolbar.

34. Use the same technique here to constrain the link arm and pin as used in a previous step. Select the inside cylindrical surface of the link arm followed by the outside surface of the pin, then click Apply. Notice how the link arm adapts to fit between the two pins as shown in the following image. Click Cancel to exit the Constraints dialog box.

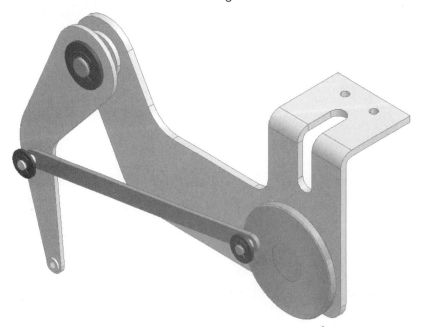

Figure 6-106

35. Use the work flow you learned in this exercise to place an assembly constraint between the front face of the link arm and the back face of the washer, using adaptivity to thicken the link arm.

36. End of exercise.

ENABLED COMPONENTS

When working in an assembly, you may want to hide a component so that you can see other components. You can turn the visibility of a component on or off by selecting its name in the Browser, or clicking on the part itself, right-clicking, and selecting Visibility from the menu. When the visibility of a component is *off* in an assembly, you can still use that component in other assemblies and open and edit it. Autodesk Inventor gives you another option, called Enabled, for controlling how components look in an assembly. By default, all components are enabled when you place or create them; their visibility is on, and they are displayed in the current display mode. When a component is enabled, it slows down the graphics regeneration speed because it has to be calculated each time a view is rotated or panned dynamically. You can disable a component. When you disable a component, it appears transparent and cannot be selected in the graphics window. A disabled component can, however, have geometry projected from it. To disable a component, select its name in the Browser, or click on the component itself, right-click, and select Enabled from the menu, as shown in the following image.

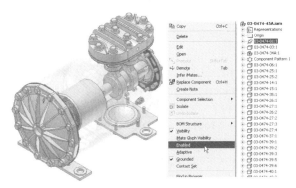

Figure 6-107

PATTERNING COMPONENTS

You can use the Pattern Component tool on the Assembly Panel Bar, shown in the following image, when you are placing multiple occurrences of the same component or subassembly that match a feature pattern on another part (component pattern) or have a set of circular or rectangular part patterns in an assembly (assembly pattern). The following sections describe component and assembly patterns.

Figure 6-108

COMPONENT PATTERNS

An associative component pattern will maintain a relationship to the feature pattern that you select. A bolt is component-patterned, for example, to a part bolt-hole pattern (circular pattern) that consists of four holes. If the feature pattern (the bolt-hole) changes to six holes, the bolts will move to the new locations and two new bolts will be added for the two new holes. To create a component pattern, there must be a feature-based rectangular or circular pattern, and the part that will be patterned should be constrained to the parent feature (that is, the original feature that was patterned in the component). Issue the Pattern Component tool from the Assembly Panel Bar, and the Pattern Component dialog box will appear, as shown in the following image.

Figure 6-109

There are three tabs in the dialog box: Associative, Rectangular, and Circular. The Associative tab is the default, and it will be used to create a component pattern. By default, the Component selection option is active. Select the component (such as the screw) or components to pattern. Next, click the Feature Pattern Select button in the dialog box and select a feature (such as a hole) that is part of the feature pattern. Do not select the parent feature. After selecting the pattern, it will highlight on the part, and the pattern name will appear in the dialog box. When done, click OK to create the component pattern and exit the operation (see the following image).

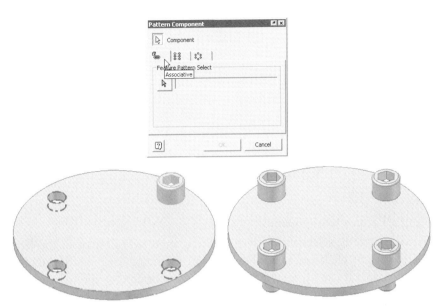

Figure 6-110

You can edit the component pattern by either selecting the pattern in the graphics window and right-clicking, or right-clicking on the pattern's name in the Browser and selecting Edit from the menu. In the Browser, the component that you patterned will be consumed into a component pattern, and the part occurrences will appear as elements, which can be expanded to see the part. You can suppress an individual pattern component by right-clicking on it and selecting Suppress from the menu, as shown in the following image.

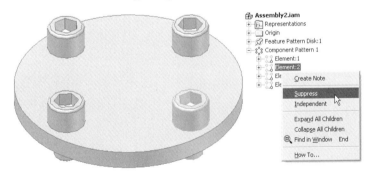

Figure 6-111

You can also break an individual part out of the pattern by right-clicking on it and selecting Independent from the menu, as shown in the following image. Once a part is independent, it no longer has a relationship with the pattern.

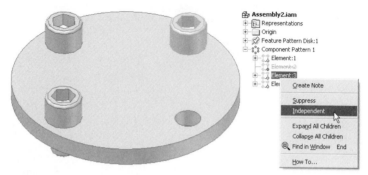

Figure 6-112

Deleting a Pattern Component

To delete a pattern, select the pattern in the graphics window or in the Browser. Then either right-click and select Delete from the menu, or press the DELETE key on the keyboard. If the Delete Component Pattern Source(s) option from the Options dialog box (under the Assembly tab) is selected, the source component will be deleted automatically. If the source component should not be deleted when the pattern elements are deleted, you should deselect this option, as shown in the following image.

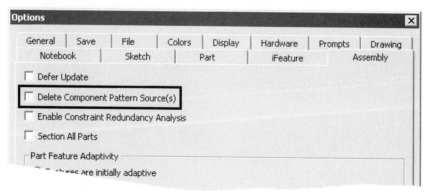

Figure 6-113

ASSEMBLY PATTERNS

You can also pattern a component in an assembly in either a rectangular or circular fashion. The resulting pattern acts like a feature pattern. After creating it, you can edit it to change its numbers, spacing, and so on. To pattern a component, you must first create it or place it in an assembly. Then issue the Pattern Component tool from the Assembly Panel Bar, and the Pattern Component dialog box will appear. Click the Rectangular or Circular tab and enter the placement values as needed.

The next two figures show examples of rectangular and circular patterns, respectively.

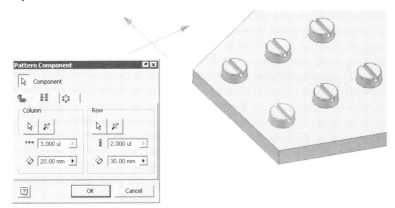

Figure 6-114

 NOTE In the rectangular pattern component example in the following image, work axes define the direction of the pattern in the X and Y directions, rather than part edges.

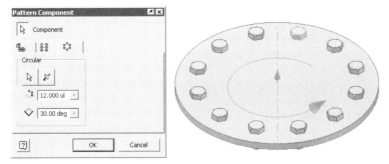

Figure 6-115

The completed pattern then acts as a single part. If one part moves, they all move. The patterned component will be consumed into a component pattern in the Browser, and each of the part occurrences will also appear as an element that you can expand. You can edit the pattern by either selecting the pattern in the graphics window and right-clicking, or right-clicking on the pattern's name in the Browser and selecting Edit from the menu.

ADDITIONAL COMPONENT PATTERN OPTIONS

Additional options that are available when editing and manipulating component patterns include replacing all occurrences in an assembly pattern in a single step, restructuring a component pattern, and controlling the component pattern visibility. The following sections describe these features.

Component Pattern Replace

When replacing a component in an assembly pattern, you can replace all occurrences in the selected assembly pattern with a newly selected component. Expand one of the elements in the Browser, as shown in the following image, and select the component to replace. With this item highlighted in the Browser, right-click and select Replace Component from the menu. The Open dialog box will appear and will enable you to select the replacement component.

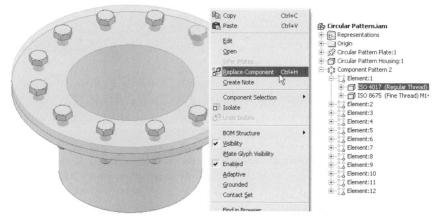

Figure 6-116

After selecting the replacement component from the Open dialog box, an Alert dialog box will appear, as shown in the following image. Component families should retain their previously placed constraints. Click OK to dismiss this dialog box.

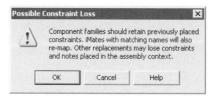

Figure 6-117

The following image illustrates the result of performing the operation to replace a pattern component. Since the use of component patterns allows for better capture of design intent, replacement of all occurrences maintains this design intent without manually replacing each component in the pattern. Overall ease of assembly use is improved and component patterns can be completed more quickly.

Figure 6-118

Component Pattern Restructure

You can restructure component patterns into or out of a subassembly. You can find these restructuring tools on the menu when you are in the assembly design environment. First, select the component pattern to restructure in the Browser. Next, right-click and select Demote from the menu. A dialog box stating that restructuring may remove assembly constraints will appear. Click OK to complete the component pattern restructuring process.

Component Pattern Visibility

Instead of expanding a pattern in the Browser and selecting each pattern instance for which to toggle off the visibility of patterned components, you can toggle the visibility of the entire component pattern off in one easy step. Select the component pattern in the Browser and right-click to open a menu, as shown in the following image. Selecting Visibility will remove the check to turn the component pattern off.

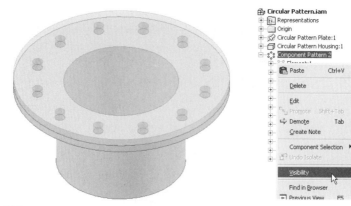

Figure 6-119

EXERCISE 6-4 Patterning Components

In this exercise, you create a fastener component pattern to match an existing hole pattern. You then replace the bolt in the pattern. You complete the exercise by demoting components to create a subassembly with the cap plate and patterned fasteners.

1. First, open an existing assembly called ESS_E06_04.iam as shown in the following image. The file contains a T pipe assembly.

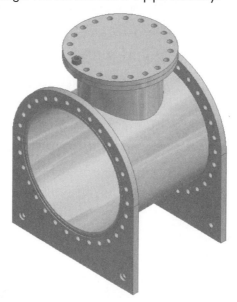

Figure 6-120

2. In the Browser, expand the parts ESS_E06_04-Cap_Bolt.ipt:1 and ESS_E06_04-Cap_Nut.ipt:1. Notice that the parts already have insert constraints applied to them as shown in the following image.

Chapter 6 Creating and Documenting Assemblies   399

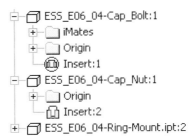

Figure 6-121

3. You will now create a pattern consisting of a collection of nuts and bolts in a circular pattern. Click the Pattern Component tool to launch the Pattern Component dialog box as shown on the right side of the following image.

4. Press CTRL and select the *ESS_E10_04-Cap_Bolt.ipt:1* and *ESS_E10_04-Cap_Nut.ipt:1* parts in the Browser as shown on the left side of the following image.

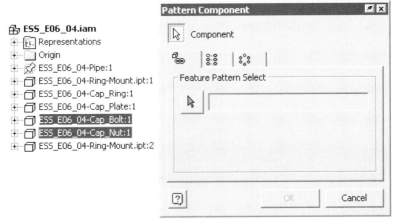

Figure 6-122

5. Next, click the Select button found under the Feature Pattern Select area of the Pattern Component dialog box. In the graphics window, point to any hole in the Cap_Plate part. When the circular pattern of holes in the cap plate is highlighted, pick the edge of one of the holes as shown on the left side of the following image.

6. You should notice a preview of the component pattern displayed as shown on the right side of the following image.

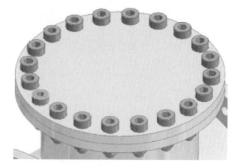

Figure 6-123

7. Click the OK button to close the Pattern Component dialog box and create the component pattern. Expand the display of the Component Pattern 1 feature in the Browser as shown in the following image.

8. In the Browser, expand Element:1 and the parts beneath it. Notice that the constraints on the original components were retained.

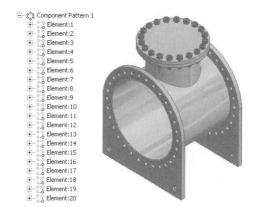

Figure 6-124

9. You will now replace the bolt in the pattern with another similar but different fastener. Under Element:1 in the Browser, right-click the *ESS_E06_04-Cap_Bolt* and select Replace Component from the menu.

10. In the Open dialog box, select *ESS_E06_04-Cap_Hex_Bolt.ipt* and click Open. A warning message box will appear to notify you that some assembly constraints may be lost.

11. Click the OK button. The pattern is updated to reflect the new bolt as shown in the following image.

Chapter 6 Creating and Documenting Assemblies 401

 NOTE The fastener in this exercise is constrained to the plate with an Insert iMate constraint. Since the replacement fastener has a similar Insert iMate, the constraint to the plate is retained.

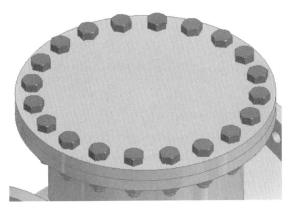

Figure 6-125

12. You will now demote a number of components. This will create a subassembly with the cap plate and component pattern.

13. Hold down CTRL while you click ESS_E06_04-Cap_Plate and Component Pattern 1 in the Browser as shown in the following image.

14. Right-click and select Demote from the menu. In the New File Name field of the Create In-Place Component dialog box, enter **Cap_Kit_Assembly.** under the New Component Name as shown in the following image.

15. When finished, click the OK button. A warning message box appears to notify you that some assembly constraints may be lost. Click Yes.

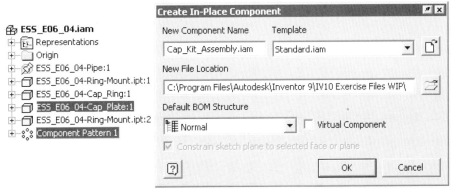

Figure 6-126

16. In the Browser, expand Cap_Kit_Assembly.iam:1. Notice that the cap plate and component pattern are now part of the new subassembly.

17. In the graphics window, drag any bolt and notice how the subassembly moves because constraints between the cap plate and cap ring are broken. Your display should appear similar to the following image.

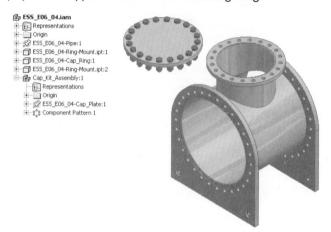

Figure 6-127

18. Place an Insert constraint between the bottom outside edge cap plate and the top outside edge of the cap ring.

19. Turn on Hidden Edge display and then zoom in and place another Insert constraint between any hole edge on the bottom of the cap plate and any hole edge on the top of the cap ring. Your display should appear similar to the following image.

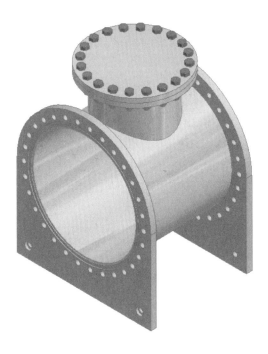

Figure 6-128

20. End of exercise.

ANALYSIS TOOLS

Various tools are available that assist in analyzing sketch, part, and assembly models. You can calculate the center of gravity of parts and assemblies, and perform an interference detection.

CENTER OF GRAVITY

At times, you may need the ability to get a visual fix on the current coordinates of the center of gravity during the design process. During the design process, the center of gravity must be kept within a particular region for the effectiveness of the overall design in which you will use the part or assembly. Having the center of gravity available will provide the output in a more real-time fashion for the center of gravity coordinates.

A center of gravity placed in an assembly could be critical to the overall design process of that assembly, whether it is used in the next assembly or in the main assembly.

Click the Center of Gravity tool on the View menu, as shown on the left side of the following image. The following image illustrates the center of gravity icon applied to an assembly model.

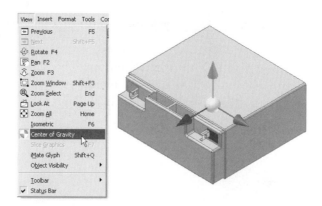

Figure 6-129

INTERFERENCE CHECKING

You can check for interference in an assembly using one or two sets of objects. To check the interference among sets of stationary components, make the assembly or subassembly in question active. Then click the Analyze Interference tool on the Tools menu, as shown on the left side of the following image. The Interference Analysis dialog box will appear, as shown on the right side of the following image.

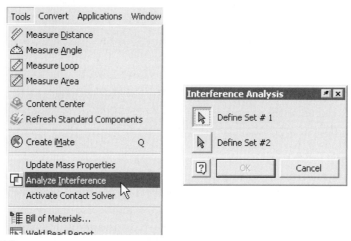

Figure 6-130

Click on Define Set #1 and select the components that will define the first set. Then click on Define Set #2 and select the components that will define the second set. A component can exist in only one set. To add or delete components from either set, first select the button that defines the set that you want to edit. Then click components to add to the set, or press the CTRL key while selecting components to remove from the set.

NOTE Use only Define Set #1 if you want to check for interference in a single group of objects.

Once you have defined the sets, click the OK button. The order in which you selected the components has no significance. If interference is found, the Interference Detected dialog box will appear, as shown in the following image.

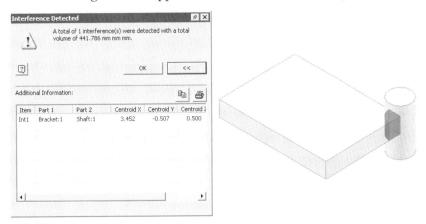

Figure 6-131

The information in the dialog box defines the X, Y, Z coordinates of the centroid of the interfering volume and lists the volume of the interference and the components which interfere with one another. A temporary solid will also be created in the graphics window that represents the interference. You can copy the interference report to the clipboard or print it from the tools in the Interference Detected dialog box. When the operation is complete, click the OK button, and the interfering solid will be removed from the screen. After analyzing and finding interference, edit the assembly or components to remove the interference.

 NOTE You can also detect interference when driving constraints.

EXERCISE 6-5 Analyzing an Assembly

In this exercise, you analyze a partially completed assembly for interference between parts and check the physical properties to verify design intent.

1. Open the assembly *ESS_E06_05.iam*. The linkage is displayed as shown in the following image.

Figure 6-132

2. Begin the process of checking for interferences in the assembly. First zoom into the linkages.

3. From the Tools menu, click Analyze Interference.

4. When the Analyze Interference dialog box displays, select the two links for Set #1 as shown in the following image.

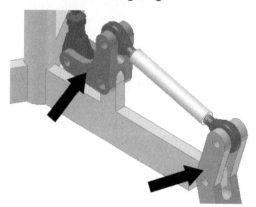

Figure 6-133

5. For Set #2, select the spyder as shown in the following image.

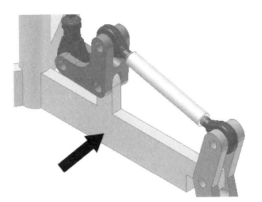

Figure 6-134

6. Click the OK button. Notice that the Interference Detected dialog box is displayed and the amount of interference displays on the parts of the assembly as shown in the following image.

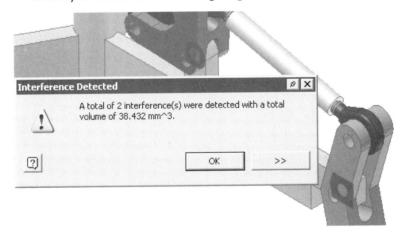

Figure 6-135

7. Continue using more features of the Interference Detected dialog box. Click the More (>>) button to expand the dialog box and display additional information.

8. Click the Copy to Clipboard button as shown in the following image. This is a quick way of sharing the interference detection information with other applications such as Microsoft Word or Notepad.

9. Then, open a text editor and paste the information into a document.

10. Exit the text editor. You do not have to save the text file.

11. Click the OK button to close the Interference Detected dialog box. Keep this assembly file open since this exercise continues.

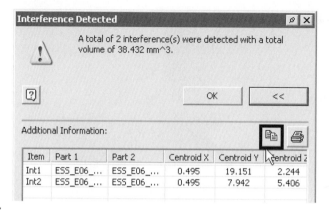

Figure 6-136

12. The design intent of the lifting mechanism relies on the correct selection of materials. This impacts the strength and mass of the assembly. You will now display the physical properties of the entire assembly.

13. View the assembly as an isometric view. Then in the Browser, right-click ESS_E06_05.iam then select iProperties as shown in the following image.

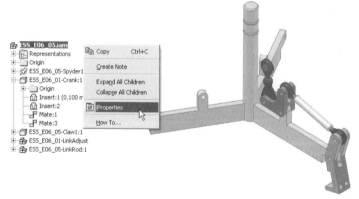

Figure 6-137

14. When the Properties dialog box displays, click the Physical tab.

15. Click the Update button. Notice the physical properties of the assembly displayed as shown in the following image.

 NOTE To get a more accurate account of the assembly properties, click on Medium under the Accuracy area and change this setting to Very High.

16. Click the OK button to dismiss this dialog box

Chapter 6 Creating and Documenting Assemblies 409

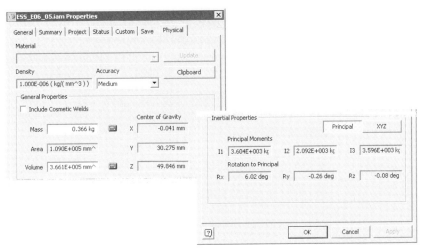

Figure 6-138

17. End of exercise.

DRIVING CONSTRAINTS

You can simulate (drive) mechanical motion using the Drive Constraint tool. To simulate motion, an angle, mate, tangent, or insert assembly constraint must exist. You can only drive one assembly constraint at a time, but you can use equations to create relationships to drive multiple assembly constraints simultaneously. To drive a constraint, right-click on the desired constraint in the Browser, as shown in the following image, and select Drive Constraint from the menu.

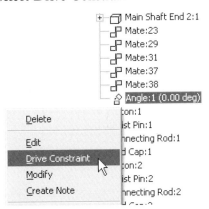

Figure 6-139

The Drive Constraint dialog box will appear, as shown in the following image. Depending on the constraint that you are driving, the units may be different. Enter a Start value; the default value is the angle or offset for the constraint. Enter a value for End and a value for Pause Delay if you want a dwell time between the steps. In the top half of the dialog box you can also choose to create an animation (AVI) file that will record the assembly motion. You can replay the AVI file without having Autodesk Inventor installed.

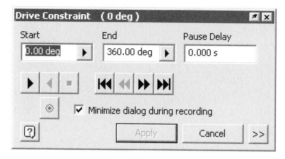

Figure 6-140

The following are descriptions of the Start, End, and Pause Delay controls.

Start   Set the start position of the offset or angle.

End   Set the end position of the offset or angle.

Pause Delay   Set the delay between steps. The default delay units are seconds.

Use the following Motion and AVI control buttons to control the motion and to create an AVI file.

▶	Forward	Drives the constraint forward, from the start position to the end position.
◀	Reverse	Drives the constraint in reverse, from the end position to the start position.
■	Pause	Temporarily stops the playback of the constraint drive sequence.
⏮	Minimum	Returns the constraint to the starting value and resets the constraint driver. This button is not available unless the constraint driver has been run.
⏪	Reverse Step	Reverses the constraint driver one step in the sequence. This button is not available unless the drive sequence has been paused.
⏩	Forward Step	Advances the constraint driver one step in the sequence. This button is not available unless the drive sequence has been paused.
⏭	Maximum	Advances the constraint sequence to the end value.

	Record	Begins capturing frames at the specified rate for inclusion in an animation file named and stored to a location that you define.

To set more details on how the motion will behave, click the More (>>) button. This will display the expanded Drive Constraint dialog box, as shown in the following image. The following sections describe the options in the dialog box in detail.

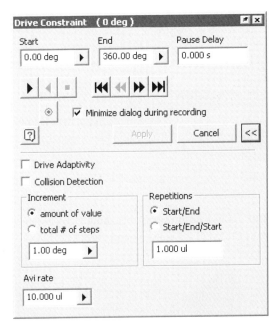

Figure 6-141

Drive Adaptivity   Click this option to adapt the component while the constraint is driven. It only applies to assembly components that have adaptivity defined and enabled.

Collision Detection   Click this option to drive the constraint until collision is detected. When interference is detected, the drive constraint will stop and the components where the collision occurs will be highlighted. It also shows the constraint value where the collision occured.

Increment   Determine the value that the constraint will be incremented during the animation.

Amount of Value   Describes the constraint in the selected number of increments.

Total # of Steps   Drives the constraint equally per number of steps.

Repetitions   Set how the driven constraint will act when it completes a cycle and how many cycles there will be.

Start/End Drives the constraint from the start value to the end value and resets at the start value.

Start/End/Start Drives the constraint from the start value to the end value and then in reverse from the end value to the start value.

AVI Rate Specify how many frames are skipped before a screen capture is taken of the motion that will become a frame in the completed AVI file.

 NOTE If you try to drive a constraint and it fails, you may need to suppress or delete another assembly constraint to allow the required component degrees of freedom. To reduce the size of an AVI file, reduce the screen size before creating the file and use a solid background color in the graphics window.

EXERCISE 6-6 Driving Constraints

In this exercise, you drive an angle constraint to simulate assembly component motion. You then drive the constraint and use collision detection to determine if components interfere.

1. Begin by opening the existing assembly file ESS_E06_06.iam as shown in the following image.

2. Use the Rotate and Zoom tools to examine the assembly. When finished, right-click in the graphics window and select Isometric View from the menu.

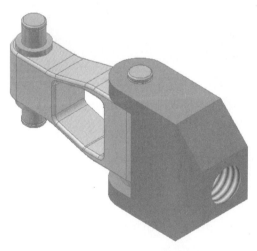

Figure 6-142

3. You will now create an angle constraint between the Pivot Base and Arm. First click the Place Constraint tool.

4. In the Place Constraint dialog box, click the Angle button.

5. Then select the Arm face followed by the Pivot Base face as shown in the following image.

6. Enter an angle of -45 degrees. Then click the OK button.

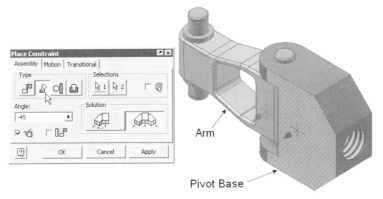

Figure 6-143

7. The assembly should appear as shown below.

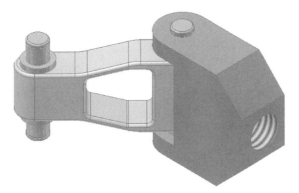

Figure 6-144

8. You will now drive the angle constraint you just placed. In the Browser, expand *ESS_E06_06-Pivot_Base:1*.

9. Right-click the Angle constraint and then select Drive Constraint from the menu.

10. In the Drive Constraint dialog box, enter **-45.00** as the Start value and **-120.00** as the End value.

11. Click the More (>>) button.

12. In the Increment section, enter **2** as amount of value as shown in the following image.

13. Click the Reverse button. Notice that the Arm assembly interferes with the Pivot Base between -75.00 and -120.00 degrees.

14. Click the Forward button.

Figure 6-145

15. You will now perform a collision detection operation. In the Drive Constraints dialog box, enter **-75.00** for the Start value.

16. Place a check the Collision Detection option.

17. Change the Increment value to .1

18. Click the Reverse button. Notice a collision is detected at -80.7 degrees and the interfering parts are highlighted as shown in the following image.

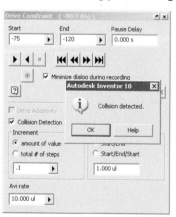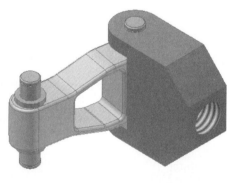

Figure 6-146

19. You will now perform a check on the full range of motion for the arm. In the Drive Constraints dialog box enter **-80.00** for the Start value and **80.00** for the End value.

20. Ensure the Collision Detection option is checked.
21. Change the Increment value to **2.00**.
22. Click the Forward button to test the full range of motion for interference. There is no collision detected in this range of motion.
23. In the Drive Constraints dialog box, click Apply. The 80.00 degrees, which represents the maximum angle before collision, is applied to the Angle constraint as shown in the following image.

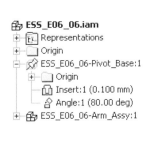

Figure 6-147

24. End of exercise.

DESIGN VIEW REPRESENTATIONS

While working in an assembly, you may want to save configurations that show the assembly in different states and from different viewing positions. Design view representations can store the following information:

- Component visibility (visible or not visible)
- Component selection status (enabled or not enabled)
- Color settings and style characteristics applied in the assembly
- Zoom magnification
- Viewing angle

You can use design view representations while working on the assembly and when creating presentation views or drawing views. You can also use them to reduce the number of items that display in an assembly. This allows large assembly files to be opened more quickly.

Once the screen orientation and part visibility is set, you can create a design view representation by clicking the Design View Representation icon at the top of the Browser, clicking the arrow next to the icon, and then clicking Other, as shown in the following image, or by clicking Design View Representations on the View menu.

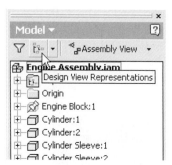 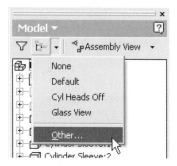

Figure 6-148

The Design View Representations dialog box will appear, as shown in the following image. In this example, two design view representations were created: Cyl Heads Off and Glass View. All names listed in the Design View Representation list box are considered public as far as the storage location is concerned. Public design view representations are saved with the assembly file and are available to all users when the assembly is accessed.

To make a design view representation current, select it from the drop-down list where you selected the Other option (or select Other to open the Design View Representations dialog box), and double-click on its name; or click it and select the Activate button. To delete a design view representation from the dialog box, select the design view representation's name from the list and then click the Delete button.

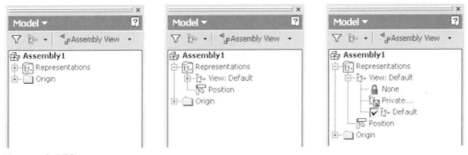

Figure 6-149

Another method for accessing design view representations is through the Assembly Browser, as shown in the following image. Each assembly file contains a folder called Representations, as shown on the left side of the figure. Expanding this folder will display the information shown in the middle of the figure. Notice the View: Default and Position listings. Expanding the View listing will display all design view representations defined in the assembly model. Notice also on the right side of the image, design view representations: None, Private, and Default. We will concentrate on this area for creating a new design view representation.

 NOTE The Position listing of the Representations folder will be discussed later in this chapter.

Figure 6-150

CREATING A NEW DESIGN VIEW REPRESENTATION

Begin the process of creating a new design view representation by expanding the Representations folder and the View: Default listing, as shown in the following image. A new design view representation will be created to show the two-cylinder engine assembly in different viewing states.

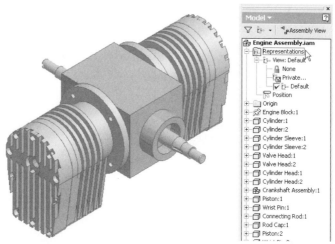

Figure 6-151

Move your cursor over the View: Default listing under the Representations folder and right-click. Then select New from the menu, as shown in on the left in the following image. This will create a new design view representation called ViewRep1, as shown in the middle of the figure. Notice also that creating this new design view representation will also make it current, as shown by the presence of the checked box. It is considered good practice to change the name of the design view representation to something more meaningful. On the right, ViewRep1 was renamed Glass View.

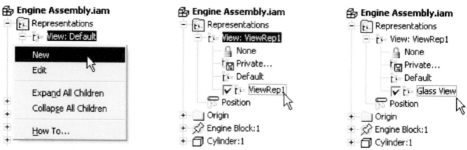

Figure 6-152

With the current design view representation set to Glass View, various parts such as the engine block, cylinders, and valve heads were changed to the glass color, as shown in the following image. This color change is saved automatically to the Glass View design view representation name. Double-clicking on the Default name will return the engine assembly back to its original representation. Double-clicking on Glass View will expose the piston and crankshaft subassemblies of the engine

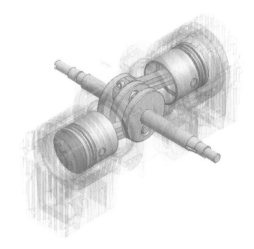

Figure 6-153

In the following image, a second design view representation was created: Cyl Heads Off. In this figure, Visibility of all outer portions of the engine assembly was turned off in the Assembly Browser and saved to this design view representation. As with all representations, double-clicking on a name or icon will make this design view representation active and update the display of the assembly model.

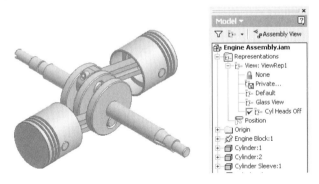

Figure 6-154

Once you are satisfied with all of the changes made to a design view representation, you can prevent any further changes to this representation by locking it. Right-clicking on the Cyl Heads Off name will display the menu shown in the following image. Selecting Lock from this menu will add a padlock icon to this design view representation.

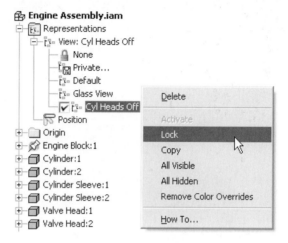

Figure 6-155

If you want to make additional changes to the Cyl Heads Off design view representation, you right-click on the name and select Unlock from the menu, as shown in the following image.

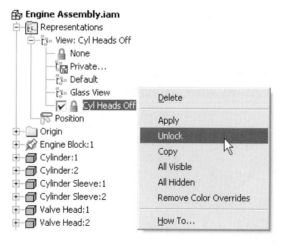

Figure 6-156

INCREASING PERFORMANCE THROUGH DESIGN VIEW REPRESENTATIONS

Through design view representations, additional control of the visibility of each subassembly is possible, allowing you to turn off the visibility of unimportant components to increase performance. The following sections describe the additional design view representation controls shown in the following image.

All Hidden Select this control so that none of the components in the assembly will be visible. This can be beneficial for managing large assembly files. Opening up a large assembly with all components hidden will allow you to manually turn on the visibility for only those components you need to work with.

All Visible Select this control to make all of the components in the assembly visible.

Remove Color Overrides Select this control to restore the original colors of an assembly. This works well if you applied an assembly level color override to any of the components in an assembly.

Figure 6-157

 NOTE The All Hidden and All Visible design view representation controls will override any existing visibility settings in the subassembly. You cannot alter or save these controls with the same name.

The design view representations applied to a subassembly from within an assembly are not associated with the design view representations created in the original subassembly. After a subassembly is placed in an assembly, changes or additional design view representations are not added automatically to the placed subassembly. You must re-import the design view representation to see the new design view representation or modifications made to a design view representation in the originating assembly.

VISIBILITY OVERRIDES

In addition to storing component visibility, zoom magnification, and viewing angle, design view representations can also store the display of origin work planes, origin work axes, origin work points, user work planes, user work axis, user work points, and sketches made within an assembly.

IMPORTING DESIGN VIEW REPRESENTATIONS FROM OTHER ASSEMBLIES

To place an assembly into another assembly with a selected design view representation, follow these steps:

1. Click the Place Component tool on the Assembly Panel Bar (as shown in the following image) and then press the hot key P or right-click and select Place Component from the menu.

Figure 6-158

The Open dialog box appears.

2. Navigate to and select the assembly that will be placed.

3. While still in the Open dialog box, click the Options button, as shown in the following image.

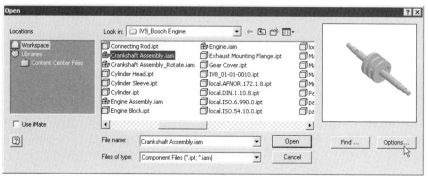

Figure 6-159

4. The File Open Options dialog box appears, showing the available design view representations (see the following image).

5. Select the design view representation to apply to the subassembly.

Figure 6-160

6. Click the OK button in the File Open Options dialog box.

7. In the Open dialog box, click the Open button.

8. In the graphics window, place the subassembly. If the Nothing Visible design view representation is selected, a wireframe box appears in the graphics window, as shown in the following image. To add assembly constraints, you will need to turn on the visibility of some components or change to a design view representation that displays the necessary components. See the next section for how to change from one design view representation to another.

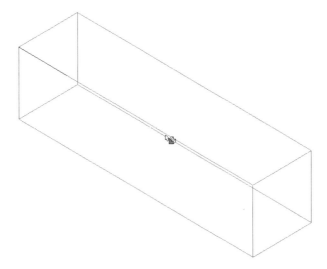

Figure 6-161

After you have placed an assembly or created an assembly as a subassembly in a parent assembly, you can change the design view representation to a different one by following these steps:

1. In the Browser, right-click on the assembly's name that you want to change (in this example, Crankshaft Assembly), or in the graphics window, move the cursor over the assembly that you want to change and then right-click. Select Representation from the menu, as shown in the following image.

Figure 6-162

The Representation dialog box appears, as shown in the following image.

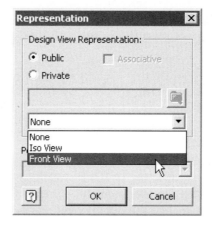

Figure 6-163

> **2.** From the list in the dialog box, select the needed Design View Representation.
>
> **3.** To complete the operation, click the OK button.

CREATING A PRIVATE DESIGN VIEW REPRESENTATION

In addition to creating design views internal to an assembly, you can also create a design view representation that is external to an assembly. This is called a private design view representation. When you create a private design view representation, the various assembly configurations are saved out to an external file with the *.idv* extension.

To create a private design view representation, expand the Representations folder in the Assembly Browser and double-click on Private, as shown in the following image.

Chapter 6 Creating and Documenting Assemblies 425

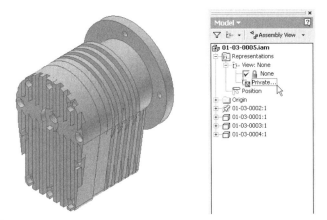

Figure 6-164

This will launch the Design View Representations dialog box. Notice in the figure below, the Storage Location is toggled automatically to Private File. You can either accept the default location to which the *.idv* file will be saved or you can select a different location.

Figure 6-165

 NOTE Private design view representations are not considered associative. This means drawing views will not reflect any changes to the settings in the private design view representation.

ASSOCIATIVE DESIGN VIEWS

You can make drawing views generated from a design view representation associative. In the following image, a design view representation called Front View, which shows the engine on the right, was saved.

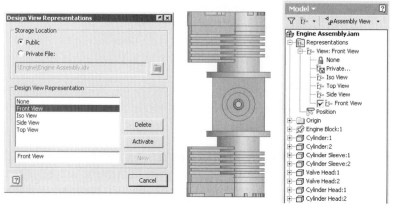

Figure 6-166

When you generate drawing views based on a design view representation, you first select the design view representation in the Drawing View dialog box, as shown in the following image. You can also check the box next to Associative. This will make the drawing view generated by the design view representation associated with the same design view representation found in the assembly file. Any changes made to the design view representation in the assembly will update the drawing view.

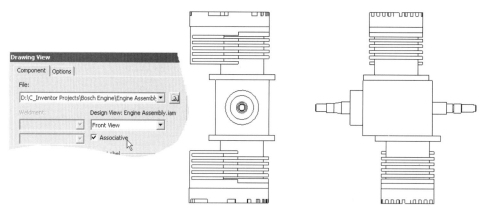

Figure 6-167

To see this in operation, the assembly file of the engine is opened. The visibility of the engine block is turned off in order to expose the internal connecting arms.

These changes are then saved to the Front View design view representation, as shown in the following image.

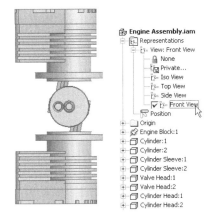

Figure 6-168

When the drawing file of the assembly is opened again, the engine block does not appear in any drawing views because this part was turned off back in the assembly file, as shown in the following image.

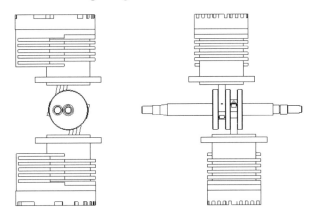

Figure 6-169

 NOTE You cannot associate private design view representations identified by the *.idv* extension with drawing views.

CREATING DRAWING VIEWS FROM DESIGN VIEW REPRESENTATIONS

A design view representation is created in an assembly file and can have information such as component visibility and viewing angle saved in it under a unique name. For example, you can turn off the visibility of a number of components and save these changes under a design view representation. When you recall this design view representation name, the components you turned off do not appear. To generate a Drawing View based on a design view representation, activate the Drawing View dialog box and click in the edit box to display all valid design views, as shown in the following image. Projected views will be based on the base view created from the design view representation, as is also shown in the following image.

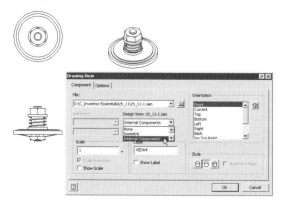

Figure 6-170

NOTE You can also make design view representations associative with the drawing view. This means that when you make a change to the design view representation contained in an assembly file, the drawing view associated with this design view representation will be updated automatically. You cannot make design views identified by an *.idv* extension (private Design Views) associative.

Once a drawing view is generated based on a design view representation, you can change the drawing view or views if you select a different design view representation. To perform this operation, first select the drawing view, and then right-click. Select Apply Design View from the menu, as shown in the following image.

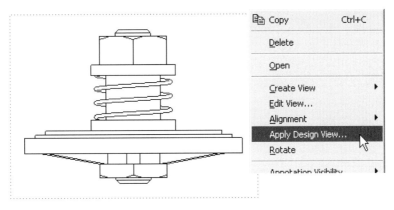

Figure 6-171

When the Apply Design View dialog box appears, click in the edit box and select a new design view representation from the list, as shown in the following image. If multiple views of the assembly appear on the sheet, set the Apply status for each view. Also set the Associative status by changing the value from No to Yes.

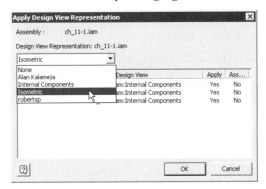

Figure 6-172

The following image illustrates the results. A design view representation called Internal Components was used to generate the first set of drawing views. After a different design view representation was selected, the drawing views were updated to reflect the new design view representation.

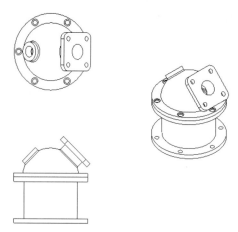

Figure 6-173

FLEXIBLE ASSEMBLIES

As already discussed, you can set the adaptivity property of a subassembly. When you place numerous instances of the same subassemby in a main assembly model, all subassemblies will expand or stretch to show the intended motion. In the following image, a typical industrial shovel is being controlled by three hydraulic cylinder subassemblies. When one of the cylinders is set to adaptive, the remaining cylinders carry the same adaptive property.

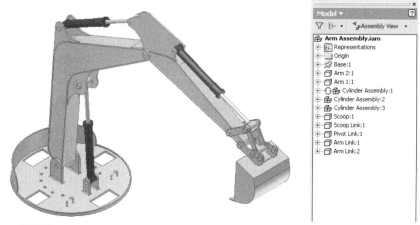

Figure 6-174

The following image illustrates the results of the adaptive cylinders. While these images may look acceptable, the problem with this solution is that when one cylinder opens or closes, the other two cylinders open or close to the same distance and direction. When working with assemblies of this nature, you want to have each cylinder move independently.

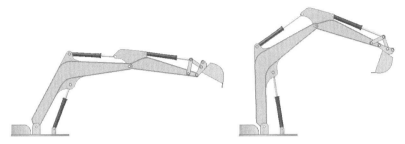

Figure 6-175

Instead of assigning the property of adaptivity that affects all cylinders, you can make each cylinder move independently by applying the Flexible property. In the following image, the cylinder assembly located in the Assembly Browser has had the Adaptivity property removed. Right-clicking on this cylinder assembly will display a menu, as shown in the figure. Click on Flexible.

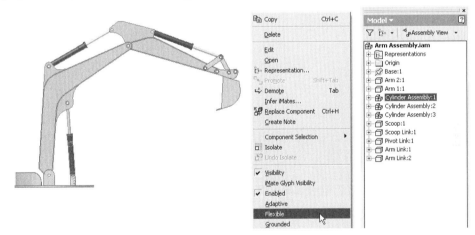

Figure 6-176

When you assign the Flexible property to the first occurrence of the cylinder assembly, an icon appears to symbolize flexibility for this subassembly, as shown in the following image. When the first cylinder moves, the others remain unaffected.

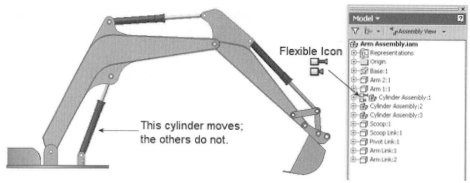

Figure 6-177

The following image shows all three cylinder assemblies set to Flexible. Now the hydraulic shovel assembly can move to any configuration because all three cylinders can move independently.

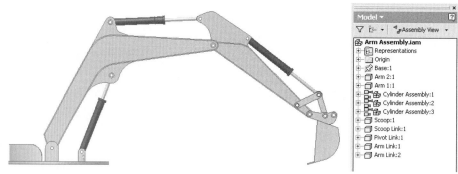

Figure 6-178

POSITIONAL REPRESENTATIONS

Use positional representations when you wish to create a motion study of an assembly model. Each motion segment or kinematical state can represent a different position of the assembly. All positional representations are saved with the assembly model.

The gripper mechanism shown in the following image will be used to demonstrate how positional representations are created and applied. One of the gripper fingers is assembled using an angle constraint. The opposite gripper finger is assembled using another angle constraint controlled by a parameter. When the angle constraints are driven, both grippers move in opposition to each other. In this example, positional representations will be used to show the gripper mechanism in its closed, middle, and open states.

To begin the process of creating a new positional representation, expand the Representations folder in the Browser, right-click on Position, and select New from the menu, as shown in the following image.

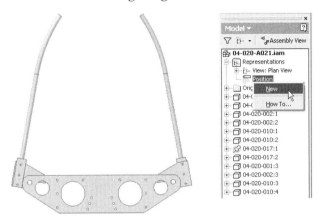

Figure 6-179

When defining a new positional representation, two are initially created: master and PositionalRep1, as shown on the left in the following image. You can think of the master positional representation as the default state of the assembly. As you create additional representations, the master representation will always return you to the default state of the assembly. You cannot edit the master positional representation. Any newly created positional representations will be based on the master.

It is considered good practice, once you have created a new positional representation, to rename it something more descriptive that better fits the intended purpose of the representation. PositionalRep1 has been renamed to Closed on the right in the following image. This name is designed to show the gripper assembly in its closed state.

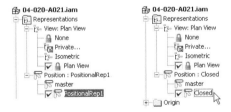

Figure 6-180

In the following image, the Angle:10 constraint appears in a suppressed state. This will allow you to drag the affected gripper arm finger freely. This suppressed state is required to configure the positional representation of the gripper in a closed position. Right-clicking on the Angle:10 constraint will display a menu; select Override.

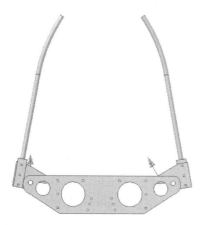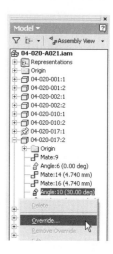

Figure 6-181

When the Override Object - Angle:10 dialog box appears, as shown in the following image, click Override under the Suppression heading. Then select Enable, as shown on the left in the figure. Next, override the angle value by clicking Override under the Value heading and entering a new value in the edit box. In the figure, 100° is the new overridden angular value.

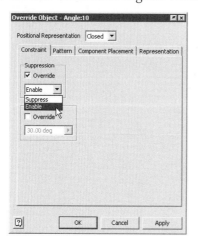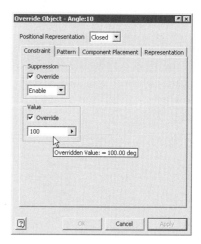

Figure 6-182

When you click the OK button, the gripper arm will appear as shown in the following image. This positional representation is in its closed position. Notice also the Angle:10 constraint appears bold in the Browser. This is to emphasize it is over-ridden in a positional representation.

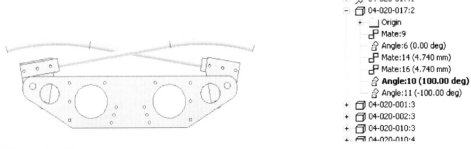

Figure 6-183

Your Browser can reflect positional representations just as it can display components as part of an assembly or standard parts found in the Library Browser. To switch the Browser to positional representations, click on the arrow next to Model and select Representations from the list, as shown on the left side of the following image. In the middle, Positional Representations have been expanded to expose the Master and current Closed representations. Expanding the Closed representation will display the constraints that were defined by the override operation, as shown on the right side of the image. This can provide a quick and easy way to review which constraints are affected by a particular positional representation override.

Figure 6-184

You may also want to show the gripper in its middle and open positions. To accomplish this, right-click on the Closed representation and select Copy from the menu, as shown on the left side of the following image. This will create a second positional representation called Closed1, as shown in the middle. On the right side of the image, a second positional representation (Closed2) has been created using the same method. All three positional representations have been expanded in order to better view their constraint information. You can override this information to reflect the middle and open positions of the gripper assembly.

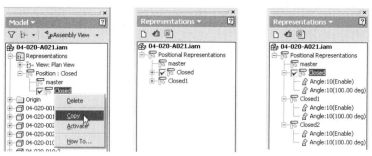

Figure 6-185

The first positional representation was renamed to Closed to represent its purpose better. It would also be advantageous to rename Closed1 to Middle and Closed2 to Open, as shown in the following image.

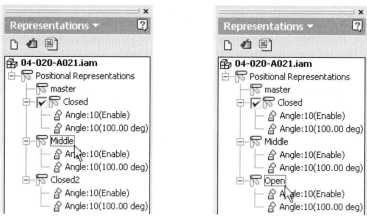

Figure 6-186

Next, the angle values of the Middle and Open representations need to be changed. A special tool is available in the Representations Browser to help accomplish this. Clicking on the Edit positional representation table button, shown on the left side of the following image, will launch an Excel spreadsheet, as shown on the right side of the image. You do not have to create this spreadsheet separately; rather, it is a function of using positional representations. This spreadsheet lists all valid positional representations and the value of the angle constraints in degrees for each.

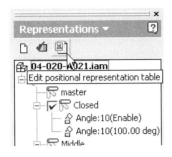

Figure 6-187

The Middle representation was changed from 100 to 40 and the Open representation from 100 to 0 (all values are in degrees in this example) as shown on the left side of the following image. To return back to the assembly and update the Representations Browser, click Close & Return to *04-020-A021.iam* on the File menu in Excel, as shown on the right side of the image.

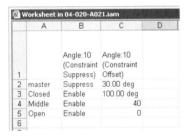 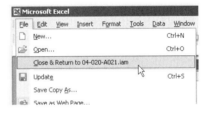

Figure 6-188

This will return you to the active assembly environment. Notice in the following image that the Middle and Open representations now reflect the changes that were made in the Excel spreadsheet.

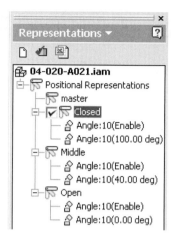

Figure 6-189

You can now double-click to activate each positional representation individually and observe the results, as shown in the following image. The left side of the image shows the closed state of the gripper (Closed representation). The middle shows the gripper half-open (Middle representation), and the right side of the image shows the gripper fully extended (Open representation)

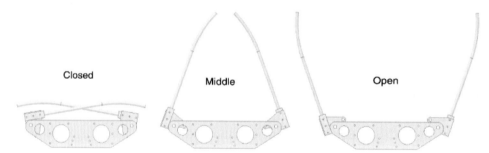

Figure 6-190

CREATING DRAWING VIEWS FROM POSITIONAL REPRESENTATIONS

When generating a base drawing view, you can specify the name of a positional representation for the view. To accomplish this, click on the Base View button on the Drawing Views Panel Bar. This will launch the Drawing View dialog box. Click the Options tab and notice the appearance of the Positional Representations area in the lower-left corner of the dialog box. Clicking on the arrow will display all positional representations defined in the assembly file. In the following image, the Closed positional representation has been selected. This will create a drawing view based on this positional representation.

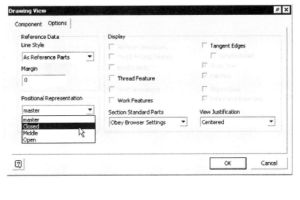

Figure 6-191

In the following image, three separate based views have been created from their corresponding positional representations: Closed, Middle, and Open.

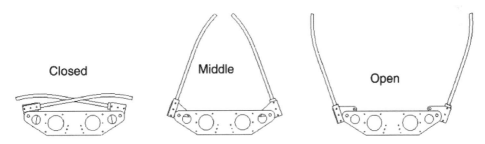

Figure 6-192

EXERCISE 6-7 Positional Representations

In this exercise, you create a few positional representations of an assembly.

1. First, open the existing assembly *ESS_E06_07.iam*.

An angle constraint has been applied between one of the plates and the bracket that holds the gripper finger. The other side of the gripper finger is driven with a similar angle constraint that has a parameter assigned with a negative angle value as shown in the following image. In this way, the gripper fingers close and open in opposing directions.

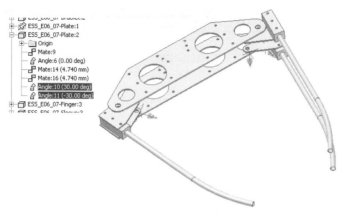

Figure 6-193

2. Begin the process of creating a new positional representation by following the next series of steps. In the Browser, expand the Representations folder and right-click on the Position node.

3. When the menu appears, click New as shown in the following image.

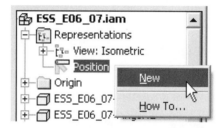

Figure 6-194

4. Expand the Position node to view the new positional representation. Creating a new positional representation for the first time creates a default representation called master. This positional representation displays the current position of the assembly.

5. The new positional representation is called by default, PositionalRep1. Rename this new positional representation to Closed as shown in the following image.

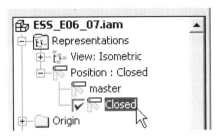

Figure 6-195

6. Next, you locate the constraint in the Browser that will be affected by the positional representation. Expand the part file ESS_E06_07-Plate:2.

7. Identify the constraint Angle:10(30 deg) as shown on the left side of the following image. This is the constraint that will be affected by the positional representation.

8. In the Browser, right-click on the Angle:10(30 deg) constraint and select Override... from the menu as shown on the right side of the following image. The Override tool is used to modify the solved state of the positional representation.

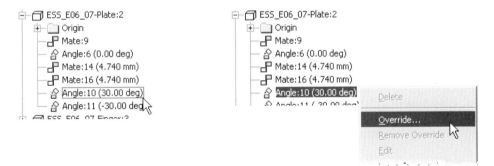

Figure 6-196

9. Clicking on Override... from the menu in the previous image will launch the Override Object dialog box as shown in the following image. You will need to make two changes in order for the positional representation to function. First, the constraint needs to be Enabled. Second, an offset value needs to be set.

10. While in the Override Object dialog box, click on the Constraint tab (this should be the default).

11. In the Suppression area, place a check in the box next to Override as shown in the following image.

12. Verify that Enabled is the current setting in the down list box.

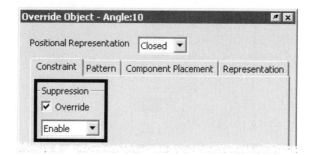

Figure 6-197

13. While still in the Override Object dialog box, identify the Value area.
14. Place a check in the box next to Override.
15. Change the value to 100 degrees in the down list box as shown in the following image.

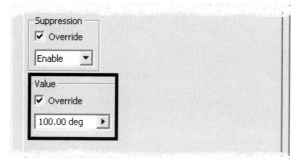

Figure 6-198

16. When finished, click the OK button.

 NOTE The default master positional representation has the Angle:10 constraint set to 30 degrees. This master state displays the gripper somewhat open. Changing the angle to 100 degrees in the previous image will close the gripper arm.

17. The Angle:10 constraint has been overriden. This action closes the gripper as shown in the following image. Notice in the Browser that the Angle:10 text constraint is bold, emphasizing this change.

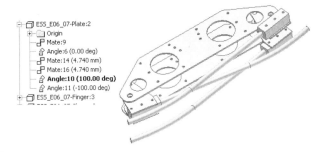

Figure 6-199

 NOTE In the Browser, expand the Representations folder and the Position node. Check the functionality of the positional representations by double-clicking on master and Closed. The gripper assembly should update to reflect the active positional representation.

18. Switch from the Model Browser to the Representations Browser by clicking on the arrow next to Model and selecting Representations from the list as shown on the left side of the following image.

19. The Representations Browser is displayed. Expand all items in this area of the Browser as shown on the right side of the following image.

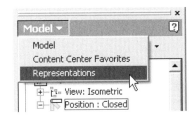 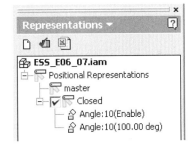

Figure 6-200

 NOTE The Representations Browser allows you to examine all positional representations associated with an assembly. It also shows the overrides for each positional representation.

20. Two additional positional representations will now be created to show the gripper in its middle and fully opened positions. Existing positional representations can be copied and then modified in order to display other versions of an assembly. In the Representations Browser, check to see that the Positional Representations mode is expanded.

21. Right-click on the Closed positional representation and select Copy from the menu as shown on the left side of the following image.

22. Notice on the right side of the following image, a new positional representation called Closed1 has been created from the existing positional representation (Closed).

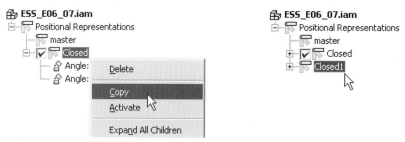

Figure 6-201

23. In the Representations Browser, expand both positional representations. Notice that all overrides are copied into the new positional representation (Closed1) as shown in the following image.

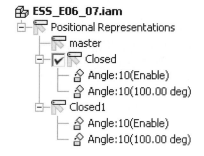

Figure 6-202

24. Create another positional reference copy. Base this copy on the existing Closed positional representation. When all copies have been made, three positional representations should be visible in the Representations Browser as shown on the left side of the following image; namely Closed, Closed1, and Closed2.

25. The positional representation copies will now be renamed to more meaningful names. In the Representations Browser, rename Closed1 to Middle and Closed2 to Open. Your Representations Browser should appear similar to the right side of the following image.

 NOTE To rename a positional representation, first single-click on its name, then single-click a second time. This will enter the rename mode where you can change its name.

Chapter 6 Creating and Documenting Assemblies   445

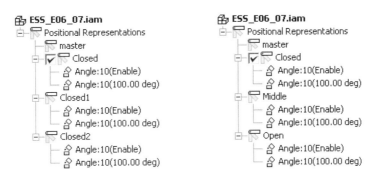

Figure 6-203

26. With the positional representations renamed, their values must now be modified in order to show different representations of the assembly. This can be accomplished using an Excel spreadsheet. To activate the Excel spreadsheet, click on the Edit positional representation table button found at the top of the Representations Browser as shown on the left side of the following image.

27. Clicking on the Edit positional representation table button in the previous image will launch a Microsoft Excel table similar to the right side of the following image. Notice that the master and all positional representations are displayed as a series of rows and columns. Notice also the degree values present in a separate column.

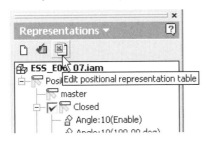

	A	B	C
1		Angle:10 (Constraint Suppress)	Angle:10 (Constraint Offset)
2	master	Enable	30.00 deg
3	Closed	Enable	100.00 deg
4	Middle	Enable	100.00 deg
5	Open	Enable	100.00 deg

Figure 6-204

28. Make the following modifications to this table.
    - For the Middle positional representation, change 100 degrees to 40 degrees.
    - For the Open positional representation, change 100 degrees to 0 degrees.

Your table should appear similar to the left side of the following image.

29. The changes you made in the Microsoft Excel spreadsheet need to be saved before you return back to the gripper assembly. From the Excel File menu, click Close & Return to ESS_E06_07 as shown on the right side of the following image.

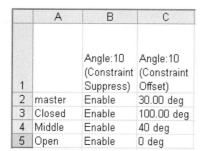 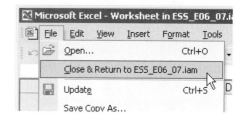

Figure 6-205

30. Saving the changes you made in the Microsoft Excel spreadsheet will update each angle value for the positional representations as shown in the following image.

31. Each positional representation can be activated by double-clicking on its name in the Representations Browser. Each should show a different state of the assembly as shown in the following image.

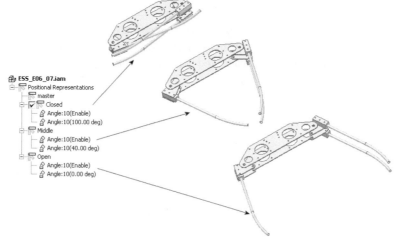

Figure 6-206

32. End of exercise.

CREATING OVERLAY VIEWS

To document assembly motion, overlay drawing views use positional representations to show an assembly in multiple positions in a single view. Overlays are available for unbroken base, projected, and auxiliary views. Each overlay can reference a design view representation independent of the parent view. Before creating an overlay view, create as many positional representations as you need to show your assembly in various positions. It would also be considered good practice to also create a number of design view representations to focus on certain components and reduce potential clutter in the drawing view.

Use the following steps for creating overlay drawing views:

1. On the Drawing Views panel bar, click the Base View tool and position this view in your drawing. The following image illustrates a hydraulic cylinder set to its closed position. It would be advantageous to show the cylinder in its fully opened position also in this base view.

Figure 6-207

2. On the Drawing Views panel bar, click the Overlay button in the following image and click the base view you positioned in the previous image.

Figure 6-208

3. When the Overlay View dialog box displays in the following image, click in the Positional Representation box and select the Open representation. You can also choose pre-defined design view representations in addition to showing a label, displaying tangent edges, and changing the style and layer of the positional representation. When finished making changes to this dialog box, click the OK button.

 NOTE A positional representation can be used only once per parent view.

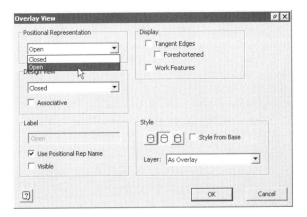

Figure 6-209

4. The new positional representation is added to the base view and is represented as a series of hidden or dashed lines as shown in the following image. Dimensions can be added between the parent and overlay views to show the distance the component has traveled.

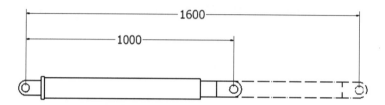

Figure 6-210

DOCUMENTING ASSEMBLIES USING PRESENTATION FILES

After creating an assembly, you can create drawing views based on how the parts are assembled. If you need to show the components in different positions (like an exploded view) hide specific components, or create an animation that shows how to assemble and disassemble the components, you need to create a presentation file. A presentation file is separate from an assembly and has a file extension of *.ipn*. The presentation file is associated with the assembly file, and changes made to the assembly file will be reflected in the presentation file. You cannot create components in a presentation file. To create a new presentation file, click the New icon in What To Do and then select the *Standard.ipn* icon on the Default tab, as shown in the following image.

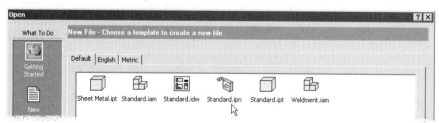

Figure 6-211

By default, the Panel Bar will show the presentation tools available to you (see the following image). The following sections explain these tools.

Figure 6-212

Create View Click to create views of the assembly. Created views are based on the same assembly model. Once created, a view will be listed as an explosion in the Browser.

Tweak Components Click to move and/or rotate parts in the view.

Precise View Rotation Click to rotate the view by a specified angle and direction using a dialog box.

Animation Click to create an animation of the assembly tweaks and saved camera views; an AVI file can be output.

The main steps for creating a presentation view are to reposition (tweak) the parts in specific directions and then create an animation if needed. The following sections discussed these steps.

CREATING PRESENTATION VIEWS

The first step in creating a presentation is to create a presentation view. Issue the Create View tool on the Presentation Panel Bar or right-click and select Create View from the menu. The Select Assembly dialog box will appear, as shown in the following image. The dialog box is divided into two sections: Assembly and Explosion Method.

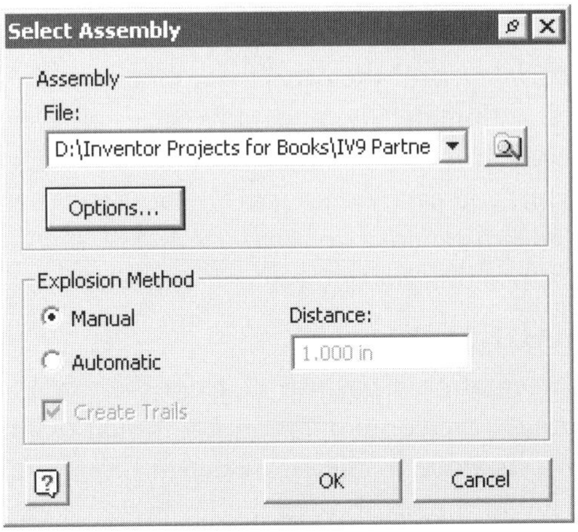

Figure 6-213

Assembly

In this section, you will determine on which assembly and design view to base the presentation view.

File   Enter or select an assembly file on which to base the presentation view.

Options   Click to display the File Open Options dialog box. Use this dialog box to select Design View Representations (covered in Chapter 10) and Positional Representations (covered in Chapter 10).

Explosion Method

In this section you can choose to explode the parts (separate the parts a given distance) manually or automatically.

Manual   Click so that the components will not be exploded automatically. After the presentation view is created, you can add tweaks that will move or rotate the parts.

Automatic   Click so that the part will be exploded automatically with a given value.

Create Trails   If you clicked Automatic, click to have visible trails (lines that show how the parts are being exploded).

Distance   If Automatic is selected, enter a value that the parts will be exploded.

After making your selection, click the OK button to create a presentation view. The components will appear in the graphics window, and a presentation view will appear in the Browser. If you clicked the Automatic option, the parts will be exploded automatically. To determine how the parts will be exploded, Autodesk Inventor analyzes the assembly constraints. If you constrained parts using the mate plane option and the normals (arrows perpendicular to the plane) for both parts were pointing outward, they will be exploded away from each other the defined distance. Think of the mating parts as two magnets that want to push themselves in opposite directions. The grounded component in the Browser will be the component that stays stationary, and the other components will move away from it. If you are creating trails automatically, generally they will come from the center of the part, not necessarily from the center of the holes. After expanding all of the children in the Browser, you can see numerically how the parts exploded. Once the parts are exploded, the distance is referred to as a *tweak*. The number by each tweak reflects the distance that the component is moved from the base component.

TWEAKING COMPONENTS

After creating the presentation view, you may choose to edit the tweak of a component or move or rotate the component an additional distance. To edit an individual tweak, click on its value in the Browser, as shown in the following image. Then enter a new value in the text box in the lower-left corner of the Browser and press ENTER to use the new value.

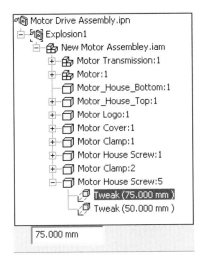

Figure 6-214

To extend all of the tweaks the same distance, right-click on the assembly's name in the Browser and select Auto Explode from the menu, as shown in the following image. Then enter a value and click the OK button, and all of the tweaks will be extended the same distance. You cannot use negative values with the Auto Explode method.

Figure 6-215

Another method for repositioning the components manually is to tweak them using the Tweak tool. To manually tweak a component, click the Tweak Component tool on the Presentation Panel Bar, press the hot key T, or right-click and select Tweak Component from the menu. The Tweak Component dialog box will appear, as shown in the following image. The Tweak Component dialog box has two sections: Create Tweak and Transformations.

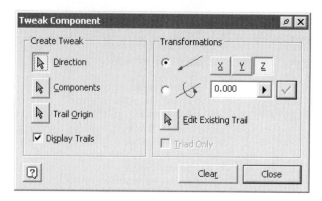

Figure 6-216

Create Tweak

In this section, you will select the components to tweak, set the direction and origin of the tweak, and control trail visibility.

Direction Determine the direction or axis of rotation for the tweak. After clicking the Direction button, select an edge, face, or feature of any component in the graphics window to set the direction triad (X, Y, and Z) for the tweak. The edge, face, or feature that you selected does not need to be on the components you are tweaking.

Components Select the components to tweak. Click the Components button and then click the components in the graphics window or Browser to tweak. If you selected a component when you started the Create Tweak operation, it will be included in the components automatically. To remove a component from the group, press and hold the CTRL key and click the component.

Trail Origin Set the origin for the trail. Click the Trail Origin button and then click in the graphics window to set the origin point. If you do not specify the trail origin, it will be placed at the center of mass for the part.

Display Trails Click if you want to see the tweak trails for the selected components. Clear the check box to hide the trails.

Transformations

In this section, you will set the type and value of a tweak.

Linear Click the button next to the arrow and line to move the component in a linear fashion.

Rotation Click the button next to the arrow and arc to rotate the component.

X, Y, Z Click the X, Y, or Z coordinate button to determine the direction for a linear tweak or the axis for a rotational tweak, or select the arrow on the triad that represents the X, Y, or Z direction.

Text Box   Enter a positive or negative value for the tweak distance or rotation angle, or click a point in the graphics window and move the cursor with the mouse button pressed down to set the distance.

Apply   After making all of the selections, click the Apply button to complete the tweak.

Edit Existing Trail   To edit an existing tweak, click the Edit Existing Trail button. Select the tweak in the graphics window and change the desired settings.

Triad Only   Click the Triad Only option to rotate the direction triad without rotating selected components. Enter the angle of rotation, and then click the Apply button. After you rotate the triad direction, you can use it to define tweaks.

Clear   Click the Clear button in the dialog box to remove all of the settings to set up for another tweak.

To tweak a component, follow these steps:

1. Issue the Tweak Component tool.
2. Determine the direction or axis of rotation for the tweak by clicking the Direction button and selecting an edge, face, or feature.
3. Select the components to tweak by clicking the Components button and clicking the components in the graphics window or Browser that will be tweaked.
4. Select a trail origin point.
5. Determine whether you want trails visible.
6. Set the type of tweak to linear or rotation.
7. Click the X, Y, or Z coordinate button to determine the direction for a linear tweak or the axis for a rotational tweak.
8. Enter a value for the tweak in the text box or select a point on the screen and drag the part into its new position.
9. Click the Apply button in the Tweak Component dialog box.

ANIMATION

After you have repositioned the components, you can animate the components to show how they assemble or disassemble. To create an animation, click the Animate tool on the Presentation Panel Bar or right-click and select Animate from the menu. The Animation dialog box will appear, as shown in the following image. The Animation dialog box has three sections: Parameters, Motion, and Animation Sequence (under the More (>>) button).

Figure 6-217

Parameters

In this section you will specify the playback speed and the number of repetitions for the animation.

Interval   Set this value for the playback speed of the animation in frames. The higher the number, the greater the time delay between frames. A smaller number will speed up the animation.

Repetitions   Set the number of times to repeat the playback. Enter the desired number of repetitions or use the up or down arrow to select the number.

 NOTE To change the number of repetitions, click the Reset button on the dialog box and then enter a new value.

Motion

In this section, you will play the animation for the active view.

Forward by Tweak   Click to drive the animation forward one tweak at a time.

Forward by Interval   Click to drive the animation forward one interval at a time.

Reverse by Tweak   Click to drive the animation in reverse one tweak at a time.

Reverse by Interval   Click to drive the animation in reverse one interval at a time.

Play Forward   Click to play the animation forward for the specified number of repetitions. Before each repetition, the view is set back to its starting position.

Auto Reverse   Click to play the animation for the specified number of repetitions. Each repetition plays start to finish, then in reverse.

Play Reverse   Click to play the animation in reverse for the specified number of repetitions. Before each repetition, the view is set back to its ending position.

Pause   Click to pause the animation playback.

Record   Click to record the specified animation to a file so that you can play it back later.

To animate components, follow these steps:

1. Issue the Animate tool.
2. Set the number of repetitions.
3. Adjust the tweaks as needed.
4. Click one of the play buttons to view the animation in the graphics window.
5. To record the animation to a file, click the Record button, and then click one of the Play buttons to start recording.

CHANGING THE ANIMATION SEQUENCE

In the Animation Sequence section, you can change the sequence in which the tweaks happen, select the tweak, and then select the needed operation. Expanding the Animation dialog box displays the following image.

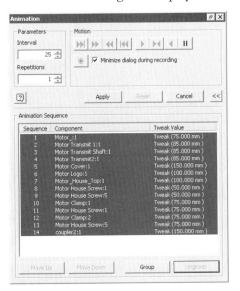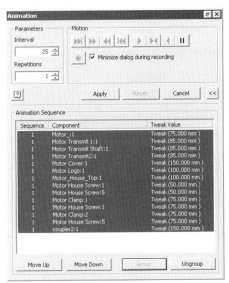

Figure 6-218

Move Up Click to move the selected tweak up one place in the list.

Move Down Click to move the selected tweak down one place in the list.

Group Select a number of tweaks and then click the Group button. When tweaks are grouped, all of the tweaks in the group will move together as you change the sequence. The group assumes the sequence order of the lowest tweak number.

Ungroup After selecting a tweak that belongs to a group, you can click the Ungroup button, and it can then be moved individually in the list. The first tweak in the group assumes a number that is one higher than the group. The remaining tweaks are numbered sequentially following the first.

PRESENTATION HIGHLIGHTING

When you click the More (>>) button in the Animation dialog box within a Presentation file (*.ipn*), as shown in the following image, the animation sequence scheme that controls the order of the animation appears. As the animation plays, each sequence will highlight in the dialog box to match the animation playback in the graphics window. This highlighting makes it easier to match the sequence number and name with the animation currently playing in the graphics window.

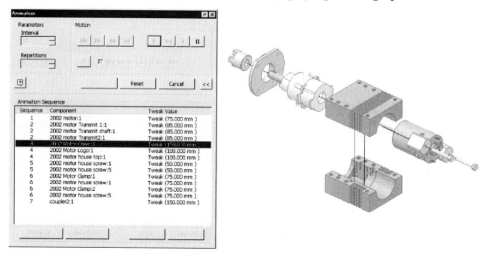

Figure 6-219

SETTING UNITS IN A PRESENTATION FILE

You can set the units of measure for use in a presentation file. You can therefore set the units in which tweaks will be created. The specified units can be different from the units set in the parent assembly. To set the units in a presentation file, create a presentation view, click Tools > Document Settings, and, on the Units tab, set the desired units from the Length drop-down list, as shown in the following image.

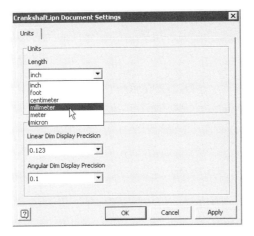

Figure 6-220

You can also edit an existing tweak and apply a unit other than the default unit set in the Document Settings dialog box. To edit a tweak's value, click on the name or icon of the tweak that you want to edit in the Browser, enter a new value in the tweak edit box (displayed at the bottom of the Browser), and press the ENTER key. The following image shows a tweak with a value of 30.000 mm being edited to 1.181 in.

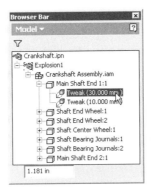

Figure 6-221

Another method used to edit a tweak is to move the cursor over the tweak in the graphics window until it is highlighted, right-click, and select Edit from the menu, as shown on the left side of the following image. The Tweak Component dialog box appears. Enter a new value in the Transformations area of the dialog box, as shown on the right side of the following image.

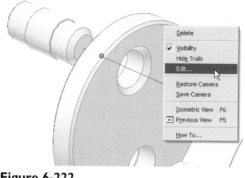 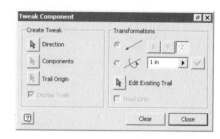

Figure 6-222

EXERCISE 6-8 Creating Presentation Views

In this exercise, you create a presentation view of an existing assembly. You add tweaks to the assembly components to create an exploded view. To navigate to the exercise in the *Electronic Student Workbook* do the following:

1. Begin the process of creating a presentation by clicking the New tool.

2. Select the Metric tab and then double-click *Standard(mm).ipn*.

3. While in the presentation graphics screen, click the Create View tool in the panel bar.

4. In the Select Assembly dialog box, click the Explore button next to the File field then open *ESS_E06_08.iam* and click the Options button.

5. In the Design View Representations field, select Internal Components, then click OK.

6. Verify that Manual is selected in the Explosion Method field.

7. Click OK to create the view. Your display should appear similar to the following image.

Figure 6-223

8. You will now set the tweak direction used to create an exploded presentation. In the Browser, click the plus sign in front of Explosion1, and the plus sign in front of ESS_E06_08.iam to expand their display. The assembly components are displayed so you can select components in the graphics screen or in the Browser.

9. Click the Tweak Components tool in the panel bar.

10. In the Tweak Component dialog box, verify that Direction is selected and Display Trails is checked.

11. Position your cursor over a cylindrical face until the temporary Z axis aligns with the axis of the assembly, as shown in the following image.

12. Click to accept the axis orientation.

Figure 6-224

13. Click on the nut at the base of the assembly to select it. In the Browser, observe that ESS_E06-NutB is highlighted.

14. In the Tweak Components dialog box, verify that the Z button is selected then enter
    -40 in the Transformations field.
15. Click Apply (green checkmark) to create the tweak. Your display should be similar to the following image.

Figure 6-225

TIP After you define the tweak axes, you can select a component or group of components then drag them to create a tweak. This way, you can quickly arrange your components visually. Later you can edit the values of the tweaks from the Browser to define their final positions.

16. Next click the valve plate to select it, then hold down the CTRL key and click the nut to unselect it. Observe that the valve plate is the only entry highlighted in the Browser.
17. Enter **-20** in the Transformations field of the dialog box.
18. Click Apply to create the tweak. Your display should be similar to the following image.

Chapter 6 Creating and Documenting Assemblies 461

Figure 6-226

 TIP You can also select components by clicking on them directly in the Browser. This is useful when trying to select hidden components.

19. The next component to tweak is the top nut. In the Browser, click *ESS_E06-NutA* to select it. Notice that the valve plate is no longer selected.

20. Enter **115** in the Transformations field then click Apply to create the tweak. Your display should be similar to the following image.

Figure 6-227

TIP Examine the Browser. Expand the components that have a plus sign in front of them to view the tweaks you have applied.

21. Complete this phase of the exercise by adding the following tweaks:

Part Name	Tweak
ESS_E06-WasherB.ipt:1	105
ESS_E06-Pin.ipt:1	85
ESS_E06-Spring.ipt:1	45
ESS_E06-WasherA.ipt:1	30
ESS_E06-Diaphragm.ipt:1	15

22. Close the Tweak Component dialog box when you are finished. Your display should be similar to the following image.

Figure 6-228

23. You will now adjust an existing tweak already created. In the Browser, expand *ESS_E06-Diaphragm.ipt:1* and *ESS_E06-WasherA.ipt:1*. The hierarchy displays tweaks nested below each component definition.

24. Right-click each of the tweaks and select Visibility to turn off the trail display for the tweak.

25. Click Tweak (15.000 mm) under the *ESS_E06-Diaphragm*.

26. Below the Browser, enter in a new value **10** in the Offset field then press ENTER. Observe that the name of the Tweak reflects the change, as shown on the left side of the following image; and the diaphragm moves to its new position, as shown on the right side of the following image.

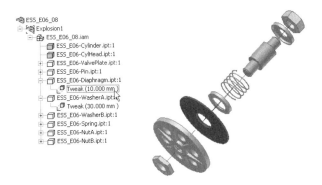

Figure 6-229

27. End of exercise.

CREATING DRAWING VIEWS FROM ASSEMBLIES AND PRESENTATION FILES

After creating an assembly or presentation, you may want to create drawings that utilize the data. You can create drawing views from assemblies using information from an assembly or presentation file. From them you can select a specific design view or presentation view. To create drawing views based on assembly data, start a new drawing file or open an existing drawing file. Then use the Base View tool and select the IAM or IPN file from which to create the drawing. If needed, specify the design view or presentation view in the dialog box. The following image shows a presentation view being selected after a presentation file was selected.

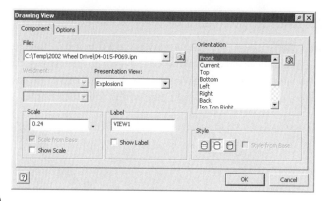

Figure 6-230

CREATING BALLOONS

After you have created a drawing view, you can add balloons to the parts and/or subassemblies. You can add balloons individually to components in a drawing or use the Auto Balloon tool to automatically add balloons to all components.

To add individual balloons to a drawing, follow these steps:

1. Activate the Drawing Annotation Panel Bar by clicking Drawing Views on the Panel Bar and selecting Drawing Annotation from the menu.

2. On the Drawing Annotation Panel Bar, click the arrow next to the Balloon tool, as shown in the following image. Two icons will appear–the first for ballooning components one at a time and the second for automatic ballooning components in a view in a single operation.

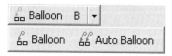

Figure 6-231

3. To balloon a single component, click the Balloon tool or use the hot key B. A preview image of the balloon will appear attached to your cursor.

4. Select a component to balloon.

5. The Item Numbering dialog box will appear, as shown in the following image.

6. Select which Source, BOM View, and Delimiter to use as shown in the following image and select OK.

 NOTE BOM refers to a Bill of Material. This consists of a table that contains information about the individual components that make up the assembly. Typical information includes but is not limited to quantity, part number, description, item number, and cost. Information held in the quantity category for example is updated whenever a component is added or removed from the assembly.

7. Position the cursor to place the second point for the leader, and then press the left mouse button.

8. Continue to select points to add segments to the balloon's leader.

9. When done adding segments to the leader, right-click and select Continue from the menu to create the balloon.

10. Select Done from the menu to cancel the operation without creating an additional balloon, or select the Back option to undo the last step that was created for the balloon.

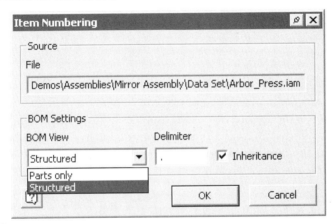

Figure 6-232

 NOTE The same dialog box will be used to create a parts list.

ITEM NUMBERING DIALOG BOX OPTIONS

Source   Specify the source file the bill of materials (BOM) will be based on.

BOM View

Structured   A structured list refers to the top-level components of an assembly in the selected view. Subassemblies and parts that belong to the subassembly will be ballooned, but parts that are in a subassembly will not be.

Parts Only View   Click to balloon only the parts of the assembly in the selected view. Subassemblies will not be ballooned, but the parts in the subassemblies will be.

NOTE In a Parts-only list, components that are assemblies are not presented in the list unless they are considered Inseparable or purchased.

Delimiter   Sets a delimiter when using the structured item numbering system..

Inheritance   Check this box if you want nested items to inherit changes made to the parent item value.

AUTO BALLOONING

In complex assembly drawings, it will become necessary to balloon a number of parts. Rather than add a balloon manually to each part, you can use the Auto Balloon tool to perform this operation on a number of parts in a single operation automatically. Clicking on the Auto Balloon button on the Drawing Annotation Panel Bar will display the Auto Balloon dialog box, as shown in the following image. A number of areas are available in this dialog box to control how you place balloons.

Selection

This area requires you to select where to apply the balloons. With the view selected, you then add components or remove components to balloon. A feature available in this area is the Ignore Multiple Instances option. When the option's box is checked, multiple instances of the same component will not be ballooned. This greatly reduces the number of balloon callouts in the drawing. If your application requires multiple instances to each have a balloon, remove the check from this box.

Placement

This area allows you to display the balloons either along a horizontal axis, along a vertical axis, or around the view being ballooned.

BOM Settings

This area allows you to control if balloons are applied to structured (first-level) components or parts only.

Style Overrides

The Style overrides area allows you to change the shape of the balloon. It is here that you can also assign a user-defined balloon shape. Click Balloon Shape to enable balloon shape style overrides.

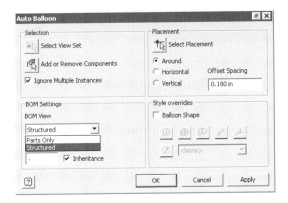

Figure 6-233

To create an auto balloon, follow these steps and refer to the corresponding figures:

1. Select the view to which to add the balloons, as shown in the following image.

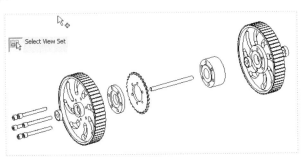

Figure 6-234

2. Select the components to which to add balloons. In the following image, all parts in the isometric drawing were selected using a window box. You will notice the color of all selected objects change. Balloons will be applied to these selected objects. To remove components, hold down either the SHIFT or CTRL key and click a highlighted component. This will deselect the component.

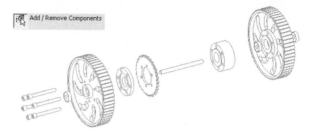

Figure 6-235

3. Select one of the three placement modes for the balloons. The following image shows an example using the Horizontal Placement mode. You can further locate this balloon group by moving your cursor around the display screen. This balloon arrangement will retain its horizontal order.

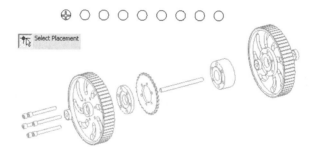

Figure 6-236

The following image shows an example using the Vertical Placement mode. You can further locate this balloon group by moving your cursor around the display screen. This balloon arrangement will retain its vertical order.

Figure 6-237

The following image shows an example using the Around Placement mode. Moving your cursor around the display screen will readjust the balloons to different locations. In this example, the balloons are located in various positions outside of the view box. After you have placed these types of balloons, you may want to drag the balloons and arrow terminators to fine-tune individual balloon locations.

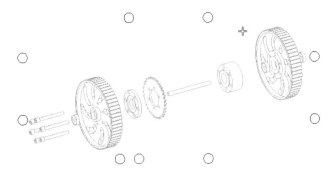

Figure 6-238

SPLIT BALLOONS

You can change a balloon's appearance to a number of predefined shapes. The following image shows areas in the Styles and Standards Editor dialog box. On the left side of the following image, the balloon shape has been changed to Circular - 2 Entries. This will allow you to place two sets of information in the balloon. This information comes from the Balloon Formatting area, as shown on the right side of the following image. Notice two entries found under the Property Display: ITEM and PART NUMBER. Property Display entries are arranged by clicking on the Property Chooser button.

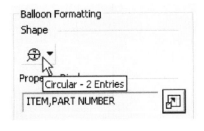 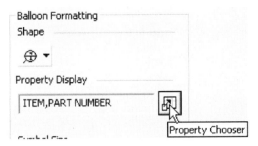

Figure 6-239

When the Property Chooser dialog box appears, you can add or remove properties. In the following image, the PART NUMBER property was added in the Selected Properties column. This property was also moved up in order for it to appear between ITEM and QTY.

Figure 6-240

The following image illustrates the results. The left side of the following image shows the default balloon with a circular shape and one entry. The right side of the following image shows the modified balloon arrangement consisting of two entries. Notice how the circle balloon expands to guarantee that all information appears. In the previous image, the Property Chooser dialog box displayed a third property, QTY. However, since the balloon shows only two entries, the QTY property was ignored. If you want this property to appear in the balloon, you would have to move it in the Selected Properties column under ITEM.

NOTE Making changes to an existing balloon style in the Styles and Standards Editor dialog box will change all balloons under that style globally in a drawing. You can also change balloons individually through the Edit Balloon dialog box, which you activate by right-clicking a balloon and selecting Edit Balloon from the menu.

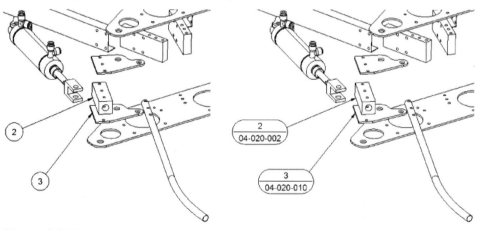

Figure 6-241

CREATING BALLOONS WITH USER-DEFINED SYMBOLS

You can also control a balloon's shape by first defining your own shape in the form of a sketched symbol. In the following image, two sketched symbols have been created: Pentagon and Octagon.

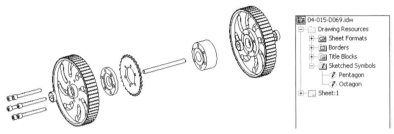

Figure 6-242

The following image illustrates the placement of automatic balloons. The Style overrides area is in the Auto Balloon dialog box. Click Balloon Shape and then click on the User Defined button. This will display all sketched symbols defined in the drawing. In this example, the Octagon sketch symbol is being used. Notice how all balloons take on the octagonal shape, as shown in the figure.

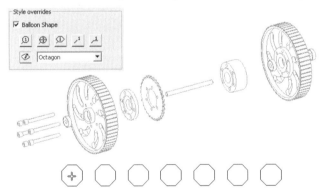

Figure 6-243

EDITING BALLOONS

You can edit balloons by right-clicking on a balloon. This will activate a menu, shown in the following image.

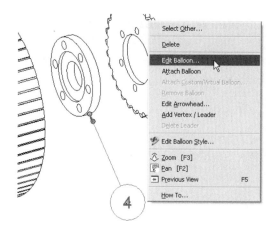

Figure 6-244

Selecting Edit Balloon from the menu will activate the Edit Balloon dialog box, as shown in the following image.

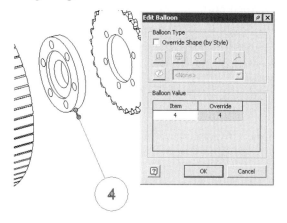

Figure 6-245

Use this dialog box to make changes to a balloon on an individual basis. The following items may be edited: Balloon Type and Balloon Value.

Editing the Balloon Type

To activate the other balloon types, click Override Shape (by Style). You can then select a new balloon type from the group supplied. Changes made to the balloon type affect the individual balloon you are editing.

To make changes to all of the balloon types globally, click on the balloon style in the Styles and Standards Editor dialog box as shown in the following image. Then click the down arrow on the Shape option in the Balloon Formatting area to expose other balloon types.

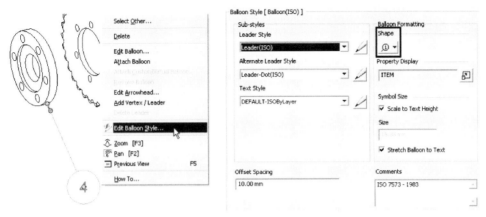

Figure 6-246

Editing the Balloon Value

Two columns exist in the Balloon Value area of the Edit Balloon dialog box: Item and Override. When you edit the Item value, changes will be made to the balloon and will be reflected in the parts list. In the following image, Item was changed to 2A. Notice in the example on the left side of the following image that both the balloon and parts list have updated to this new value.

When you make a change in the Override column of the Edit Balloon dialog box, the balloon will update to reflect this change; however, the parts list remains unchanged. The example on the right side of the following image illustrates this. The balloon was overridden with a value of 2A, but the parts list has this item (Rod Cap) listed as 2.

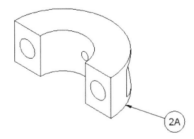 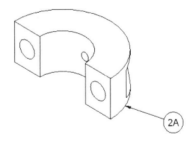

			Parts
ITEM	QTY	PART NUMBER	
1	1	Connecting Rod	
2A	1	Rod Cap	
3	2	Rod Cap Screw	
4	1	Piston	
5	1	Wrist Pin	

			Parts
ITEM	QTY	PART NUMBER	
1	1	Connecting Rod	
2	1	Rod Cap	
3	2	Rod Cap Screw	
4	1	Piston	
5	1	Wrist Pin	

Figure 6-247

NOTE Any changes performed with the Balloon Edit dialog box will be reflected in all parts lists associated with that view.

Custom Value Input Override

A user may enter a value in a balloon that is not reflected in the parts list. This is the purpose of a *custom value override*. This option is only available through the Edit Balloon dialog box.

Aligning Balloons Vertically or Horizontally

You can align balloons horizontally or vertically. Select a number of balloons to align. The first balloon selected will act as an anchor point for the alignment of the other balloons (see the following image).

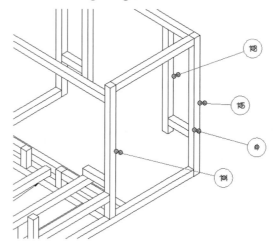

Figure 6-248

Once you have selected multiple balloons, right-click on any of them to display the menu and the Align option, as shown in the following image.

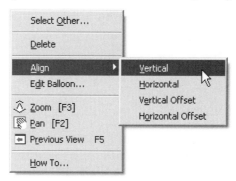

Figure 6-249

Select either the Horizontal or Vertical options to perform the alignment. If you select Horizontal, the balloons will be aligned along the *X* axis. If you select Vertical, the balloons will be aligned along the *Y* axis, as shown in the following image.

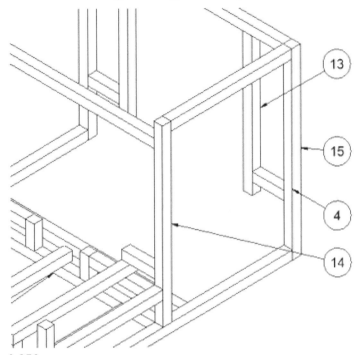

Figure 6-250

You can also add an offset spacing value between balloons when performing alignment operations. The same rules apply for balloon offsets as for balloon alignment; namely, the first balloon selected will be the point or anchor for the other balloons. Right-click on any balloon to display the menu, as shown in the following image.

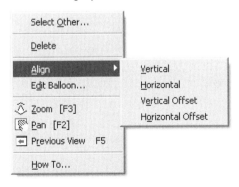

Figure 6-251

Select the Horizontal Offset or Vertical Offset option to create a uniform space between balloons. The offset distance is based on the diameter of the balloons placed. The offset distance is a multiplier found in the Offset Spacing option in the Balloon category of the Styles and Standards Edit dialog box, as shown in the following image. Notice the you can also change the text style of balloons in this dialog box.

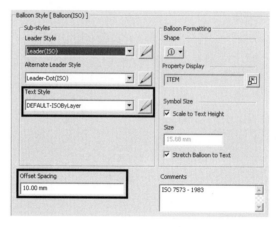

Figure 6-252

Attaching a Custom Balloon

If you have predefined a custom part, you can select it from a list of custom parts for custom balloon content. To attach a custom balloon, right-click on the balloon and select Attach Custom Balloon, as shown in the following image. If there is no custom part available, this menu item will be grayed out.

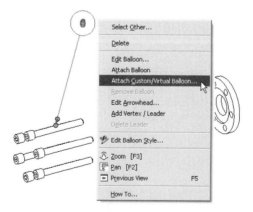

Figure 6-253

The dialog box shown in the following image will prompt you for your selection of custom balloons. You can only select one custom item at a time.

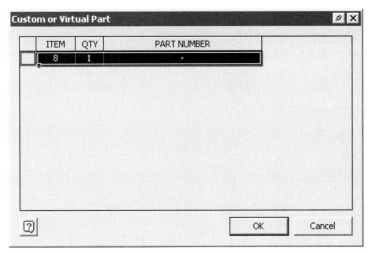

Figure 6-254

After selecting the custom part, click the OK button to return to the drawing. Drag the cursor around for placement of the attached balloon.

Click to place the balloon in the desired location, as illustrated in the following image.

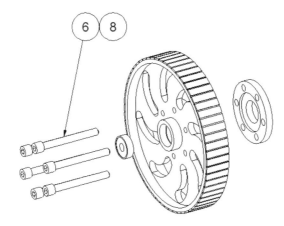

Figure 6-255

The drawing can contain as many drawing views, based on different IAM and IPN files, as needed. The same rules apply for creating drawing views from assemblies or presentation files as those which apply to part (IPT) files.

PARTS LISTS

After you have created the drawing views and placed balloons, you can also create a parts list, as shown in the following image. As noted earlier in this section, components do not need to be ballooned before creating a parts list.

	Parts List		
ITEM	QTY	PART NUMBER	DESCRIPTION
1	1	BASE	ARBOR BASE
2	1	FACE PLATE	FACE PLATE
3	1	PINION SHAFT	PINION SHAFT
4	1	LEVER ARM	LEVER ARM
5	1	THUMB SCREW	THUMB SCREW
6	1	TABLE PLATE	TABLE PLATE
7	1	RAM	RAM
8	2	HANDLE CAP.ipt	HANDLE CAP
9	1	COLLAR	COLLAR
10	1	GIB PLATE	GIB PLATE
11	1	GROOVE PIN	GROOVE PIN
12	4	ANSI B18.3 - 1/4 - 20 - 7/8	Hexagon Socket Head Cap Screw
13	4	ANSI B18.3 - 10-32 UNF x 0.58	Hexagon Socket Set Screw - Flat Point
14	1	ANSI B18.6.2 - 10-32 UNF - 0.1875	Slotted Headless Set Screw - Flat Point

Figure 6-256

To create a parts list, click the Parts List tool (see the following image) on the Drawing Annotation Panel Bar. Select a view on which to base the parts list. If balloons were not created, the Parts List Item Numbering dialog box will appear as was previously discussed in the Creating Balloons section of this chapter. After specifying your options, click the OK button and a parts list will appear. If the components were ballooned already, the list will appear without displaying the Parts List Item Numbering dialog box. Click a point in the graphics window to place the parts list. The information in the list comes from the properties of each component. You can also split a long parts list into multiple columns.

Figure 6-257

To split a parts list placed on the drawing already, follow these steps:

1. Right-click on the parts list and select Edit Parts List from the menu.

2. In the Edit Parts List dialog box, select the row after which the split will occur.

3. Right-click and select Wrap Table at Row from the menu, as shown in the following image.

Chapter 6 Creating and Documenting Assemblies 479

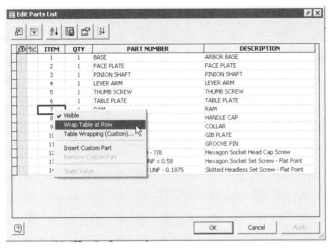

Figure 6-258

4. To complete the operation, click the OK button. The following image illustrates the results of performing a column split.

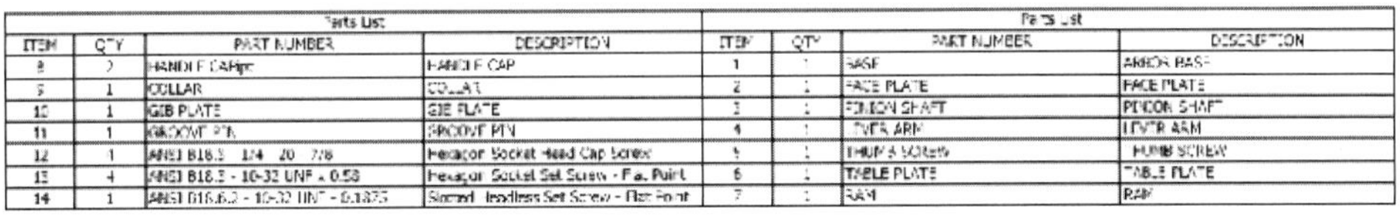

[table image]

Figure 6-259

EDITING BOM DATA You can open a Bill of Materials dialog box directly from the drawing environment, and edit the assembly BOM. All changes are saved in the assembly and corresponding component files. Right-click on the parts list in the Drawing Browser and then select Bill of Materials from the menu as shown in the following image.

Figure 6-260

This will launch the Bill of Materials dialog box based on a specific assembly file. This dialog box provides a convenient location to edit the iProperties and Bill of Material properties for all components in the assembly. You can even perform edits on multiple components at once. Notice in the following image that the Bill of Materials Editor displays the BOM Structure. In this Bill of Materials dialog box, perform the desired edits. When finished, click Done to close the Bill of Materials dialog box. All changes will be updated in the parts list and balloons.

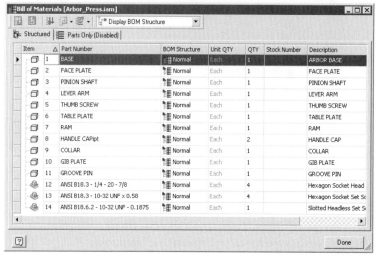

Figure 6-261

PARTS LIST TOOLS

The information supplied earlier in this chapter developed the basis of inserting and manipulating a parts list. This section will expand on the capabilities of the parts list and will cover the following topics:

- A detailed account of the Parts List Operations tools available at the top of the Edit Parts List dialog box.
- The function of the spreadsheet view.
- Additional sections of the Edit Parts List dialog box.
- Expanded capabilities of the Edit Parts List dialog box.
- Nested parts lists.

PARTS LIST OPERATIONS (TOP ICONS)

Once you have placed a parts list in the drawing, various controls are available to assist with the modification of the parts list. Right-click on the parts list in a drawing and select Edit Parts List from the menu, as shown in the following image, or double-click on the parts list in the drawing.

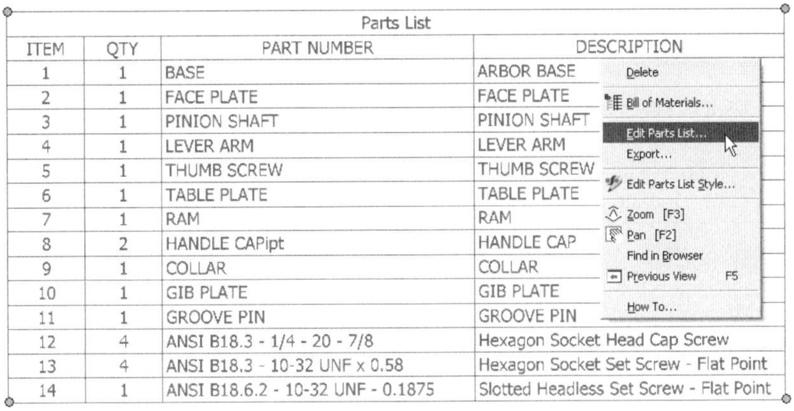

Figure 6-262

This displays the Edit Parts List dialog box, as shown in the following image. The row of icons along the top of this dialog box consists of numerous features that allow you to change the parts list.

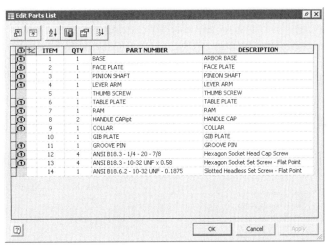

Figure 6-263

The Parts List table lists detailed explanations of all parts list icons:

Parts List Button	Function	Explanation
	Column Chooser	Opens the Parts List Column Chooser dialog box, where you can add, remove, or change the order of the columns for the selected parts list. Data for these columns is populated from the properties of the component files.

	Group Settings	Opens the Group Settings dialog box, where you select parts list columns to be used as a grouping key, and group different components into one parts list row. This button is active only when generating a parts list using the Structured mode.
	Sort	Opens the Sort Parts List dialog box, where you can change the sort order for items in the selected parts list.
	Export	Opens the Export Parts List dialog box so that you can save the selected parts list to an external file of the file type you choose.
	Table Layout	Opens the Parts List Table Layout dialog box where you can change the title text, spacing or heading location for the selected parts list.
	Renumber	Renumbers item numbers of parts in the parts list consecutively.

CREATING CUSTOM PARTS

Custom parts are useful for displaying parts in the parts list that are not components or graphical data such as paint or a finishing process. To add a custom part, right click on an existing part in the Edit Parts List dialog box as shown in the following image and click Insert Custom Part. When a custom part no longer needs to be documented in the parts list, you can right click in the row of the custom part and click Remove Custom Part from the menu as shown in the following image.

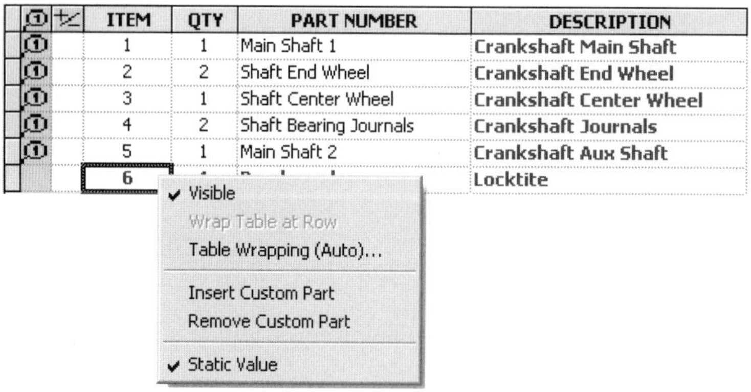

Figure 6-264

SPREADSHEET VIEW

This section is in the middle of the Edit Parts List dialog box, where the contents of the selected parts list appear. As you make changes to the parts list properties, the changes are reflected in the spreadsheet view. Items that have corresponding balloons in the drawing file are marked with a balloon symbol in the left-hand column, as shown in the following image.

ITEM	QTY	PART NUMBER	DESCRIPTION
1	1	BASE	ARBOR BASE
2	1	FACE PLATE	FACE PLATE
3	1	PINION SHAFT	PINION SHAFT
4	1	LEVER ARM	LEVER ARM
5	1	THUMB SCREW	THUMB SCREW
6	1	TABLE PLATE	TABLE PLATE
7	1	RAM	RAM
8	2	HANDLE CAPipt	HANDLE CAP
9	1	COLLAR	COLLAR
10	1	GIB PLATE	GIB PLATE
11	1	GROOVE PIN	GROOVE PIN
12	4	ANSI B18.3 - 1/4 - 20 - 7/8	Hexagon Socket Head Cap Screw
13	4	ANSI B18.3 - 10-32 UNF x 0.58	Hexagon Socket Set Screw - Flat Point
14	1	ANSI B18.6.2 - 10-32 UNF - 0.1875	Slotted Headless Set Screw - Flat Point

Figure 6-265

NESTED PARTS LISTS

An added function of a parts list is the ability to distinguish individual parts from subassemblies. In the case of subassemblies, a visual representation in the form of a +/- button appears along the left column. Clicking on a + button will expand the subassembly and display a list of its component parts. In the example shown on the left side of the following image, the Crankshaft Assembly is first identified with a + and the item number 6. When expanding the Crankshaft Assembly, the individual components of the nested items appear with an additional number, such as the Main Shaft 1 part, which is listed as item 6.1 (see right side of the following image).

ITEM	QTY	PART NUMBER
1	1	Engine Block
2	2	Cylinder
3	2	Cylinder Sleeve
4	2	Cylinder Head
5	2	Valve Head
6	1	Crankshaft Assembly
7	2	Piston
8	2	Wrist Pin
9	2	Connecting Rod
10	2	Rod Cap
11	4	Rod Cap Screw

ITEM	QTY	PART NUMBER
1	1	Engine Block
2	2	Cylinder
3	2	Cylinder Sleeve
4	2	Cylinder Head
5	2	Valve Head
6	1	Crankshaft Assembly
6.1	1	Main Shaft 1
6.2	2	Shaft End Wheel
6.3	1	Shaft Center Wheel
6.4	2	Shaft Bearing Journals
6.5	1	Main Shaft 2
7	2	Piston

Figure 6-266

If you applied balloons to a drawing view, as shown in the following image, the balloons will expand to hold the nested parts list items.

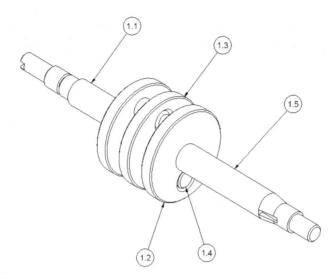

Figure 6-267

EXERCISE 6-9 Creating Assembly Drawings

In this exercise, you create drawing views of an assembly and then add balloons and a parts list.

1. This exercise begins with the formatting of a drawing sheet. Open the drawing file *ESS_E06_09.idw*. This drawing consists of one blank A1 sheet with border and title block.

2. In the Browser, right-click Sheet1 and select Edit Sheet.

3. In the Edit Sheet dialog box, in the Name field, type Assembly to rename the sheet.

4. In the Size list, select A2. When Finished, click the OK button.

5. The next phase of this exercise involves updating the file properties in order to update the title block fields. On the File menu, click iProperties.

6. Click on the Summary tab and make the following changes:

- In Title, enter **Pump Assembly**.
- In Author, enter your name.

7. Next click on the Project tab and make the following changes:

- In Part number, enter **123-456-789**.
- In Creation Date, click the down arrow, and then select today's date.

8. When finished, click the OK button. Zoom in to the title block and notice that the entries have updated as shown in the following image.

DRAWN AJK	3/11/2005			
CHECKED		TITLE		
QA				
MFG		Pump Assembly		
APPROVED				
		SIZE A2	DWG NO 123-456-789	REV
		SCALE		SHEET 1 OF 1

Figure 6-268

9. You will now create drawing views. Zoom out to view the entire sheet.

10. Click the Base View tool in the panel bar to display the Drawing View dialog box.

11. Click the Explore Directories button next to the File list, select Assembly Files from the Files of Type drop-down list and double-click *ESS_E06_09.iam* as the view source.

12. From the Orientation list, select Top.

13. Set the Style to Hidden Line (do not select Hidden Line Removed).

14. Set the Scale to 1.

15. Move the view to the upper left corner of the sheet then click to place the view. The drawing view should appear similar to the following image.

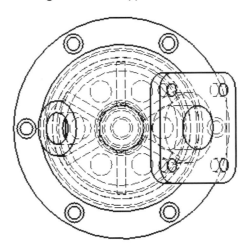

Figure 6-269

16. You will now begin the process of generating a section view from the existing top view.

17. Before you create the section view, you will need to turn off the Section property for some components so they are not sectioned by following the next series of directions:

   • In the Browser, expand the Assembly sheet.

   • In the Browser, expand View1:*ESS_E06_09.iam*. (Your view number may be different)

   • Then expand *ESS_E06_09.iam*.

   • Select *ESS_E06-Pin.ipt*, *ESS_E06-Spring.ipt*, *ESS_E06-NutA.ipt*, and *ESS_E06-NutB.ipt*.

 TIP To select multiple items, hold down the CTRL key while clicking the items from the Browser.

   • Right-click any of the highlighted parts in the Browser and turn off Section (remove the check mark) from the menu as shown in the following image.

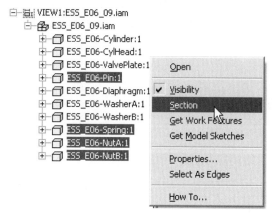

Figure 6-270

18. Now create the section view by clicking the Section View tool in the panel bar.

19. Select the top view of the assembly and draw a horizontal section line through the center of the top view as shown in the following image.

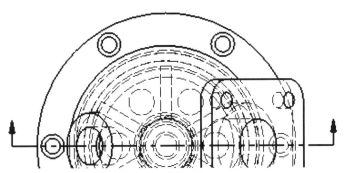

Figure 6-271

20. Right-click and select Continue.

21. Move the preview below the top view and click to place the view.

22. Notice that the nuts, spring, and pin are not sectioned as shown on the left side of the following image. Notice also the presence of the section label and drawing view scale.

23. You will now edit the section view and turn off the label and scale. On the sheet, select the sectioned assembly view, right-click, and select Edit View from the menu to display the Drawing View dialog box.

24. Clear the check marks from the box next to visible in the Label and Scale sections then click OK. The Label and Scale no longer display at the bottom of the view as shown on the right side of the following image.

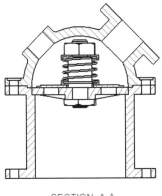 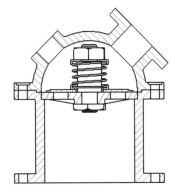

Figure 6-272

25. You will now hide certain components from displaying in the top view. In the Browser, under the expanded component listing for *ESS_E06_06.iam*, select both *ESS_E06-CylHead.ipt* and *ESS_E06-Spring.ipt*.

26. Right-click and clear the check mark from Visibility. The cylinder head and spring do not display in the top view as shown in the following image.

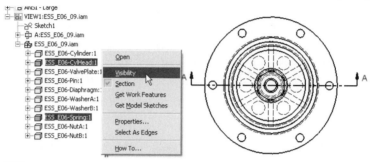

Figure 6-273

27. Next add a text note to the top view stating the cylinder head and spring are not visible. Click the Text tool in the Drawing Annotation panel bar.

 NOTE To change panel bars, click on the title in the panel bar then select Drawing Annotation Panel from the context menu.

28. Select a point to the left of the top assembly view.

29. In the text entry area, type **CYLINDER HEAD AND**, press ENTER, type **SPRING REMOVED** then click OK. The text will be added to the top view as shown in the following image. If necessary, select and drag the text to a better location.

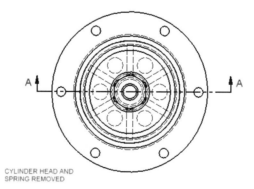

Figure 6-274

30. You will now create an isometric view that will be based on the section view. Click the Projected View tool in the Drawing Views panel bar.

31. Select the sectioned assembly view as the base view.

32. Move the preview up and to the right, and then click the sheet.

33. When the view positioned, right-click and select Create from the menu. The isometric pictorial view is created as shown in the following image.

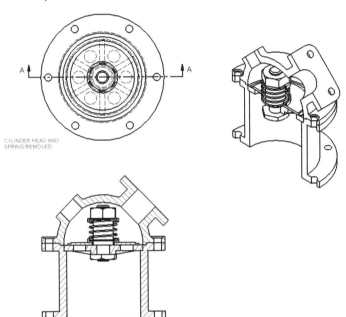

Figure 6-275

34. You will now create a parts list to identify the item number, quantity, part number, and description. Begin by clicking the Parts List tool in the Drawing Annotation panel bar.

35. The Parts List dialog box will display. Select the sectioned assembly view.

36. Verify that the BOM Settings area is set to Structured components and click the OK button.

37. Move the parts list until it joins the title block and border then click to place it. Zoom in to display the parts list as shown in the following image.

Parts List			
ITEM	QTY	PART NUMBER	DESCRIPTION
1	1	Cylinder	CYLINDER
2	1	Cyl Head	CYLINDER HEAD
3	1	Valve plate	VALVE PLATE
4	1	Pin	THREADED PIN
5	1	Diaphragm	DIAPHRAGM
6	1	ESS_E09_03-Washer A	RETAINING WASHER
7	1	ESS_E09_03-Washer B	RETAINING WASHER
8	1	Spring	SPRING - Ø1 X Ø22 X 25
9	1	Nut A	FLAT NUT, REG - M16 X 1.5
10	1	Nut B	FLAT NUT, THIN - M16 X 1.5

Figure 6-276

38. In the steps that follow, you will edit the parts list to reorder the columns and adjust the column widths. Begin by first selecting the parts list in the graphics window or from the Browser.

39. Right-click and select Edit Parts List from the menu to display the Edit Parts List dialog box.

40. Click the Column Chooser button to display the Parts List Column Chooser dialog box and make the following changes:

- In the Available Properties list, select MATERIAL.
- Click Add.
- In the Selected Properties list, select DESCRIPTION.
- Click Move Up two times.

41. When finished, click the OK button to close the Parts List Column Chooser dialog box. The properties are reordered so that the Description column is next to the Item column as shown in the following image

42. You will now edit the column widths. Once again launch the Edit Parts List dialog box, right-click on the DESCRIPTION header, and choose Column Width from the menu. Make the following column width changes:

- Enter **65** in the Column Width dialog box to change the column's display width.
- Change the QTY field's width to 20.
- Change the MATERIAL field's width to 42.

43. Click the OK button to close the dialog box.

44. Move the parts list so it is flush with the border and the title block. Your display should appear similar to the following image.

\multicolumn{5}{c}{Parts List}
ITEM	DESCRIPTION	QTY	PART NUMBER	MATERIAL
1	CYLINDER	1	Cylinder	Cast Iron
2	CYLINDER HEAD	1	Cyl Head	Cast Iron
3	VALVE PLATE	1	Valve plate	Steel, Mild
4	THREADED PIN	1	Pin	Steel, Mild
5	DIAPHRAGM	1	Diaphragm	Stainless Steel
6	RETAINING WASHER	1	ESS_E09_03-Washer A	Steel, Mild
7	RETAINING WASHER	1	ESS_E09_03-Washer B	Steel, Mild
8	SPRING - Ø1 X Ø22 X 25	1	Spring	Stainless Steel
9	FLAT NUT, REG - M16 X 1.5	1	Nut A	Steel, Mild
10	FLAT NUT, THIN - M16 X 1.5	1	Nut B	Steel, Mild

Figure 6-277

45. Next, you will edit the properties in the individual part model files and update the parts list. First, open the assembly file *ESS_E06_09.iam*.

46. In the Browser, right-click *ESS_E06-NutA.ipt:1* and select Properties to display the Properties dialog box.

47. Click the Project tab and in the Part Number field, enter PURCHASE. This part does not require a design drawing for manufacture so the drawing file name need not be listed.

48. Click OK to close the dialog box.

49. Repeat the previous steps for *ESS_E06-NutB.ipt:1*. Then return to the pump assembly drawing.

50. Select the parts list in the graphics window or from the Browser.

51. Right-click and select Edit Parts List from the menu to display the Edit Parts List dialog box.

52. In the Edit Parts List dialog box, the part numbers that have been changed are highlighted.

53. Right-click in one of the highlighted cells and select Update All. The part numbers will update with the new values.

54. Click OK to update the parts list as shown in the following image.

Parts List				
ITEM	DESCRIPTION	QTY	PART NUMBER	MATERIAL
1	CYLINDER	1	Cylinder	Cast Iron
2	CYLINDER HEAD	1	Cyl Head	Cast Iron
3	VALVE PLATE	1	Valve plate	Steel, Mild
4	THREADED PIN	1	Pin	Steel, Mild
5	DIAPHRAGM	1	Diaphragm	Stainless Steel
6	RETAINING WASHER	1	ESS_E09_03-Washer A	Steel, Mild
7	RETAINING WASHER	1	ESS_E09_03-Washer B	Steel, Mild
8	SPRING - Ø1 X Ø22 X 25	1	Spring	Stainless Steel
9	FLAT NUT, REG - M16 X 1.5	1	PURCHASE	Steel, Mild
10	FLAT NUT, THIN - M16 X 1.5	1	PURCHASE	Steel, Mild

Figure 6-278

55. You now place balloons in the assembly drawing. Balloons are used to identify parts in an assembly. The item number in the balloon corresponds to the item number in the parts list.

56. Pan and zoom to display the sectioned assembly view.

57. Change to the Drawing Annotation Panel bar and click the Balloon tool.

58. Select the edge of the component shown in the following image as the start of the leader.

59. Click a point on the sheet to define the end of the first leader segment.

60. Right-click and select Continue to place the balloon as shown in the following image.

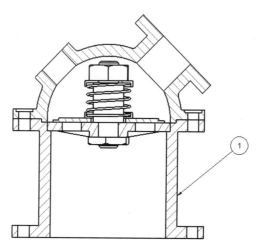

Figure 6-279

61. Repeat the previous steps to place the remaining balloons as shown in the following image.

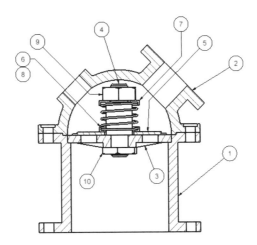

Figure 6-280

 TIP To add a balloon for the spring, place the balloon for the washer (6) then turn off the Balloon tool. Right-click balloon 6 and select Attach Balloon. Click the spring, position the new balloon, and then click to place the balloon.

62. End of exercise.

CHAPTER SUMMARY

To	Do This	Tool
Place a component in an assembly	Click the Place Component tool on the Assembly Panel Bar or on the Assembly toolbar.	
Create a component in an assembly	Click the Create Component tool on the Assembly Panel Bar or on the Assembly toolbar.	
Place an assembly constraint	Click the Constraint tool on the Assembly Panel Bar or on the Assembly toolbar.	
Edit an assembly constraint	Double-click the constraint in the Browser or right-click and select Edit from the menu.	
Edit a component in place	Double-click the component in the graphics window or Browser.	
Control component visibility	Right-click the component in the graphics window or Browser and select Visibility from the menu.	✔ Visibility
Check component interference	Click Tools > Analyze Interference.	Analyze Interference
Move an assembly through a range of motion	Drag an underconstrained component.	

To	Do This	Tool
Create a presentation file	Create a new file with the *.ipn* extension and click the Create View button.	Standard.ipn
Create a balloon	Click the Balloon tool on the Drawing Annotation Panel Bar.	
Create a parts list	Click the Parts List tool on the Drawing Annotation Panel Bar.	

Applying Your Skills

SKILL EXERCISE 6-1

In this exercise, you create a new component for a charge pump and then assemble the pump.

1. Open the assembly file ESS_E06_10.iam as shown in the following image. This exercise uses the skills you have learned in previous exercises to assemble a pump and create a new part in place.

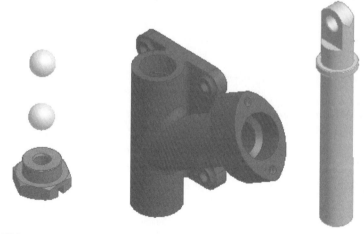

Figure 6-281

2. Place the following predefined components into the assembly:
   - 1 occurrence of ESS_E06_10-Union.ipt
   - 1 occurrence of ESS_E06_10-Seal.ipt
   - 2 occurrences of ESS_E06_10-M8x30.ipt

3. Next, create the gland in place. Project edges from the pump body to define the flange. See the following image for the necessary dimensions needed to complete the gland.

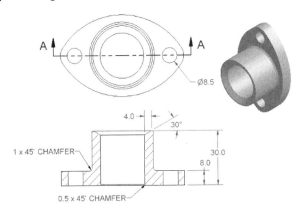

Figure 6-282

4. Use assembly constraints to build the model as shown in the following image.

ASSEMBLY TIPS:

To assemble the balls, mate the center point of the ball with the center line of the body then place a tangent constraint between the ball's surface and the sloped surface of the seat.

To assemble the seal into the body, first apply a mate constraint between the centerline of the seal and the centerline of the bore in the body. Next, apply a mate constraint between the angled surface of the seal and the angled surface of the seal's seat in the body. Apply similar constraints between the gland and seal.

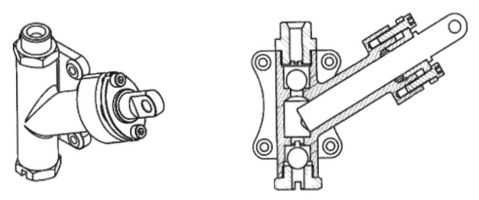

Figure 6-283

5. The completed assembly model should appear similar to the following image.

Figure 6-284

    6. End of exercise.

SKILL EXERCISE 6-2

In this exercise, you create an assembly drawing, a parts list, and balloons for a charge pump.

    1. In this exercise, you create an assembly drawing, parts list, and balloons for a charge pump.

The charge pump assembly is shown in the following figure.

Figure 6-285

    2. Create a new drawing with a single A2 sheet named Assembly.

    3. Insert a top view of *ESS_E06_11.iam* with a scale of 1:1.

4. Using the top view as the base view, create a sectioned front view. Exclude the ram, valve balls, and machine screws from sectioning.

5. Use the Create View tool to create an independent top-right isometric view of the pump assembly.

6. Insert the parts list then modify its format according to the column widths as shown in the following table.

Column	Width
Item	15
Qty	15
Description	57
Part Number	50
Material	60

7. The results should appear similar to the following image.

Figure 6-286

8. Add balloons to identify the components.

9. Complete the title block to include the title and author (your name) as shown in the following image.

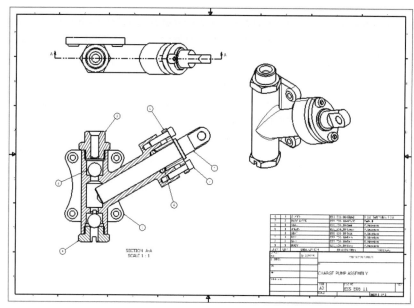

Figure 6-287

10. End of exercise.

CHECKING YOUR SKILLS

Use these questions to test your knowledge of the material covered in this chapter.

1. **True__ False__** The only way you can create an assembly is by placing existing parts into it.

2. Explain top-down and bottom-up assembly techniques.

_____

_____

3. **True__ False__** An occurrence is a copy of an existing component.

4. **True__ False__** Only one component can be grounded in an assembly.

5. **True__ False__** Autodesk Inventor does not require components in an assembly to be fully constrained.

6. **True__ False__** A sketch must be fully constrained to adapt.

7. What is the purpose of creating a presentation file?

_____

_____

8. **True__ False__**   A presentation file is associated to the assembly file on which it is based.

9. **True__ False__**   When creating drawing views from an assembly, you can create views from multiple presentation views or design views.

CHAPTER 7

Advanced Sketching and Constraining Techniques

In the first few chapters you learned how to create sketches using basic techniques. That experience has given you a good foundation of knowledge on creating basic sketches. In this chapter you will learn how to create more complex sketches, and learn to use tools that will reduce the number of steps needed to create a constrained sketch, use construction geometry to better control the sketch, set up relationships between dimensions, and create parts that are driven from a table.

CHAPTER OBJECTIVES

After completing this chapter, you will be able to

- Use construction geometry to help constrain sketches
- Create and constrain an ellipse
- Create a 2D spline
- Create a pattern of sketch geometry
- Share a sketch
- Utilize both the symmetry constraint and mirror tool
- Slice the graphics window
- Sketch on another parts face
- Change the display of dimensions
- Create relationships between dimensions
- Create parameters
- Create a part that is driven from a Microsoft Excel spreadsheet

CONSTRUCTION GEOMETRY

Construction geometry can help you create sketches that would be difficult to create without it. You can constrain and dimension construction geometry like normal geometry, but you will not see the construction geometry in the part when you turn the sketch into a feature. When you sketch, the sketches by default have a normal geometry style, meaning that the sketch geometry is visible in the feature.

Construction geometry can reduce the number of constraints and dimensions required to constrain a sketch fully and can help define the sketch. A construction circle inside a hexagon can drive the size of the hexagon, for example. Without construction geometry, the hexagon would require six constraints and dimensions. It would require only three constraints and dimensions with construction geometry; the circle would have tangent or coincident constraints applied to it and the hexagon. You create construction geometry by changing the line style before or after you sketch geometry in one of the following two ways:

- Before sketching, click the Construction icon on the standard toolbar, as shown in the following image.
- After creating the sketch, select the geometry that you want to create and click the Construction icon on the standard toolbar.

Figure 7-1

After turning the sketch into a feature, the construction geometry will disappear or be consumed. When you edit a feature's sketch that you created with construction geometry, the construction geometry will reappear during editing and disappear when the part is updated. You can add or delete construction geometry to or from a sketch just like any geometry that has a normal style. In the graphics window, construction geometry will be lighter in color and thinner in width than normal geometry. The image on the left in the following image shows a sketch with a construction line for the angled line. The angled line has a coincident constraint applied to it at every point that touches it. The image on the right shows the sketch after it has been extruded. Note that the construction line was not extruded.

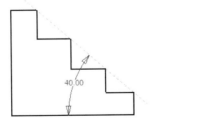 

Figure 7-2

ELLIPSES

You can create ellipses by clicking the Ellipse tool under the Circle tool on the Sketch Panel Bar, as shown in the following image.

Figure 7-3

To create an ellipse, follow these steps:

1. Click the Ellipse tool.
2. Click a point that will define the center of the ellipse and move the cursor to select a point that defines the orientation and length of the first axis of the ellipse.
3. Move the cursor until the desired shape of the ellipse is found that defines the second axis of the ellipse.
4. Click the point to create the ellipse.

You can trim, extend, and dimension the ellipse. To dimension an ellipse, follow these steps:

1. Click the General Dimension tool.
2. Click on the ellipse.
3. Move the cursor and click a point to either define the major or minor distances.
4. Repeat the process to dimension the other major or minor distance.

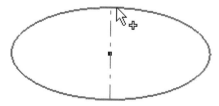

Figure 7-4

You can also offset an ellipse. To offset an ellipse, start the Offset tool, click on the ellipse, move the cursor to where you want to create the new offsetted ellipse, and click that point. This creates a concentric ellipse. A concentric ellipse will have the same center point (a coincident constraint) as the ellipse from which it was offset.

2D SPLINES

A spline is geometry that you sketch as free-flowing in shape and that you can use to create complex shapes. Points are used to define the shape of the spline; you can constrain these points as needed. The points that define a spline will be visible when you create or select a spline. To create a spline, follow these steps.

1. Click the Spline tool under the Line tool on the Sketch Panel Bar, as shown in the following image.

2. Either click arbitrary points in the graphics window or click existing points in the sketch.

3. To exit the command and keep the spline, press ENTER, or right-click and select Continue from the menu.

Figure 7-5

Once you have created a spline, you can edit and control it by right-clicking on a point that lies on the spline or in between the points on the spline. You will only get menu options that pertain to a selected point. When you right-click on the spline, the spline options will appear, as shown in the following image. The following sections describe the options. Many of these shape controls can be described precisely using geometric sketch constraints and sketch dimensions.

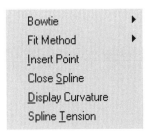

Figure 7-6

BOWTIE

The Bowtie option, shown in the following image, allows you to modify precisely the spline's shape at the shape point. There are three options under Bowtie: Handle, Curvature, and Flat. These options control the curvature of the spline.

Figure 7-7

Handle

The Handle option allows you to control the length of the handle. The handle (also referred to as a handlebar) is a line segment that has nodes at each end and is always tangent to the spline. Control the handle's visibility by clicking on a point on the spline, right-clicking, and selecting Bowtie > Handle. If you clicked the Handle option and selected the spline, the handle will be visible. Once the handle is visible, you can click a node and either rotate or change the handle's length. Rotating the handle will change the tangent angle at that point. The longer the handle length, the longer the spline will be tangent to it. You can control handlebar lengths with dimensions. Handlebar lengths have no units; they can be thought of as a weight. A length (or weight) of 1 is the most typical or "natural." In a loose sense (depending on the actual geometry), a length of 2 causes the spline to hug the curve twice as long, while a length of 1/2 causes it to hug the curve half as long. These are not hard and fast rules. If the spline is a straight line, for example, changing the length of the tangent has little or no effect, except that if it is too large, the spline will overshoot the endpoint and double back on itself.

Figure 7-8

Curvature

The Curvature option allows you to control the radius of a spline at the point where the curvature exists. The curvature is represented by an arc. Control the curvature's visibility by clicking on a point on the spline, right-clicking, and selecting Bowtie > Curvature. The curvature arc will then appear. If you clicked the Curvature option and the spline is selected, the curvature and handle will be visible. To adjust the curvature, click and drag one of the curvature nodes (as shown in the following image) or constrain it to existing geometry. If, after changing the radius of curvature, the spline appears to overshoot the intended direction, you can reduce the length of the handle to compensate.

You can create second-order contact between a spline and a circle (or a spline and another spline) by constraining all three elements of the spline (point-coincident, handlebar-tangent, curvature bar-concentric) to the other geometry.

Figure 7-9

Flat

The Flat option allows you to control the length of the flat. When you click the Flat option, handle and curvature nodes are not visible, and a line with two nodes representing the flat will appear. Clicking one of the nodes and dragging it to a new location can adjust the length of the flat, as shown in the following image.

Figure 7-10

FIT METHOD

The Fit Method option allows you to control how the spline will be generated between the points. There are three methods from which to choose: Smooth, Sweet, and AutoCAD (as shown in the following image).

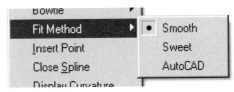

Figure 7-11

Smooth This method creates smooth splines–lightweight representations of the splines. It approximates the properties of the sweet splines with fewer control points, leading to smaller file sizes and faster recompute speeds. You cannot control the tension of smooth splines (they will be converted automatically to sweet splines if you try). Compared with sweet splines on the same set of points, smooth splines tend to look more aesthetically pleasing, hence the use of the term "smooth" (an industrial designer's term for nice-looking curves).

Sweet This method creates sweet splines, also called minimum energy splines, which are formulated to seek a balance between tension and curvature. They are loosely modeled on the bending and stretching properties of materials. A very stiff material such as a steel plate bends slowly, for example, distributing its curvature over its entire length. A flexible piece of rubber, on the other hand, localizes its bending to the positions where the bend is forced, and stays somewhat flat elsewhere. While sweet splines have nice geometric and aesthetic properties, you should use them with caution. Models with many surfaces generated from sweet splines may have larger file sizes, longer load times, and longer recompute times.

AutoCAD This method creates AutoCAD splines, which allow compatibility with AutoCAD. These splines are most appropriate for two-way interaction with AutoCAD files, but they have two primary drawbacks. First, they have reduced continuity (visible with curvature combs), which may cause surfaces derived from them to have poor reflective qualities (these reflective qualities can be seen using zebra stripes). Second, they do not support bowties.

To change the spline method, right-click on the spline and select Smooth, Sweet, or AutoCAD from Fit Method. the next three images show the same spline generated using Smooth, Sweet, and AutoCAD methods, respectively.

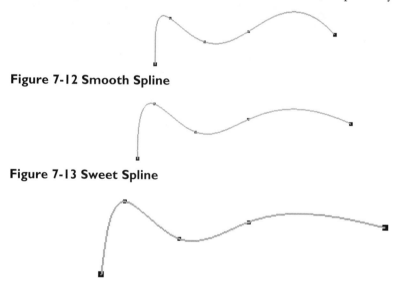

Figure 7-12 Smooth Spline

Figure 7-13 Sweet Spline

Figure 7-14 AutoCad Spline

INSERT POINT

The Insert Point option allows you to insert a point into the spline. You can control and edit this new point like any other point on the spline. To insert a point, right-click on the spline and select Insert Point from the menu. Move the cursor where you want to locate the point and click. The following image shows a point being added to the right side of the spline.

Figure 7-15

CLOSE SPLINE

The Close Spline option allows you to close the first and last point of an open spline. To close an open spline, right-click on the spline and select Close Spline from the menu. The following image shows an open spline on the left and the same spline closed on the right.

Figure 7-16

DISPLAY CURVATURE

The Display Curvature option allows you to better see the curvature of the spline by turning on the visibility of curvature lines. You cannot use these display curvature lines to edit the spline. As the spline has more curvature, the lines will be longer. To display the curvature of a spline, right-click on the spline and select Display Curvature from the menu. The main value of displaying the curvature comb is to visualize the continuity of the curve and to get an idea of the quality of surfaces generated from such a curve. Curvature combs magnify discontinuities in a curve. Sudden changes in curvature, for example, show up as sharp changes in the path of the teeth of the curvature comb.

The following image shows a spline with the display curvature lines visible.

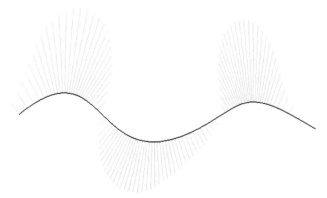

Figure 7-17

SPLINE TENSION

The Spline Tension option offers a method for controlling a spline by adjusting its tension. You can adjust the tension of a spline between 0 and 100. The lower the number is, the more curvature it will have, and the higher the number, the straighter it will be between the points. The default for new splines is set to 0. As you adjust the tension, a preview image will appear in a lighter color. To adjust the tension of a spline, right-click on the spline, select Spline Tension from the menu, and slide the pointer to the desired value. The following image shows a spline with the Spline Tension dialog box.

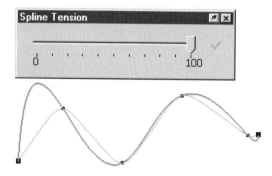

Figure 7-18

CONSTRAIN AND DIMENSION A SPLINE

To add constraints to a spline, you need to add them to any visible handlebars, curvature arc, or flat of any point. You can add the following constraints: concentric, equal, collinear, horizontal, perpendicular, parallel, tangent, and vertical. You can add dimensions to any visible handlebars, curvature, flat, or any point. The following image shows a spline that has dimensions added to the handlebar and curvature arc and constraints shown on the handlebar.

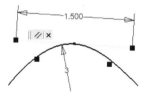

Figure 7-19

EXERCISE 7-1 Complex Sketching

In this exercise, you work with spline curve controls to create the body of a toaster. The profile of the toaster is based on a free form spline curve and the slots on top of the toaster are created by projecting the base as a closed loop.

1. Open *ESS_E07_01.ipt*.

2. Use the Look At and Zoom tools to adjust the view as shown in the following image.

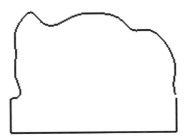

Figure 7-20

3. Edit the sketch.

4. Zoom in to the right side of the sketch and notice that the endpoints between the vertical lines and the spline do not meet.

5. Click the Coincident constraint tool.

6. Apply a coincident constraint between the endpoint of the line and spline as shown in the following image.

Figure 7-21

7. Click the Tangent constraint tool.

8. Apply a tangent constraint between the line and spline.

9. Pan to the left side of the sketch and apply a tangent constraint between the line and the spline.

Figure 7-22

10. Use the zoom and pan tools to view the top of the sketch.

11. Right-click on the highest control point on the right side of the sketch.

12. Click Bowtie then click Flat. The handlebar and curvature are displayed.

Figure 7-23

13. Apply a 1 unit dimension to the handlebar.

14. Apply a horizontal constraint to the handlebar.

15. Right-click on the highest control point on the left side of the sketch.

16. Click Bowtie then select Handle. The handlebar is displayed.

17. Apply a horizontal constraint to the handlebar.

18. Apply a collinear constraint between the left and right handlebars.

Figure 7-24

19. Apply the five dimensions as shown in the following image.

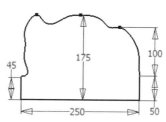

Figure 7-25

20. In the next steps you add constraints and dimensions to control the free form 2D spline. Use the zoom and pan tools to view the lower left side of the sketch.

21. Right-click the control point as shown in the following image.

Figure 7-26

22. Click Bowtie then click Curvature. The curvature bar and handlebar are displayed.
23. Add a vertical constraint to the handlebar.
24. Add a colinear constraint to the handlebar and the vertical line.
25. Apply a **40** mm dimension to the curvature bar.
26. Apply a **40** mm dimension to the control point and the end point of the vertical line.

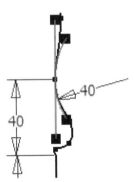

Figure 7-27

27. View the sketch as an isometric.
28. Extrude the sketch using a depth of **150** mm.
29. Rotate to an isometric view as shown.

Figure 7-28

30. In the next steps you shell the body for the toaster then create an offset work plane above the toaster. This plane will be used to create the openings on top of the toaster.
31. Click the Shell tool.
32. Remove the bottom face of the toaster from the shell.
33. Create a shell with a **2** mm wall thickness.
34. Click the Work Plane tool.
35. Create a work plane offset **225** mm from the XZ plane in the origin folder.

Figure 7-29

36. Click the 2D Sketch tool and click the workplane.
37. Click the Project Geometry tool.
38. Move the cursor on the bottom outside edge, then cycle through the edges and select the base loop.

Figure 7-30

39. Expand the sketch entry in the Browser. The projected geometry is listed as *Projected Loop1*.
40. Use the Look At tool to view the sketch.
41. Click the Two Point Rectangle tool.
42. Create a rectangle on the sketch.

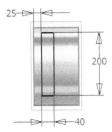

Figure 7-31

43. Click the Vertical constraint tool to constrain the rectangle using the midpoint option.

44. Click the midpoint of the top edge of the rectangle and the midpoint of the top edge of the base loop.

Figure 7-32

45. Change to an isometric view.

46. Click the Extrude tool and click the rectangle as the profile, change the Extents to TO, click the inside face of the shell feature, change the operation to Cut and then click OK.

47. Turn off the visibility of the work plane and your model should resemble the following image.

Figure 7-33

48. Close the file. Do not save changes.

PATTERN SKETCHES

When creating a sketch that will have a rectangular or a polar pattern, you can pattern the geometry in the sketch instead of creating multiple sketches or waiting to pattern the feature. The rectangular pattern does not need to be oriented in a horizontal or a vertical alignment. To pattern a sketch, follow these steps:

1. Click either the Rectangular Pattern or Circular Pattern tool on the Sketch Panel Bar, as shown in the following image.

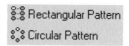

Figure 7-34

2. The Rectangular or Circular Pattern dialog box will appear. The following image shows the Rectangular Pattern dialog box on the left and the Circular Pattern dialog box on the right.

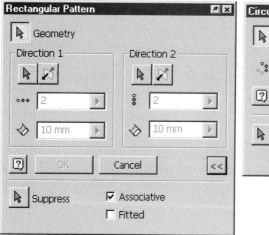
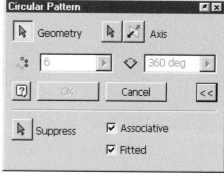

Figure 7-35

3. Click the geometry button and select the geometry you want to pattern. You can select multiple objects.

4. Define the direction (1 or 2) or the axis by clicking the arrow and then selecting an object. For a rectangular pattern, the selected object must be a line in the active sketch.

5. If needed, click the Reverse Direction button, shown in the following image to change the direction of the pattern.

Figure 7-36

    6. Enter the information for the count, spacing, and angle.

    7. For a rectangular pattern you can select a second direction, if needed.

    8. Click the OK button to create the sketch pattern.

In the bottom pane of the Pattern dialog boxes are the following options: Suppress, Associative, and Fitted.

Suppress   Click to keep the selected object(s) from appearing in the pattern. When you suppress a patterned occurrence, it will be represented in a hidden linetype in the graphics window.

Associative   Click to have the pattern reflect any changes made to the geometry that you patterned. If the option's box is clear, any changes made to the geometry that you patterned will not be reflected in the patterned objects.

Fitted   Click to have the objects being patterned equally spaced within the specified angle or distance. If the option's box is clear, the pattern spacing will specify the angle or distance between the patterned occurrences.

Once you have created the pattern sketch, the pattern will be grouped together. The patterned objects are referred to as elements. You can edit the individual elements or delete the entire pattern. To edit the patterned sketch, move the cursor over an element and right-click. A menu will appear that gives you the option to suppress or unsuppress elements (depending upon the element the cursor is over when you right-click) and either delete or edit the pattern. The following image shows the edit pattern options while the cursor is over an unsuppressed occurrence of a feature.

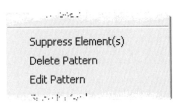

Figure 7-37

- Click Suppress Element(s) and then select the object(s) to suppress. To unsuppress suppressed objects, reselect them.

- Click Delete Pattern to delete the entire pattern. The original object(s) that were patterned will not be deleted.

- Click Edit Pattern to change the count, spacing, angle, suppress/unsuppress occurrences, and toggle the Associative and Fitted options.

SHARED SKETCHES

When creating a feature that may use the same sketch geometry, same dimensions, and will lie on the same face of the sketch of an existing feature, you can share the sketch instead of creating a new one. A shared sketch is an associated copy of the original sketch and is placed above the original feature in the Browser. The shared sketch has the same name as the original sketch, and any modifications to an original sketch will be updated to all the features that use the shared sketch. Once you have shared a sketch, you can select the shared sketch as the profile for additional sketched features. There is no limit to the number of features that can use the shared sketch. The visibility of the shared sketch is on by default; you turn it on and off as needed by right-clicking on the shared sketch in the Browser and selecting Visibility from the menu. To share a sketch, follow these steps:

1. Locate the feature in the Browser that contains the sketch you want to share.
2. Click the plus sign to the left of that feature to expand the sketch.
3. Right-click on the sketch name and select Share Sketch from the menu, as shown in the following image.
4. Issue the feature tool that you need (to extrude, revolve, etc.) and then select the shared sketch as the profile for the feature. The right side of the following image shows a shared sketch that has been extruded down to form a rectangle and the shared sketch (rectangle) has also been extruded up with face draft.

 NOTE A shared sketch that is consumed by a feature cannot be deleted.

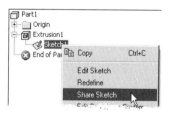
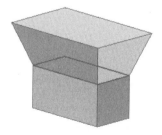

Figure 7-38

MIRROR SKETCHES AND SYMMETRY CONSTRAINT

When you need to create a symmetrical part, you can use the Mirror sketch tool and the symmetry constraint to reduce the number of constraints and dimensions required to fully constrain the sketch. The Mirror tool will apply the symmetry constraint(s) between the selected geometry and the mirrored geometry automatically. As the geometry on one half changes, so will the mirrored geometry on the other half. Follow these guidelines to create a mirrored sketch:

1. Analyze what the finished sketch will look like and determine where the line of symmetry will be.

2. Add constraints and dimension to the sketch either before or after the mirror operation. Draw half of the finished sketch and the line of symmetry. The line of symmetry must be a single line, but its style can be normal, construction, or centerline. The following image shows a sketch that will be mirrored upon the left vertical construction line.

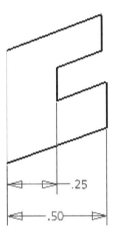

Figure 7-39

3. Click the Mirror tool (shown in the following image) on the Sketch Panel Bar.

Figure 7-40

4. The Mirror dialog box will appear as shown in the following image.

Figure 7-41

5. Select the objects in the graphics window to be mirrored; by default, the Select option is active.

6. Click the Mirror line button in the Mirror dialog box and select the line in the graphics window that will be the line of symmetry.

7. To complete the operation, click the Apply button.

8. A symmetry constraint will be applied between both halves automatically. Any change to a dimension or constraint on the first half will be reflected on the mirrored side. You can delete the symmetry constraint like any other geometric constraint. The following image shows the geometry on the left mirrored across the vertical construction line with the constraints shown.

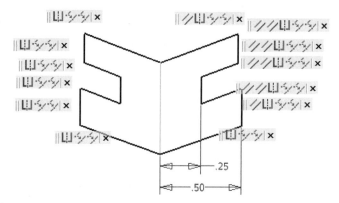

Figure 7-42

You can also add a symmetry constraint between two sets of objects manually. The object(s) must be the same object type (line, arc, circle, etc.) and must lie on opposite sides of a line of symmetry. After you have drawn the geometries, add the Symmetry constraint from the Sketch panel bar under the sketch constraints menu, shown in the following image. Select the two objects that will be symmetrical and then select the line of symmetry.

Chapter 7 Advanced Sketching and Constraining Techniques 519

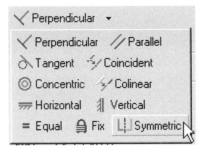

Figure 7-43

SLICE GRAPHICS

While creating parts, you may need to sketch on a plane that is difficult to see because other features are obscuring the view. The Slice Graphics option will temporarily slice away the portion of the model that obscures the active sketch plane on which you want to sketch. The following image shows a part that is shelled out (in the middle). The image on the left is shown in a normal state, and the image on the right is a menu shown with the Slice Graphics option. To temporarily slice the graphics screen, follow these steps:

1. Make a plane that the graphics will be sliced through the active sketch.

2. Rotate the model so the correct side will be sliced (the side of the model that faces you will be sliced away).

3. While editing the sketch, right-click and select Slice Graphics from the menu (as shown in the following image), press the F7 key, or click Slice Graphics on the View menu. The model will be sliced on the active sketch plane.

4. Use sketch tools from the Sketch toolbar to create geometry on the active sketch.

5. To restore the sliced graphics, right-click and select Slice Graphics, click Slice Graphics on the View menu, or click the Sketch or Return button on the Command bar to leave sketch mode.

Figure 7-44

 NOTE When working in an assembly, there are different slice graphics tools (Assembly Section Views) available from the Assembly toolbar.

SKETCH ON ANOTHER PART'S FACE

In Chapter 6 you learned that when you create a new part in an assembly, you have the option to create the first sketch on any planar face or plane on any part and the option of having a flush assembly constraint automatically applied between the selected face on the other part and the first sketch. This relationship constrains the sketch to the selected face; if the face moves on the selected part, so will the sketch. The option for creating a sketch on a face or plane of another part is not limited to the first sketch. Any time you create a sketch, you can place it on a face or plane of another part. This sketch will also be associated to the selected face or plane. When creating a sketch, click the planar face or plane on the part on which you want to create the sketch. This will create a work plane and constrain it to the selected face or plane of the other part, and it will create a sketch on the new work plane. The work plane will be "adaptive," meaning that if the other part's face or plane moves, so will the sketch that is tied to the work plane. For more information on adaptivity, refer to Chapters 6. The right side of the following image shows a sketch that was created on the angle part. Notice that the work plane in the Browser is adaptive.

Figure 7-45

EXERCISE 7-2 Projecting Edges and Sketching On Another Part's Face

1. In this exercise, you create a cover for an existing part. You define the shape of the cover by projecting geometry from another part. You also examine how the cover updates when the underlying part geometry is modified..
   Open *ESS_E07_02.iam*.

2. Use the Zoom and Rotate tools to examine the part.

3. Right-click on the graphics window and select Isometric View.

4. First, you create a new cover plate for the ring. Click the Create Component tool. In the Create In-Place Component dialog box:
   Enter **Cover** in the New File Name edit box.
   Click the Browse button to the right of the Template edit box.
   Click the Metric tab in the Open Template dialog box.
   Select the *Standard(mm).ipt* template and click OK.
   Ensure that Constrain sketch plane to selected face or plane is checked.
   Click OK.

5. Click the face as shown in the following image.

Figure 7-46

6. Click the Project Geometry tool.

7. Move the cursor over the ring face as shown below. Click when all loops on the face are highlighted as shown in the following image. Do not project individual edges.

Figure 7-47

8. Right-click in the graphics window and click Done.
9. Click the Return tool to exit the sketch.
10. Click the Extrude tool.
11. For the profile click inside the inner circle and in between both circles and enter a distance of **10**mm and your sreen should resemble the following image.

Figure 7-48

12. Use the Rotate tool to adjust the view as shown in the following image with the flat of the first part facing you.
13. Click the Sketch tool, and then select the flat face of the first part as shown below.

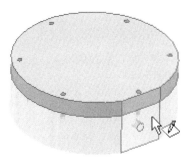

Figure 7-49

14. Click the Project Geometry tool.

15. Project the outside edge of the Cover and the circular edge of the hole from the Ring, as shown below.

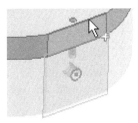

Figure 7-50

16. Use the Look At tool to look directly at the new sketch.

17. Draw, constrain and dimension two lines and an arc as shown in the following image.

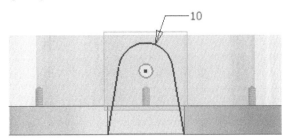

Figure 7-51

18. Change to an isometric view so you can see the new sketch.

19. Extrude the sketch **5**mm as shown in the following image.

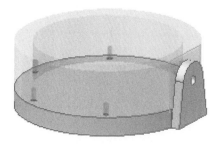

Figure 7-52

20. Click the Return tool to return to the assembly environment.
21. Notice the adaptive symbol on the component Cover:1.
22. Make the component ESS_E07_02-Ring:1 the active component.
23. Edit the sketch of the Base Extrusion and change the 100mm diameter dimension to 150mm.
24. Update the part.
25. Click the Return tool to return to the assembly environment and your assembly should resemble the following image.

Figure 7-53

26. Close the file. Do not save changes.

DIMENSION DISPLAY, RELATIONSHIPS, AND EQUATIONS

In Chapter 2 you learned how to create independent dimensions that had no relationship to other dimensions. When creating part features, you may want to set up relationships between fetaures and/or sketch dimensions . The length of a part may need to be twice that of its width, for example, or a hole may always need to be in the middle of the part. In Autodesk Inventor, there are a few different methods you can use to set up relationships between dimensions. The following sections will cover these methods.

DIMENSION DISPLAY

When you create each dimension, it is automatically tagged with a label (parameter name) that starts with the letter "d" and a number, for example "d0" or "d27." The first dimension created for each part is given the label "d0." Each dimension that you place for subsequent part sketches and features sequences goes up one number at a time. If you erase a dimension, the next dimension does not go back and reuse the erased value. Instead, it keeps sequencing from the last value on the last dimension created. When creating dimensional relationships, you may want to view a dimension's display style to see the underlying parameter label of the dimension. There are five options for displaying a dimension's display style.

Display as value   This is the default dimension display style. Use to display the actual value of the dimensions on the screen.

Display as name   Use to display the dimensions on the screen as the dimension label or actual parameter name (i.e., d12 or Length).

Display as expression   Use to display the dimensions on the screen in the format of label = value, showing each actual value (i.e., d7 = 20 mm or Length = 50 mm).

Display as tolerance   Use to display the dimensions on the screen that have a tolerance style as the tolerance (i.e., 40 ± .3).

Display precise value   Use to display the dimensions on the screen that have a tolerance style to its exact value with the tolerance applied (i.e., 40.3).

To change the dimension display style, right-click in the graphics window and click Dimension Display as shown in the following image, or click Tools > Document Settings and, on the Units tab, change the display style. After you select a dimension display style, all visible dimensions that you did not change individually to a display style will change to that style. As you create dimensions, they will reflect the current dimension display style.

Figure 7-54

To change an individual dimension's display style, right-click on a dimension in the graphics window and select Dimension Properties on the menu. On the Document Settings tab, you may then select the style from the Modeling Dimension Display drop-down list, as shown in the following image.

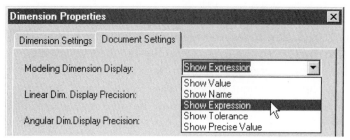

Figure 7-55

DIMENSION RELATIONSHIPS

Setting a dimensional relationship between two dimensions requires setting a relationship between the dimension you are creating and an existing dimension. When entering text in the Edit Dimension dialog box, enter the dimensions label (d#) of the other dimension, or click the dimension with which you want to set the relationship in the graphics window. The left side of the following image shows the Edit Dimension dialog box after selecting the 10 mm dimension to which the new dimension will be related. After establishing a relationship to another dimension, the dimension will have a prefix of fx: as shown on the right side of the following image.

Figure 7-56

EQUATIONS

You can also use equations whenever a value is required. A couple of examples would be: (d9/4)*2, or 50 mm + 19 mm. When creating equations, Autodesk Inventor allows prefixes, precedence, operators, functions, syntax, and units. To see a complete listing of valid options, use the Help system and navigate to the topic of equations and the title Edit box reference. You can enter numbers with or without units; when no unit is entered, the default unit will be assumed. As you enter an equation, Autodesk Inventor calculates it. An invalid expression will appear in red and a correct expression will appear in black. For best results while using equations, include units for every term and factor in the equation.

To create an equation in any edit box, follow these steps:

1. Click in the edit field.

2. Enter any valid combination of numbers, parameters, operators, or built-in functions. The following image shows an example of an equation that uses both millimeters and inches for the units.

3. Press ENTER or click the green check mark to accept the expression.

 NOTE Use ul (unitless) where a number does not have a unit; for example, use a unitless number when dividing, multiplying, or specifying values for a pattern count.

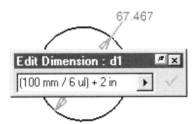

Figure 7-57

PARAMETERS

Another method of setting up relationships between dimensions is to use parameters. A parameter is a user-defined name that is assigned a numeric value, either explicitly or through equations. You can use multiple parameters in an equation, and you can use parameters to define another parameter, such as depth = length - width. You can use a parameter anywhere a value is required. There are four types of parameters: model parameters, user parameters, reference parameters, and linked parameters.

Model Parameters   This type is created automatically and assigned a name when you create a sketch dimension, feature parameter (such as extrusion distance, draft angle, or coil pitch), and the offset, depth, or angle value of assembly constraint. Autodesk Inventor assigns a default name to each model parameter as you create it. The default name format is a "d" followed by an integer incremented for each new parameter. You can rename model parameters via the Parameters dialog box.

User Parameters   This type is created manually from an entry in the Parameters dialog box.

Reference Parameters   This type is created automatically when you create a driven dimension. Autodesk Inventor assigns a default name to each reference parameter as you create it. The default name format is a "d" followed by an integer incremented for each new parameter. You can rename reference parameters via the Parameters dialog box.

Linked Parameters   This type is created via a Microsoft Excel spreadsheet and linked into a part or assembly file.

You create and/or edit parameters by clicking the Parameters tool, shown in the following image, on the Sketch, Part Features, or Assembly panel bars.

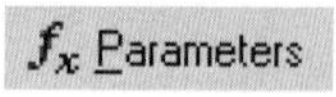

f_x Parameters

Figure 7-58

After you click the Parameters tool, the Parameters dialog box will appear. The following image shows an example with a few model, user, and reference parameters created. The Parameters dialog box is divided into two sections: Model Parameters and User Parameters. A third section named Reference Parameters will appear if driven dimensions exist. The values from dimensions or assembly constraints used in the active document automatically fill the Model Parameters section. The User Parameters section will be defined manually. You can change the names and equations of both types of parameters, and you can add comments by double-clicking in the cell and entering the new information. The column names for Model and User Parameters are the same; the following sections define them.

Parameter Name   The name of the parameter will appear in this cell. To change the name of an existing parameter, click in the box and enter a new name. When creating a new user parameter, enter a new name after clicking the Add button. When you update the model, all dependent parameters update to reflect the new name.

Units   Enter a new unit of measurement for the parameter in this cell. With Autodesk Inventor you can build equations that include parameters of any unit type. All length parameters are stored internally in centimeters; angular parameters are stored internally in radians. This becomes important when you combine parameters having different units in equations.

Equation   The equation will appear in this cell; it will determine the value of the parameter. If the parameter is a discrete value, the value appears in rounded form to match the precision setting for the document. To change the equation, click on the existing equation and enter the new equation.

Nominal Value   The nominal tolerance result of the equation will appear in this cell, and it can only be modified by editing the equation.

Tol.   From the drop-down list, select a tolerance condition: upper, median, nominal, or lower. Tolerances will be covered later in this chapter.

Model Value   The actual calculated model value of the equation in full precision will appear in this cell.

Export Parameters Column   Click to export the parameter to the Custom tab of the Properties dialog box. The parameter will also be available in the bill of materials and parts list Column Chooser dialog boxes.

Comment   You may choose to enter a comment for the parameter in this cell. To add a comment, click in the cell and enter the comment.

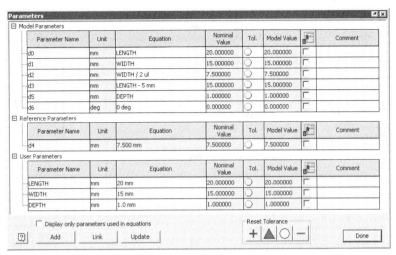

Figure 7-59

USER PARAMETERS

User parameters are parameters that you define in a part or an assembly file. Parameters defined in one environment are not directly accessible in the other environment. If parameters are to be used in both environments, use a linked parameter via a common Microsoft Excel spreadsheet. You can use a parameter any time a numeric value is required. When creating parameters, follow these guidelines:

- Assign meaningful names to parameters; other designers may edit the part file and will need to understand your thought process.

- The parameter name cannot include spaces, mathematical symbols, or special characters. You can use these to define the equation.

- The parameter name cannot consist of only numbers. It must include at least one alphabetic character, and the alphabetic character must appear first. W1 or Width1, for example, would be valid. The names 123 and 1W would be invalid parameter names.

- Autodesk Inventor detects capital letters and uses them as unique characters. Length, length, and LENGTH are three different parameter names.

- When entering a parameter name where a value is requested, the upper and lower case of the letters must match the parameter name.

- When defining a parameter equation, you cannot use the parameter name to define itself (i.e., Length = Length/2 would be invalid).

- If you use the same user parameter name in multiple part files, it should be defined in a template file. When you create new parts based on that template, the parameter will already be defined.

- Duplicate parameter names are not allowed. Model-, User-, and Spreadsheet-driven parameters must have unique names.

To create and use a User parameter, follow these steps:

1. Click the Parameters tool on the Sketch, Part Features, or Assembly Panel Bars.
2. Click the Add button at the bottom of the Parameters dialog box.
3. Enter the information for each of the cells.
4. After creating the parameter(s), you can enter the parameters' names anywhere a value is required. When editing a dimension, click the arrow on the right and select List Parameters from the menu, as shown on the left side of the following image. All the available user parameters will appear in a list similar to that on the right side of the image. Click the desired parameter from the list.

 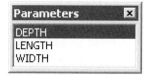

Figure 7-60

LINKED PARAMETERS

If you want to use the same parameter name and value for multiple parts, you can create a spreadsheet using Microsoft Excel. You can then either embed or link the spreadsheet into a part or assembly file through the Parameters dialog box. When you embed a Microsoft Excel spreadsheet, there is no link between the spreadsheet and the parameters in the Autodesk Inventor file, and any changes to the spreadsheet will not be reflected in the Autodesk Inventor file. When you link a Microsoft Excel spreadsheet to an Autodesk Inventor part or assembly file, any changes in the spreadsheet will also update the parameters in the Autodesk Inventor file. You can link more than one spreadsheet to an Autodesk Inventor file, and you can link each spreadsheet to multiple part and assembly files. By linking a spreadsheet to an assembly file and to the part files that comprise the assembly, you can drive parameters from both environments from the same spreadsheet. There is no limit to the number of part or assembly files that can reference the same spreadsheet. Each linked spreadsheet appears in the Browser under the 3rd Party folder.

When creating a Microsoft Excel spreadsheet with parameters, follow these guidelines:

- The data in the spreadsheet can start in any cell, but they must be specified when you link or embed the spreadsheet.

- The data can be in rows or columns, but they must be in this order: parameter name, value or equations, unit of measurement, and (if needed) a comment.

- The parameter name and value are required, but the other items are optional.

- The parameter name cannot include spaces, mathematical symbols, or special characters. You can use these to define the equation.

- Parameters in the spreadsheet must be in a continuous list. A blank row or column between parameter names eliminates all parameters after the break.

- If you do not specify a unit of measurement for a parameter, the default units for the document will be assigned when the parameter is used. To create a parameter without units, enter ul (unitless) in the Units cell.

- Only those parameters defined on the first worksheet of the spreadsheet become linked to the Autodesk Inventor file.

- You can include column or row headings or other information in the spreadsheet, but they must be outside the block of cells that contains the parameter definitions.

On the left side of the following image is an example of three parameters that were created in rows with a name, equation, unit, and comment. The right side of the image shows the same parameters created in columns.

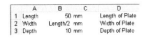 

Figure 7-61

After you have created and saved the spreadsheet, you can create parameters from it by following these steps:

1. Click the Parameters tool on the Sketch, Part Features, or Assembly Panel Bars.

2. Click Link on the bottom of the Parameters dialog box.

3. The Open dialog box will appear similar to that shown in the following image.

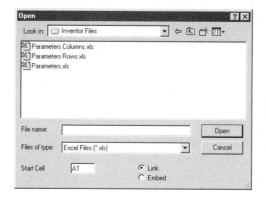

Figure 7-62

4. Navigate to and select the Microsoft Excel file to use.

5. In the lower-left corner of the Open dialog box, enter the start cell for the parameter data.

6. Select whether the spreadsheet will be linked or embedded.

7. Click the Open button (a section showing the parameters is added to the Parameters dialog box, as shown in the following image). If you embedded the spreadsheet, the new section will be titled Embedding #.

8. To complete the operation, click Done.

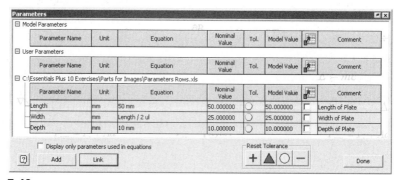

Figure 7-63

To edit the parameters that are linked or embedded, follow these steps:

1. Open the Microsoft Excel file.

2. Make the required changes.

3. Save the Microsoft Excel file.

4. Open the Autodesk Inventor part or assembly file that uses the spreadsheet.

5. Click the Update tool on the standard toolbar.

You can also follow these steps:

1. Open the Autodesk Inventor part or assembly file that uses the spreadsheet.

2. Expand the 3rd Party folder in the Browser.

3. Either double-click the spreadsheet or right-click and select Edit from the menu, as shown in the following image.

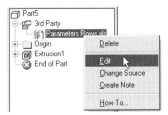

Figure 7-64

1. The Microsoft Excel spreadsheet will open in a new window for editing.

2. Make the required changes.

3. Save the Microsoft Excel file.

4. Activate the Autodesk Inventor part or assembly file that uses the spreadsheet.

5. Click the Update tool from the standard toolbar.

 NOTE If you embedded the spreadsheet, the changes will not be saved back to the original file, but will only be saved internally to the Autodesk Inventor file.

EXERCISE 7-3 Auto Dimension, Relationships, and Parameters

In this exercise you create a sketch and dimension it using the Auto Dimension tool. You then set up relationships between dimensions and define user parameters in both the model and an external spreadsheet.

1. Start a new file based on the default *Standard (mm).ipt* file.

2. Draw the geometry shown in the following figure. The size should be roughly 25 mm wide by 40 mm high.

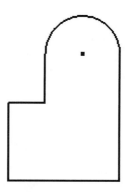

Figure 7-65

3. Add a fix constraint to the lower left corner and ensure that the arc is tangent to both vertical lines.

4. Add the four dimensions as shown in the following image.

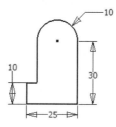

Figure 7-66

5. Edit the 10 mm radius dimension of the arc and for its value select the 25 mm horizontal dimension as shown in the following figure.

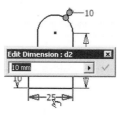

Figure 7-67

6. In the Edit Dimension dialog box, type **/4** after the d#. For example, d1/4. The 1 may be a different number depending upon the order your geometry was dimension.

7. Edit the 10 mm vertical dimension and for its value select the 30 mm vertical dimension. Then type in **/2** after the d#.

8. Change the value of the 25 mm horizontal dimension to 50 mm and the 30 mm vertical dimension to 40 mm. When done your sketch should

resemble the following figure. The fx: before the two dimensions denotes they are equation-driven or a reference to a parameter.

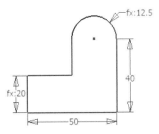

Figure 7-68

9. Next, you create parameters and drive the sketch from them. Click the Parameters tool from the 2D Sketch Panel bar.

10. In the Parameter dialog box, create a User parameter by clicking the Add button and then type in the following information:
    Parameter Name = Length
    Units = mm
    Equation = 75
    Comment = Bottom length

11. Create another User parameter by selecting the Add button and type in the following information. When done the User Parameter area should resemble the following figure.
    Parameter Name = Height
    Units = mm
    Equation = Length/2
    Comment = Height is half of the length

Parameter Name	Unit	Equation	Nominal Value	Tol.	Model Value		Comment
Length	mm	75 mm	75.000000		75.000000		Bottom length
Height	mm	Length / 2 ul	37.500000		37.500000		Height is half of the length

Figure 7-69

12. Close the Parameter dialog box by clicking Done.

13. Change the dimension display by right-clicking the graphics window and click Dimension Display > Expression.

14. Edit the bottom horizontal dimension and change its value to the parameter 'Length.' Double-click on the dimension and in the Edit Dimension dialog box, click the arrow; from the menu click List Parameters and then click Length from the list.

15. Then edit the right vertical dimension and change its value to Height. When done your screen should resemble the following figure.

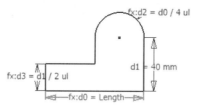

Figure 7-70

16. Click the Parameters tool. From the User parameters section double-click on the equation cell of the parameter name Length and change its value to 35 mm.

17. Repeat the same steps for the parameter name Height and change its value to 50 mm and the comment field to 'Height of the right side'.. When done making the changes, the User Parameter area should resemble the following figure.

Parameter Name	Unit	Equation	Nominal Value	Tol.	Model Value		Comment
Length	mm	35 mm	35.000000		35.000000		Bottom length
Height	mm	50 mm	50.000000		50.000000		Height of the right side

Figure 7-71

18. To complete the changes click the Done button.

19. Click the Update tool and the sketch should update to reflect the changes.

20. Save the file as **ESS_E07_03.ipt**

21. Now you create a spreadsheet that has two parameters. Create an Excel spreadsheet that has the column names Parameter Name, Equation, Unit, and Comment along with the following data.
Parameter Name = BaseExtrusion
Equation = 30
Units = mm
Comment = Extrusion distance for the base feature
Parameter Name = Draft
Equation = 5
Units = deg
Comment = Draft for all features

	A	B	C	D
1	Parameter Name	Equation	Unit	Comment
2	BaseExtrusion	30	mm	Extrusion distance for the base feature
3	Draft	5	deg	Draft for all features

Figure 7-72

22. Save the Excel spreadsheet as *ESS_07_Parameters.xls*.

23. Make Autodesk Inventor the current application and then click the Parameters tool.

24. In the Parameter dialog box click the Link button and then select the *ESS_07_Parameters.xls* file (but do not click Open yet).

25. For the Start Cell enter A2. If you fail to do this, no parameters will be found.

26. Click the Open button and a new spreadsheet area will appear in the Parameters dialog as shown in the following figure. Then click the Done button to complete the operation.

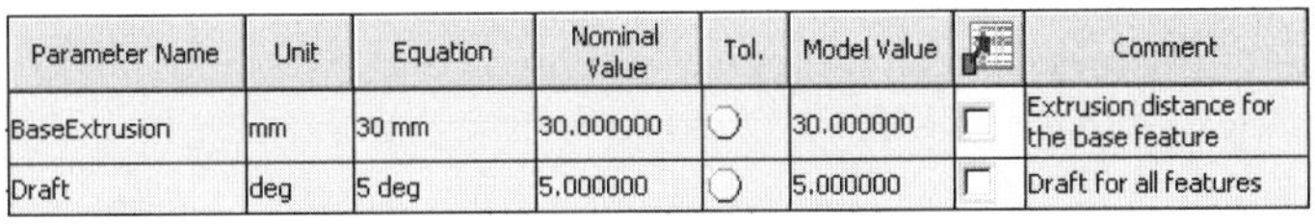

Parameter Name	Unit	Equation	Nominal Value	Tol.	Model Value		Comment
BaseExtrusion	mm	30 mm	30.000000	○	30.000000	☐	Extrusion distance for the base feature
Draft	deg	5 deg	5.000000	○	5.000000	☐	Draft for all features

Figure 7-73

27. Change to an isometric view.

28. Click the Extrude tool and then click the More tab in the Extrude dialog box. For the Taper's value enter Draft or click the arrow from the menu click List Parameters and then click the parameter Draft.

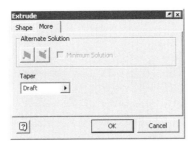

Figure 7-74

29. Click the Shape tab. For the Distance value use the parameter BaseExtrusion as shown in the following image.

Figure 7-75

30. To complete the operation, click the OK button and your model should resemble the following image.

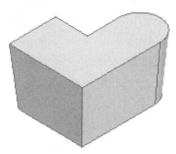

Figure 7-76

31. In the Browser expand the 3rd Party icon and either double-click or right-click on the name *ESS_07_Parameters.xls* and select Edit from the menu.

32. In the Excel spreadsheet make the following changes:
    For the Parameter Name 'BaseExtrusion' change the Equation to 50mm.
    For the Parameter Name 'Draft' change the Equation to –3 degrees.

33. Save the spreadsheet and then close Excel.

34. Make Autodesk Inventor the current application and click the Update tool from the Standard Toolbar. When done your model should resemble the following image.

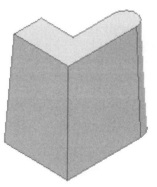

Figure 7-77

35. Close the file and do NOT save changes.

Chapter 7 Advanced Sketching and Constraining Techniques

CHAPTER SUMMARY

To	Do This	Tool
Create construction geometry	Change the line style to Construction in the Style area on the standard toolbar.	Normal / Construction
Create splines	Click the Spline tool on the 2D Sketch Panel Bar.	
Pattern a sketch object	Click either the Rectangular Pattern tool or Circular Pattern tool on the 2D Sketch Panel Bar.	
Share a sketch	Right-click on the sketch name in the Browser and select Share Sketch from the menu.	
Mirror a sketch	Click the Mirror tool on the 2D Sketch Panel Bar.	
Temporarily slice away a portion of the model that obscures the plane	After making a sketch active, right-click and select Slice Graphics from the menu.	Insert Image
Project selected edges, vertices, work features, curves, or silhouette edges	Click the Project Geometry tool on the 2D Sketch Panel Bar.	
Sketch on a plane of another part	Click the 2D Sketch tool on the standard toolbar and click on any planar face or plane of another part.	Sketch
Create sketch constraints and dimensions automatically	Click the Auto Dimension tool on the 2D Sketch Panel Bar.	
Create parameters	Click the Parameters tool on the Sketch, Part Features, or the Assembly Panel Bars.	f_x

Applying Your Skills

SKILL EXERCISE 7-1

In this exercise, you create a new polygon, pattern sketch objects, and use sketch editing and constraint controls in a 2D layout of an electronic remote control.

1. Open *ESS_E07_04.ipt*.
2. Use the Look At and Zoom tools to adjust the view as shown below.

Figure 7-78

3. Activate Sketch1.
4. Start the Polygon tool.
   In the Polygon dialog box, change the number of sides to **4** and use the Inscribed method.
   Place the center point coincident with the intersection of the short diagonal construction line and construction arc, and then place the corner point coincident with the arc (above). Do not close the Polygon dialog box.

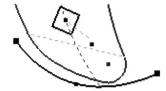

Figure 7-79

5. In the Polygon dialog box, change the number of sides to **3**, and create a three-sided polygon with the center and corner points coincident with the diagonal construction line.

Figure 7-80

6. Use parallel and equal constraints and two dimensions to control the size and position of both polygons.

Chapter 7 Advanced Sketching and Constraining Techniques 541

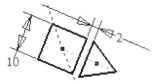

Figure 7-81

7. Zoom in on the intersecting ellipses.

8. Click the General Dimension tool.

9. Select each ellipse between quadrant points and place the following dimensions.

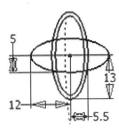

Figure 7-82

10. Start the Trim tool and trim ellipse segments as necessary to complete the outline of the control buttons, and then use the Fix constraint tool to lock the trimmed button curves. The dimensions will be deleted as you trim.

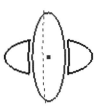

Figure 7-83

11. Click the spline and drag any shape point (small square) to another location. Notice how both shape points move, but the fix points (large squares on each end) do not.

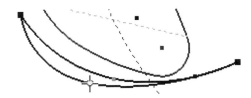

Figure 7-84

12. Undo the move.

13. Drag the spline control points (large square) individually for a smoother curve. Experiment with different positions to see how the curvature changes. You can also drag the entire spline to reposition it, if necessary.

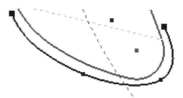

Figure 7-85

14. Drag the endpoints of the spline to make them coincident with the endpoints of the construction line.

Figure 7-86

15. Click the General Dimension and dimension the spline control points as shown.

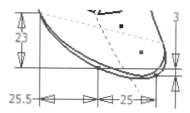

Figure 7-87

16. Delete the four arc segments from the original profile, and then place tangent constraints between the outside spline segments and arcs.

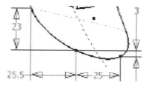

Figure 7-88

17. Click the Rectangular Pattern tool.

18. Click Zoom All (move the dialog box if necessary), and then select all four sides of the rectangular sketch object (above the polygons).

19. Click the Direction 1 arrow button and then select the bottom side of the rectangle (flip the arrow if necessary so it points to the right). Repeat the procedure and select the left side of the rectangle and make sure the Direction 2 arrow points up.

20. In the Rectangular Pattern dialog box:
    Change Direction 1 Count value to **3** and the Spacing value to **12** mm, and then change Direction 2 Count value to **5** and the Spacing value to **10** mm. Expand the dialog box. Make sure the Associative check box is checked, and then click the Suppress arrow button.
    In the pattern preview, select the five pattern instances that interfere with the elliptical buttons.
    Click OK to create the sketch pattern.

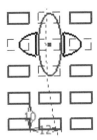

Figure 7-89

21. Click the Circular Pattern tool.

22. Move the Circular Pattern dialog box if necessary, and then click all three sides of the triangular polygon you created earlier in the exercise.

23. Click the Axis arrow button, and then click the center point of the four-sided polygon.

24. In the Circular Pattern dialog:
    Change the Count value to **4**.
    Make sure the Angle value is **360** deg.
    Expand the dialog box and make sure the Associative and Fitted check boxes are checked.
    Click OK to create the sketch pattern.

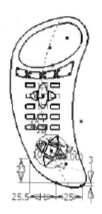

Figure 7-90

25. Close the file. Do not save changes.

CHECKING YOUR SKILLS

Use these questions to test your knowledge of the material covered in this chapter.

1. **True__ False__** You can dimension to geometry that uses the construction style.

2. **True__ False__** Splines cannot have geometric constraints applied between them and other geometry.

3. **True__ False__** Modifications to a shared sketch will update all the features that use that shared sketch.

4. **True__ False__** Slice Graphics will permanently slice away a portion of the model.

5. **True__ False__** The Project tool can project vertices, work features, curves, or silhouette edges of another part in an assembly to the active sketch.

6. **True__ False__** When creating parameters in a spreadsheet, the data items must be in the following order: parameter name, value or equations, unit of measurement, and, if needed, a comment.

7. Explain how to suppress a patterned occurrence.

8. What is the difference between a model parameter and a user parameter?

   _____

   _____

9. What is a reference parameter?

   _____

CHAPTER 8

Advanced Part Modeling Techniques

In this chapter, you will learn how to use advanced modeling techniques. Using advanced modeling techniques, you can create transitions between parts that would otherwise be difficult to create. You can edit advanced features like other sketched and placed features, but typically their creation requires more than one unconsumed sketch. This chapter also introduces you to techniques in this chapter that will help you be more productive in your modeling.

CHAPTER OBJECTIVES

After completing this chapter, you will be able to

- Extrude an open profile
- Create ribs, webs, and rib networks
- Emboss text and profiles
- Create sweep features
- Create coil features
- Create loft features
- Split a part or split faces of a part
- Copy features within a part
- Reorder part features
- Mirror model features
- Suppress features of a part
- Work with file properties
- Create and place standard iParts
- Create and place custom iParts
- Create iFeatures
- Create table-driven iFeatures
- Publish parts
- Publish featuers

USING OPEN PROFILES

In Chapter 3 you learned how to extrude a closed profile. In this section you will learn how to extrude an open profile. An open profile is a sketch that does not form a closed area or boundary. You extrude an open profile bidirectionally–in a positive or negative direction normal to the profile plane and in a fill direction. A fill direction will extend the profile until it touches a termination face (in all directions), creates a closed area, and encloses any existing feature. The following image shows the original part on the left with a line, based on a work plane, in the middle of the sketch. The part in the middle shows the open profile (the line) extruded to the left (the Match shape option was used and the cylindrical boss was enclosed, respecting the hole through the boss). The part on the right shows the open profile extruded to the right.

Open Profile   Open Profile Extruded to Left   Open Profile Extruded to Right

Figure 8-1

When working with an open profile, you can only use the Extrude tool. The Revolve and Sweep tools can use an open profile; however, the result will be a surface. The open profile can be consumed inside or extend beyond the part, or a combination of the two. When you extrude the open profile, it will either be extended or trimmed back to be enclosed with the part. You can constrain and dimension open profile like any other profile.

To extrude an open profile, follow these steps:

1. Make a sketch active.
2. Draw an open profile.
3. Add geometric constraints and dimensions as needed.
4. Issue the Extrude tool.
5. Click the open profile on the part.

6. Click on the side of the part that will be filled in. The following image shows the left side of the part selected as the fill side. Figure 8-3 shows the right side selected.

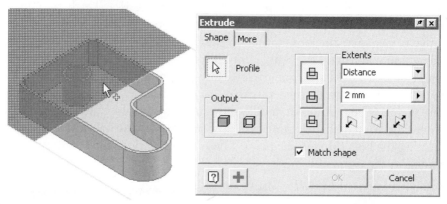

Figure 8-2

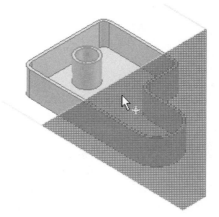

Figure 8-3

7. Change the Extents option as needed: Distance, To Next, To, From To, and All. Then specify the distance and the extrusion direction if required.

8. Click the Match Shape option if you want the extrude to leave islands as voids. The image on the left side in the following image shows the extrude open profile with the Match Shape option selected. The image on the right side shows it without the option selected.

9. To complete the operation, click OK.

 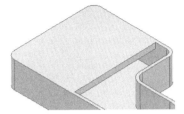

Figure 8-4 (Left) Extrude open profile with **Match** shape checked. **(Right)** Extrude open profile with **Match** shape not checked.

RIB AND WEB FEATURES

Ribs and webs are used primarily to reinforce or strengthen features in mold and cast parts, but you can also use them in machined parts and in other cases where additional support and minimal weight are required. The following image shows a part with a sketch and with the sketch used to create a rib and a web.

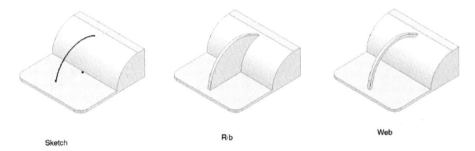

Figure 8-5

Using the Rib tool located on the Part Features Panel Bar, as shown in the following image, you can create ribs, webs, and rib networks.

- A rib is a thin-walled feature that is typically closed.
- A web is similar in width but usually has an open shape.
- A rib network consists of a series of thin-walled support features.

Figure 8-6

A rib or web feature is defined by a single, open, unconsumed profile that is then refined using the options in the Rib dialog box. If there is no unconsumed sketch in the part file, Autodesk Inventor will warn you with the message: "No unconsumed visible sketches." After starting the Rib tool, the Rib dialog box will appear as shown in the following image. The following sections describe the options in the dialog box.

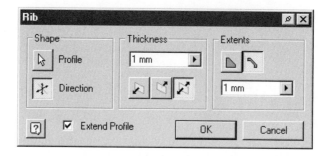

Figure 8-7

SHAPE

Profile — Select this button to choose the sketch to extrude. The sketch can be an open profile, or you can select multiple-intersecting or nonintersecting profiles to define a rib or web network.

Direction — Select this button, then in the graphics window position the cursor around the open profile to specify whether the rib will extend parallel or perpendicular and in which direction relative to the sketch geometry. Once the correct direction is displayed, click in the graphics window. The following image shows a rib in all four directions.

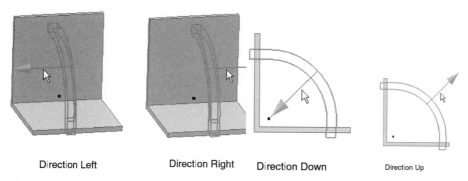

Direction Left   Direction Right   Direction Down   Direction Up

Figure 8-8

THICKNESS

In this section you specify the thickness and thickness direction of the rib.

Edit Box — Enter the width of the rib feature using this edit box.

Flip Buttons — Select the flip buttons to specify which side of the profile to apply the thickness value to or to add the same amount of material to both sides of the profile.

EXTENTS

In this section you specify a rib (To Next) or a web (Finite).

To Next  This button will extend the ends of the open profile and the area between the profile and the next available set of faces along the rib direction.

 Finite  The Finite button enables an edit box in which to specify an offset distance from the sketch geometry. By enabling the Finite button and entering a distance, you can create a web feature. Note the Edit box for Distance is not available with the To Next option.

Extend Profile   The Extend Profile checkbox specifies whether to extend the endpoints of the sketch to the next available face or to leave the ends of the open profile as determined by the end of the rib feature. If you click the Extend Profile option, the ends are extended; if you leave the option's box clear, the ends cap at the end of the sketched profile. The Extend Profile option is always available for a finite web, but is only active for the To Next option when the direction and face(s) on the model meet appropriate conditions.

CREATING RIBS AND WEBS

To create a rib or web, follow these steps:

1. Create an active sketch in the location where you will place the rib or web.
2. Sketch an open profile that defines the basic shape of one edge of the rib or web.
3. Add constraints and dimensions to the sketch as needed.
4. Click the Rib tool located on the Part Features Panel Bar.
5. Specify the direction of the rib or web. Click the Direction button in the Rib dialog box. Then in the graphics window, position the cursor around the open profile to specify whether the rib or web will extend parallel or

perpendicular and in which direction relative to the sketch geometry. Once the correct direction is previewed, click in the graphics window.

6. Enter a value for the rib or web in the Thickness edit box. This will specify the width of the rib or web.

7. Also in the Thickness section, use the Flip buttons to choose which side of the profile to apply the thickness value or to add the same amount of material to both sides of the profile.

8. Specify the depth of the profile by clicking either the To Next or Finite buttons in the Extents area of the Rib dialog box.

9. If you use the Finite option, enter an offset distance and click Extend Profile if the endpoints are to be extended to the next available face.

10. To complete the operation, click the OK button.

RIB NETWORKS

You can also create rib networks using the Rib tool. You can use multiple intersecting or nonintersecting sketch objects within a single profile to create a rib network. The creation process is the same for creating a single rib, except the thickness is applied to all objects within the profile. When you select the profile objects, you will need to select them individually. If the rib network is to have equal spacing between the objects, use the 2D Rectangular Pattern or 2D Circular Pattern tool to create the profile. The following image shows a part with multiple intersecting lines defined in a single profile, that same profile used to create a rib network, and again as a web network.

Rib Networks as a Profile   Rib Networks as Ribs   Rib Network as Webs

Figure 8-9

EXERCISE 8-1 Creating Ribs and Webs

In this exercise, you sketch an open profile and then use the Rib tool to create a rib. Then you edit the rib feature to change the rib to a web. You complete the exercise by sketching overlapping lines and creating a rib network.

1. Open *ESS_E08_01.ipt*.

2. Use the Zoom and Rotate tools to examine the entire part.
3. Make the work plane the active sketch.
4. Click the Look At tool and then click the work plane.
5. Turn off the visibility of the work plane.
6. Use the Line and Three point arc tool to sketch the line and arc as shown in the following image. Make sure that the arc you sketch, if extended to the right, would intersect the top of the part. Dimensions could be added as required.

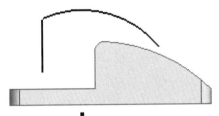

Figure 8-10

7. Click the Return tool to exit the sketch mode.
8. Change to the isometric view as shown in the following image.

Figure 8-11

9. Click the Rib tool. The Direction button is active. Move the cursor down and click a point to get the preview as shown below.

Figure 8-12

10. In the Rib dialog box Enter **8 mm** in the Thickness edit box and then click OK. When complete the rib should resemble the following image.

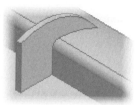

Figure 8-13

11. In the Browser, right-click the Rib feature then click Edit Feature.

12. In the Rib dialog box click the Finite button, enter **25 mm** in the Extents edit box and then click OK as shown in the following image.

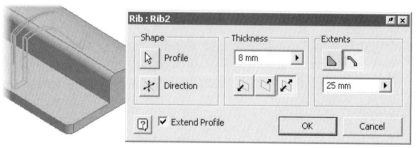

Figure 8-14

13. When complete the rib should resemble the following image.

Figure 8-15

14. Use the Rotate tool to view the bottom of the part.

15. Make the inside flat face the active sketch as shown in the following image.

Figure 8-16

16. Sketch three separate, overlapping line segmets as shown in the following image. The line should be parallel and perpendicular to the existing lines and each other. Dimensions could be added as required.

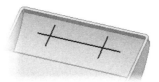

Figure 8-17

17. Click the Return tool.
18. Click the Rib tool.
19. Select the three line segments for the profile.
20. In the Rib dialog box:
    - verify that the Thickness edit box still shows a value of **8 mm**.
    - Click the Finite button.
    - Enter **6 mm** in the Extents edit box.
    - Enter **6 mm** in the Extents edit box.
21. Click the Direction button and move the cursor around your profile until the preview shows the rib direction going down into the pocket as shown in the following image, then click to accept that direction.

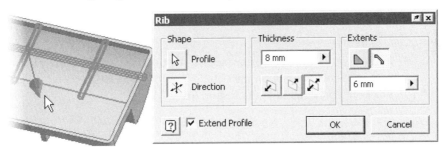

Figure 8-18

22. Click OK in the Rib dialog box and your rib network should resemble the following image.

Figure 8-19

23. Use the Zoom and Rotate tools to examine the rib network and the rest of your part.
24. Close the file and do NOT save changes.

EMBOSS TEXT AND PROFILES

To better define a part, you may need to have a shape or text either embossed (raised) or engraved (cut) into a model. In this section you will learn how you can emboss or engrave a closed shape or text onto a planar or curved face. You can define a shape using the sketch tools on the 2D Sketch Panel Bar. The shape needs to be closed. There are two steps to embossing or engraving: first, you create the shape or text and, second, you emboss the shape or text onto the part. The following sections describe the steps.

STEP 1 - CREATING TEXT

To place text onto a sketch, follow these steps:

1. Make a sketch active.
2. Click the Create Text tool on the 2D Sketch Panel Bar, as shown in the following image.

Figure 8-20

3. Drag a rectangle where you will place the text. The Format Text dialog box will appear.
4. In the Format Text dialog box, specify the text font and format style and enter the text to place on the sketch. The following image shows the dialog box with text entered in the bottom pane.

Figure 8-21

5. When you are done entering text, click the OK button.

6. When the text is placed, a rectangular set of construction lines defines the perimeter of the text. You can add dimensions or constraints to these construction lines to refine the text's location and orientation, as shown in the following image.

7. To edit the text, move the cursor over the text, right-click, and select Edit Text from the menu. The same Format Text dialog box will appear that was used to create the text.

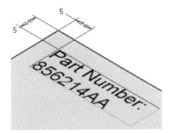

Figure 8-22

STEP 2 - EMBOSSING TEXT

To emboss a closed profile or text, click the Emboss tool on the Part Features Panel Bar, as shown on the left in the following image. The Emboss dialog box will appear as shown on the right. The Emboss dialog has the following options:

 Profile   Select a profile (closed shape or text) to emboss. You may need to use the Select Other tool to select the text.

	Depth	Enter an offset depth to emboss or engrave the profile if you selected the Emboss from Face or Engrave from Face types.
	Top Face Color	Select a color from the drop-down list to define the color of the top face of the embossed area (not its lateral sides).
	Emboss from Face	Select this option to add material to the part.
	Engrave from Face	Select this option to remove material from the part.
	Emboss/ Engrave from Plane	Select this option to add and remove material from the part by extruding both directions from the sketch plane. Direction changes at the tangent point of the profile to a curved face.
	Flip Direction	Select either of these buttons to define the direction of the feature.
	Wrap to Face	Check this box for Emboss from Face or Engrave from Face types to wrap the profile onto a curved face. Only a single non-seamed face can be selected. The profile will be slightly distorted as it is projected onto the face. The wrap stops when a perpendicular face is encountered.
	Face	Click this button after checking the Wrap option, and then select the face around which the profile will be wrapped.

Figure 8-23

To emboss a closed shape or text, follow these steps:

1. Click the Emboss tool on the Part Features Panel Bar.

2. Define the profile by selecting a closed shape or text. If needed, use the Select Other tool.

3. Select the type of emboss: Emboss from Face, Engrave from Face, or Emboss/ Engrave from Plane.

4. Specify the depth, color, direction, and face as needed. If you selected Emboss/Engrave from plane option, you can also add a taper angle to the created emboss/engrave feature.

 NOTE You can edit the embossed feature like any other feature.

EXERCISE 8-2 Creating Text and Emboss Features

In this exercise, you emboss and engrave sketched profile objects on faces of a razor handle model. You then create a sketch text object and engrave it on the handle.

1. Open *ESS_E08_02.ipt*.

2. Use the Rotate and Zoom tools to examine the part, then turn on visibility of Sketch11 and rotate the view to display the top triangular face as shown in the following image.

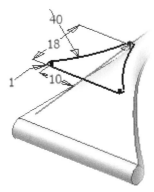

Figure 8-24

3. Click the Emboss tool to engrave a sketched profile. Notice that the only available profile is automatically selected.

4. In the Emboss dialog box, click the Engrave from Face option and change the Depth to **0.9 mm**.

5. - Click the Top Face Color button and choose Aluminum (Flat) from the Color dialog box drop down list.

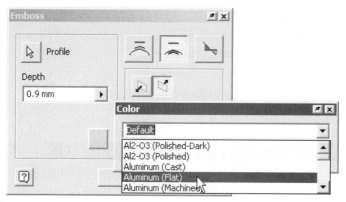

Figure 8-25

6. Click OK twice to close both dialog boxes and create the engraved feature.

Figure 8-26

7. Change to an Isometric View and turn on the visibility of Sketch10.

8. Click the Emboss tool.

9. For the profile, click the oblong and the six closed profiles. Be sure to select the top and bottom half of the first two herringbone profiles (on the left).

Figure 8-27

10. In the Emboss dialog box, verify that the Emboss from Face option is selected, and change the Depth to **0.2 mm**.

11. Click the Top Face Color button and choose Black from the Color dialog box drop down list as shown in the following image.

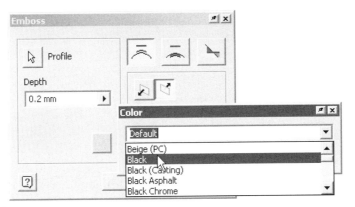

Figure 8-28

12. Click OK twice to close both dialog boxes and create the embossed feature.

13. Use the Rotate tool to examine the engraved and embossed features.

14. Next you create a text sketch object. Turn on the visibility of Sketch9.

15. Double-click Sketch9 to edit it.

16. Use the Look At tool to set up a view normal to the sketch.

17. Click the Text tool.

18. Click in an open area below the part and under the left edge of the construction rectangle to specify the insertion point and display the Format Text dialog box.

19. Click the Center and Middle Justification buttons and click the Italic option. Change the Stretch value to **120**.

20. Click in the text field and type **The SHARP EDGE.**

21. In the text field, double-click on the word 'The' and change the text size to **2.5 mm** as shown in the following image.

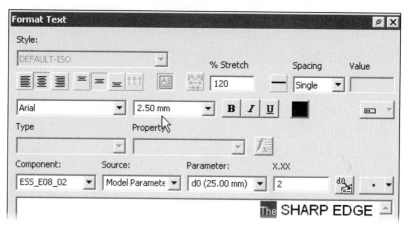

Figure 8-29

22. Click OK to place the text.
23. Draw a diagonal construction line coincident with the text bounding box corners as shown in the following image.

Figure 8-30

24. Place a coincident constraint between the midpoint of both diagonal construction lines as shown in the following image.

Figure 8-31

25. Click the Return button to finish editing the sketch and then change to the Isometric View.
26. Click the Emboss tool.
    - For the profile, select the text object.
    - In the Emboss dialog box, click the Engrave from Face option.
    - Change the Depth to **0.5 mm**.
    - For the direction click the right button.
    - Click the Top Face Color button and click Black from the Color drop down list.

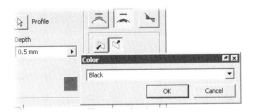

Figure 8-32

27. Click OK twice to close both dialog boxes and create the engraved feature.

28. Use the Rotate tool to examine the engraved and embossed features.

Figure 8-33

29. Close the file and do NOT save changes.

SWEEP FEATURES

A sweep feature is unlike other sketched features. A sweep feature requires two unconsumed sketches: a profile to be swept and a path that the profile will follow. The two sketches cannot lie on the same plane and cannot be parallel. The path can be an irregular shape or be based on a part edge by projecting the edges onto the active sketch. The path can be either an open or closed profile and can lie in a plane or lie in multiple planes (3D Sketch). Handles, cabling, and piping are examples of sweep features. A sweep feature can be a base or a secondary feature. To create a sweep feature, use the Sweep tool on the Part Features Panel Bar, as shown on the left in the following image. The Sweep dialog will appear as shown on the right. The following list includes descriptions of the options in the Sweep dialog box.

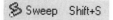
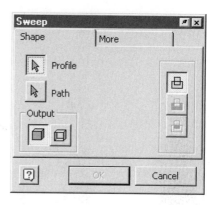

Figure 8-34

SHAPE

Profile Click to choose the sketch profile to sweep. If the Profile button is depressed, it means that you need to select a sketch or sketch area. If there are multiple closed profiles, you will need to select the profile that you want to sweep. If there is only one possible profile, Autodesk Inventor will select it for you and you can skip this step. If you selected the wrong profile or sketch area, reclick the Profile button and deselect the incorrect sketch by clicking it while holding down the CTRL key, then select the desired sketch profile.

Path Click this button to choose the path along which to sweep the profile. The path can be either an open or closed profile.

Operation buttons This is the column of buttons along the right side of the dialog box. By default, the Join operation is selected. Use the operations buttons to add or remove material from the part (using the Join or Cut options) or to keep what is common between the existing part and the completed sweep (using the Intersect option).

    Join–Adds material to the part.
    Cut–Removes material from the part.
    Intersect–Creates a new feature from the shared volume of the sweep feature and existing part volume. Material not included in the shared volume is deleted.

Output Buttons In the Output section, click Solid to create a solid feature or Surface to create the feature as a surface.

MORE

On the More tab, enter a value for the angle from which you want the profile to be drafted. By default, the taper angle is 0°, as shown in the following image.

Figure 8-35

In the next two sections, you will learn how to create sweep features using both 2D and 3D paths.

CREATING A SWEEP FEATURE

In this section you will learn how to create sweep features. You first need to have two unconsumed sketches. One sketch will be swept along the second sketch that represents the path. To create a sweep feature, follow these steps:

1. Create two unconsumed sketches: one for the profile and the other for the path. The profile and path must lie on separate nonparallel planes. It is recommended that the profile intersect the path. Use work planes to place the location of the sketches, if required. The sketch that you use for the path can be open or closed. Add dimensions and constraints to both sketches as needed.

2. Click the Sweep tool on the Part Features Panel Bar.

3. The Sweep dialog box will appear. If two unconsumed sketches do not exist, Autodesk Inventor will notify you that two unconsumed sketches are required.

4. Click the Profile button and then select the sketch that will be swept in the graphics window. If only one closed profile exists, this step is automated for you.

5. If it is not depressed already, click the Path button and then select the sketch to be used as the path in the graphics window.

6. Determine if the resulting sweep will define a solid or a surface by clicking either the Solid or Surface button in the Output area.

7. If this is a secondary feature, click the operation that will specify whether material will be added, removed, or if what is common between the existing part and the new sweep feature will be kept.

8. If you want the sweep feature to have a taper, click on the More tab and enter a value for the taper angle.

9. Click the OK button to complete the operation.

The image shows two sketches that have dimensions applied to them. The path is lying flat and the profile is drawn on a work plane that is normal to the path. The image on the right shows the completed sweep.

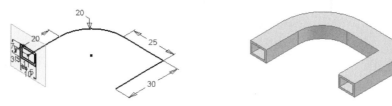

Figure 8-36

3D SKETCHING

To create a sweep feature whose path does not lie on a single plane, you need to create a 3D sketch that will be used for the path. You can use a 3D sketch to define the path for a lip or to define the routing path for an assembly component, such as a pipe or duct work that crosses multiple faces on different planes. You need to define a 3D sketch in the part environment, and you can do this within an assembly or in its own part file. You can use the Autodesk Inventor adaptive technology during 3D sketch creation to create a path that updates automatically to reflect changes to referenced assembly components. In this section you will learn strategies on how to create 3D sketches.

3D Sketch Overview

When creating a 3D sketch, you use many of the same sketching techniques that you have already learned, with the addition of a few tools. 3D sketches use work points and model edges/vertices to define the shape of the 3D sketch by creating line or spline segments between them. You can also create bends between line segments. When creating a 3D sketch, you use a combination of lines, splines, fillet features, work features, constraints, and existing edges and vertices.

3D Sketch Environment

The 3D sketch environment is used to create 3D curves with 3D or a combination of both 2D and 3D curves. Before creating a 3D sketch, change the environment to the 3D sketch environment by clicking the 3D Sketch tool on the standard toolbar under the 2D Sketch tool, as shown on the left in the following image. The Panel Bar's tools will then change to the 3D Sketch tools, as shown in the center of the figure. The tools that are unique to the 3D Sketch Panel Bar are explained throughout this section. While in the 3D Sketch environment, all features appear in the Browser with a 3D sketch name. The image on the right in the figure shows an example of a 3D Sketch in the Browser. Once a feature uses the 3D sketch, it will be consumed under the new feature in the Browser.

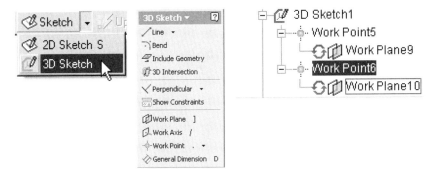

Figure 8-37

3D Path From Existing Geometry

The easiest way to create a 3D path is to use existing geometry. If you wanted to create a lip on an existing part, for example, you would use the existing edges to define the path. You can include existing geometry by projecting part edges, vertices, and geometry from visible sketches into a 3D sketch. To use existing geometry to create a 3D path, follow these steps:

1. Activate the 3D Sketch environment by clicking the 3D Sketch tool on the standard toolbar under the 2D Sketch tool.

2. Click the Include Geometry tool on the 3D Sketch Panel Bar, as shown in the following image. This tool will include existing sketch geometry and existing part edges into a 3D sketch. The projected geometry is updated to reflect changes to the original geometry.

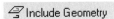

Figure 8-38

3. Click each of the model edges that you want to use for the 3D sketch. When finished, right-click and select Done from the menu, or press the ESC key on the keyboard. If you click an incorrect edge, you can delete it from the sketch manually. The following image shows an example of including outside edges of a part in a 3D sketch.

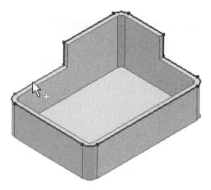

Figure 8-39

4. If a plane does not exist where you want to place the profile to be swept, create a work plane that defines the plane.

5. Make the work plane, or an existing plane, the active sketch.

 NOTE Use the 2D Sketch tool to make the plane the active sketch.

6. Sketch, constrain, and dimension the profile that will be swept. The following image shows a sketch created, constrained, and dimensioned on the work plane.

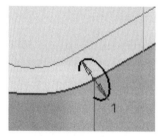

Figure 8-40

7. Click the Sweep tool on the Part Features Panel Bar.

8. If the Profile button is not depressed already in the Sweep dialog box, click it and then select the sketch that will be swept in the graphics window.

9. Click the Path button and then select the 3D sketch to use as the path in the graphics window.

10. Select the operation that will specify whether material will be added, removed, or if what is common between the existing part and the new sweep feature will be kept.

11. If you want the sweep feature to have a taper, click on the More tab and enter a value for the taper.

12. Click the OK button to complete the operation. The following image shows the completed part.

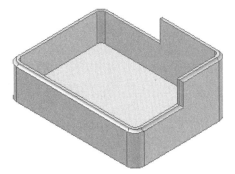

Figure 8-41

3D Sketch From Intersection Geometry

Another option to help create a 3D path is to use geometry that intersects with the part. If the intersecting geometry defines the 3D path, you can use it. The intersection can be defined by a combination of the following: a planar or nonplanar part face, a surface face or quilt, or a work plane. To create a 3D path from an intersection, follow these steps:

1. Create the intersecting features that describe the desired path.

2. Change to the 3D Sketch environment by clicking the 3D Sketch tool on the standard toolbar under the 2D Sketch tool.

3. Click the 3D Intersection tool, shown in the following image, on the 3D Sketch Panel Bar.

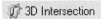

Figure 8-42

4. The 3D Intersection Curve dialog box appears, as shown on the right in the following image.

5. Select the two intersecting features.

6. To complete the operation, click OK.

The left side of the following image shows a 3D path being created on a cylindrical face at the edge defined by a work plane that intersects it at an angle.

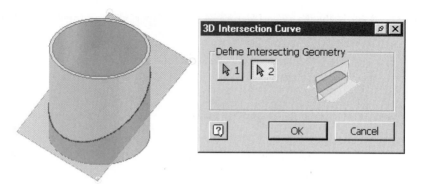

Figure 8-43

The following image shows a cylinder with a surface that was extruded to the outside face of the cylinder. A 3D sketch was then created using the 3D Intersection tool.

Figure 8-44

Constructed Paths

Another option for creating 3D paths is to define a path by creating work points at locations where the 3D path will intersect, and then connecting the points using a 3D line or spline. Because this method of constructing a 3D path depends on work points, you should review the Creating Work Points section in Chapter 4 to become comfortable with the creation and editing of work points. The following is a brief review of the methods used to create work points:

- Click an endpoint or midpoint of an edge, then click an edge or axis, and a work point is created at the intersection (or theoretical intersection) of the two.

- Click an edge and plane, and a work point is created at the intersection (or theoretical intersection) of the two.

- Click three nonparallel faces or planes, and a work point is created at their intersection (or theoretical intersection).

- A work point is also generated automatically if you select a vertex while creating a 3D line.

You can also use grounded work points, but they are not associated dynamically with the part or any other work features, including the original locating geometry. When you modify surrounding geometry, the grounded work point remains in the specified location.

To create a 3D path using work points and a 3D line, follow these steps:

1. Create work points and grounded work points, as needed, to define the 3D path.

2. Change to the 3D Sketch environment by clicking the 3D Sketch tool on the standard toolbar under the 2D Sketch tool.

3. If you want the 3D path to place a bend automatically, you can set the size of the bend. Click Tools > Document Settings, and change the 3D Sketch Auto-Bend Radius setting on the Sketch tab, as shown in the following image.

Figure 8-45

4. Click the 3D Line tool on the 3D Sketch Panel Bar.

5. Select the work points in the order that the path will follow. By default, a bend is applied automatically between 3D line segments. The automatic bend option can be toggled on and off by right-clicking while in the 3D line command and selecting or deselecting Auto-Bend on the menu, as shown in the following image.

Figure 8-46

6. To add a bend between two 3D lines manually, use the Bend tool, shown in the following image, on the 3D Sketch Panel Bar. In the 3D Sketch Bend dialog box, enter a value for the bend and then select two 3D lines. Once the bend is placed, you can edit it by double-clicking on the dimension and entering a new value.

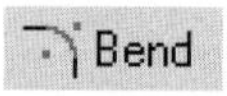

Figure 8-47

7. When you are done selecting points, right-click and select Done from the menu.

8. Next, create a new sketch that defines the desired profile that will be swept along the 3D path.

9. Click the Sweep tool and create the 3D sweep by selecting the profile and path sketches.

The following image shows a part with construction work axes, work planes, and work points; the constructed 3D path with bends; and the completed swept part.

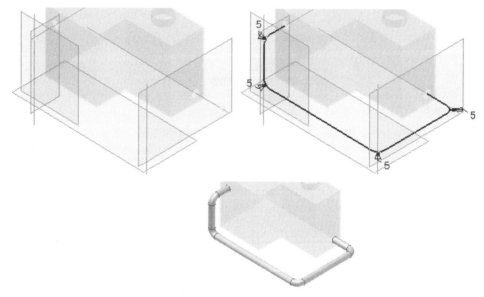

Figure 8-48

Splines

You can also create a spline between work points or vertices on a part.

To create a 3D path using work points and a 3D line, follow these steps:

1. Create work points and/or grounded work points, as needed, to define the 3D path.

2. Change to the 3D Sketch environment by clicking the 3D Sketch tool on the standard toolbar under the 2D Sketch tool.

3. Click the Spline tool on the 3D Sketch Panel Bar as shown in the following image.

Chapter 8 Advanced Part Modeling Techniques 573

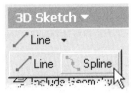

Figure 8-49

    4. Select the work points in the order that the path will follow, as shown in the following image.

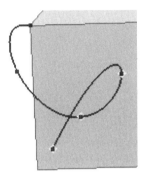

Figure 8-50

    5. To exit the spline command, right-click and select Continue from the menu, and then either press the ESC key or right-click and select Done from the menu.

    6. If desired, you can add tangent constraints between the spline and a part edge by clicking the Tangent constraint tool on the 3D Sketch Panel Bar. The following image shows a tangent constraint being added between the spline and the top edge of the part.

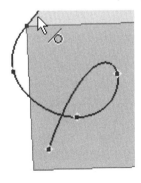
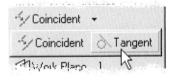

Figure 8-51

Coincident Constraints

When working with 3D lines, there are two types of 3D coincident constraints that hold the endpoints of the lines and work points together. Line segments created with the 3D Line tool inside a 3D sketch are the only geometry that work with 3D coincident constraints. Each 3D coincident constraint appears in the sketch as individual icons. The following is a definition of the two coincident constraints:

A constraint symbol (shown in the following image) near the endpoint of a line indicates a 3D coincident constraint between a line's endpoints and the underlying work point.

Figure 8-52

The following rules apply to this 3D coincident constraint:

1. Deleting the symbol removes the connection between one or more line endpoints and the underlying work point.
2. Deleting the symbol does not delete the constraint between shared line endpoints.
3. Lines joined by a bend have a separate (single) endpoint coincident constraint to the work point.
4. All other line endpoints coincident to the work point share a common coincident constraint.
5. When two lines are connected by a bend, deleting their endpoint coincident constraint does not destroy the bend. You can reattach the common endpoint by adding a new coincident constraint.

- The 3D coincident constraint symbol (shown in the following image) near a line's midpoint indicates the constraint between shared line points.

Figure 8-53

The following rules apply to this 3D coincident constraint:

1. The symbol will only appear if an endpoint is shared by more than one line.
2. Deleting a symbol breaks the sharing of the endpoint.
3. If a bend exists between two line segments, deleting the symbol destroys the bend. The two line endpoints are located at the work point but have different properties.

4. The endpoint of the line with the deleted symbol is not attached to the work point.

5. The endpoint of the other line is constrained to the end point.

6. You can reattach the free lines endpoint by adding a new coincident constraint.

The following image illustrates an example of the 3D constraints shown on 3D lines.

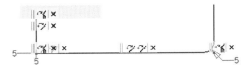

Figure 8-54

To add a 3D coincident constraint, click the Coincident tool on the 3D Sketch Panel Bar, as shown in the following image, and follow these guidelines:

Figure 8-55

- You can only add coincident constraints between an unattached line endpoint and an existing work point or vertex.

- You must first delete an existing endpoint or midline (midpoint) constraint symbol to detach a line's endpoint from an existing work point.

- You must select the line endpoint as the first selection in the coincident constraint.

- A new work point is created if the second selection is a part vertex.

EXERCISE 8-3 Creating Sweep Features

In this exercise, you construct a 3D path entirely from existing part edges and use the Sweep tool to create a lip.

1. Open *ESS_E08_03.ipt*.

2. Use the Rotate tool to rotate the part into the position shown in the following image.

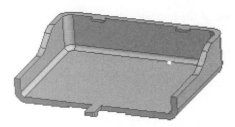

Figure 8-56

3. From the Tools menu click Application Options. In the Options dialog box:
   - Click the Sketch tab.
   - Uncheck the Autoproject edges for sketch creation and edit check box.
   - Select the OK button.

4. Right-click in the graphics window then click New 3D Sketch. You can also click the arrow next to 2D Sketch, then click 3D Sketch to enter 3D Sketch mode.

5. Next you use the Include Geometry tool to select existing part edges and define the 3D path.

6. Click the Include Geometry tool.

7. Click the 13 edges that define the top outside edges as shown in the following image (shown in wireframe display).

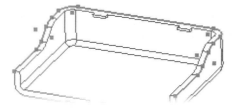

Figure 8-57

8. Right-click then click Done.
9. Click the Return tool.
10. Zoom in on the part as shown below in the following image.
11. Create a work plane. To define the work plane that is normal to the start of the 3D sketch, click the Work Plane tool.
    - Click the end point of the edge
    - Click the edge as shown in the following image.

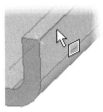

Figure 8-58

12. Right-click in the graphics window then click New Sketch.

13. Click the work plane.

14. You start creating the 2D sketch by projecting the start point of the 3D path to the sketch plane. Click the Project Geometry tool.
    - Click the start point of the 3D sketch.
    - Right-click then click Done. This point is used to orient the rectangular sketch that you create in the next step.

Figure 8-59

15. Click the Two Point Rectangle tool and create a two point rectangle as shown in the following image, start at the projected point.

Figure 8-60

16. Add horizontal and vertical dimension as desired.

17. Click the Return tool.

18. Now you sweep the 2D profile along the 3D path using the Sweep tool. Click the Sweep tool.

19. The rectangular profile is selected for you and the Path button is active, click the 3D sketch geometry.

20. Click the OK button in the Sweep dialog box.

21. Turn off the visibility of the work plane
22. Use the Zoom and Rotate tools to examine the Sweep feature.

Figure 8-61

23. From the Tools menu click Application Options. In the Options dialog box:
    - Click the Sketch tab.
    - Check the Autoproject edges for sketch creation and edit check box.
    - Click the OK button.

24. Close the file and do NOT save changes.

COIL FEATURES

Using the Coil feature, you can easily create many types of helical or coil geometry. You can create many types of springs by selecting different settings in the Coil dialog box. You can also use the Coil feature to remove or add a helical shape around the outside of a cylindrical part to represent a thread profile.

To create a coil you need to have at least one unconsumed or shared sketch available in the part. This sketch describes the profile (or shape) of the coil feature and can also describe the coil's axis of revolution. If no unconsumed sketch is available, Autodesk Inventor will prompt you with an error message that states there are "No unconsumed visible sketches on the part." After an unconsumed sketch is available, you can click the Coil tool from the Part Features Panel Bar, as shown in the following image. The Coil dialog box appears. The following sections explain the tabs.

Figure 8-62

COIL SHAPE TAB

The Coil Shape tab, as shown in the following image, allows you to specify the geometry and orientation of the coil.

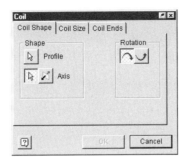

Figure 8-63

Profile Click to pick the sketch you will use as the profile shape of the coil feature. By default, the Profile button is shown depressed; this tells you that you need to select a sketch or sketch area. If there are multiple closed profiles, you will need to select the profile that you want to revolve. If there is only one possible profile, Autodesk Inventor will select it for you and you can skip this step. If you select the wrong profile or sketch area, reclick the Profile button and select a new profile or sketch area. You can only use one closed profile to create the coil feature.

Axis Click to pick the sketched line or centerline, a projected straight edge, or an axis about which to revolve the profile sketch. If selecting an edge or sketched centerline, it must be part of the sketch. If selecting a work axis, it cannot intersect the profile.

Flip Click to change the direction in which the coil will be created along the axis (see the following image). The direction will be changed on either the positive or negative X- or Y-axis, depending upon the edge or axis that you selected. You will see a preview of the direction in which the coil will be created.

Figure 8-64

Rotation Click to specify the direction in which the coil will rotate. You can choose to have the coil rotate in either a clockwise or counterclockwise direction. The operation buttons are the column of buttons along the center of the dialog box that will appear if a base feature exists, as shown in the following image. The operation buttons are only available if the Coil feature is not the first feature in the part. By default, the Join operation is selected. You can select the other operations to either add or remove material from the part using the Join or Cut options, or keep what is common between the existing part volume and the completed coil feature using the Intersect option.

Figure 8-65

Join– Adds material to the part.

Cut– Removes material from the part.

Intersect– Keeps what is common to the part and the sweep feature.

COIL SIZE TAB

The Coil Size tab, as shown in the following image, allows you to specify how the coil will be created. You are presented with various options for the type of coil that you want to create. Based on the type of coil that you select, the other parameters for Pitch, Height, Revolution, and Taper will become active or inactive. Specify two of the three parameters that are available and Autodesk Inventor will calculate the last field for you.

Figure 8-66

Type   Select the parameters that you want to specify: Pitch and Revolution, Revolution and Height, Pitch and Height, or Spiral. If you select Spiral as the Coil Type, only the Pitch and Revolution values are required.

Pitch   Type in the value for the height to which you want the helix to elevate with each revolution.

Revolution   Specify the number of revolutions for the coil. A coil cannot have zero revolutions. Fractions can be used in this field. For example, you can create a coil that contains 2.5 turns. If end conditions are specified (see the Coil Ends tab section), the end conditions are included in the number of revolutions.

Height   Specify the height of the coil. This is the total coil height as measured from the center of the profile at the start to the center of the profile at the end.

Taper   Type an angle at which you want the coil to be tapered along its axis.

NOTE A Spiral coil type cannot be tapered.

COIL ENDS TAB

The Coil Ends tab, as shown in the following image, lets you specify the end conditions for the start and end of the coil. When selecting the Flat option, the helix—not the profile that you selected for the coil—is flattened. The ends of a coil feature can have unique end conditions that are not consistent between the start and end of the coil.

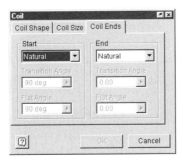

Figure 8-67

Start Select either Natural or Flat for the start of the helix. Click the down arrow to change between the two options.

End Select either Natural or Flat for the end of the helix. Click the down arrow to change between the two options.

Transition Angle This is the rotational angle (specified in degrees) in which the coil achieves the coil start or end transition. Normally it occurs in less than one revolution.

Flat Angle This is the rotational angle (specified in degrees) that describes the amount of flat coil that extends after the transition. It specifies the transition from the end of the revolved profile into a flattened end.

The following image shows a coil created as the base feature. The image on the left shows the coil in its sketch stage: the rectangle will be used as the profile and the centerline will be used as the axis of rotation. The finished part on the right shows the coil with flat ends.

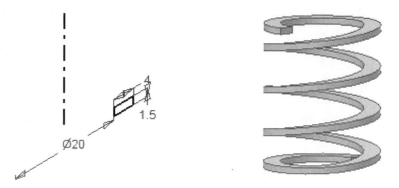

Figure 8-68

The following image shows a coil created as a secondary feature. The image on the left shows the coil in its sketch stage. The sketch is drawn on a work plane that is located on the axis of the cylinder. The rectangle will be used as the profile, and the work axis will be used as the axis of rotation. The finished part on the right shows the coil with flat ends.

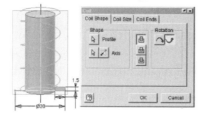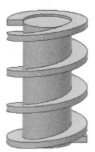

Figure 8-69

LOFT FEATURES

The Loft tool creates a feature that blends a shape between two or more planar sections (profiles) that describe different shapes. Loft features are used frequently in the creation of plastic or molded parts. Many of these types of parts have complex shapes that would be difficult to create using standard modeling techniques. You can create loft features that blend between two or more cross-section profiles that reside on different planes. You can use a rail, or multiple rails, to define a path(s) that the loft will follow. There is no limit to the number of sections or rails that you can include in a loft feature. Three types of geometry are used to create a loft: sections, rails, and points. The following sections describe these types of geometry.

Sections Define the shape(s) between which the loft will blend. The following rules apply to sections:

1. There is no limit to the number of sections that you can include in the loft feature.

2. Sections do not have to be sketched on parallel planes.

3. You can define sections by 2D sketches (planar), 3D sketches (nonplanar), or existing faces or edges on a part.

4. All sections must be either open or closed. You cannot mix open and closed profile types within the same loft operation.

Rails Define rails by the following: 2D sketches (planar), 3D sketches (nonplanar), or part edges. The following rules apply to rails:

1. There is no limit to the number of rails that you can create or include in the loft feature.

2. Rails must not cross each other and must not cross mapping curves.

3. Points defined by rail curves and mapping sets cannot be coincident on internal sections and rails.

4. Rails affect all of the sections, not just faces or sections they intersect. Section vertices without defined rails are influenced by neighboring rails.

5. All rail curves must be open or closed.

6. Closed rail curves define a closed loft, meaning that the first section is also the last section.

7. No two rails can have identical guide points, even though the curves themselves may be different.

8. Rails can extend beyond the first and last sections. Any part of a rail that comes before the first section or after the last is ignored.

9. If a rail is not an edge of the final body, the surface will be smooth across the rail.

10. You can apply a 2D or 3D sketch tangency constraint between the rail and the existing geometry on the model.

Points You can map points to help define how the sections will blend to each other. The following rules apply to mapping points:

1. A set of mapping points constitutes a mapping curve.

2. A mapping point can lie anywhere on an edge.

3. The points in a set must either follow the order of the sections (first point on first section, etc.) or follow the reverse of this order (first point on last section, etc.).

4. Mapping curves must not cross other mapping curves or rail curves.

5. Mapping curves can share their first and/or last points.

To create a loft feature, follow these steps:

1. Create the profiles that will be used as the sections of the loft. Use work features, sketches, or projected geometry to position the profiles, if required.

2. Click the Loft tool on the Part Features panel bar, as shown in the following image.

Figure 8-70

3. On the Curves tab of the Loft dialog box, the Sections option will be the default option. In the graphics window, click the sketches and/or part faces in the order in which the loft sections will blend.

4. If rails are to be used in the loft, click Click to add in the Rails section of the Curves tab and then click the rail or rails.

5. If needed, change the options for the loft on the Conditions and Transition tabs.

The following sections explain the options in the Loft dialog box and on its tabs.

CURVES TAB

The Curves tab, shown in the following image, allows you to select which sketches or part faces will be used as sections and whether or not a rail will be used, and to determine the output condition.

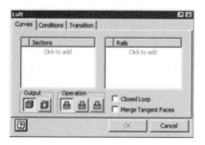 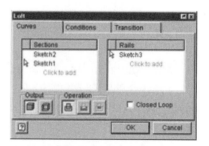

Figure 8-71

Sections   Select two or more 2D or 3D sketches or part faces that make up the loft feature. The sketches and part faces can be on any plane or face.

Rails   Select a sketch or sketches to be used as rails.

Output   Select whether the loft feature will be a solid or a surface. Open sketch profiles selected as loft sections will define a lofted surface.

Operation   Use the operations buttons to add or remove material from the part (using the Join or Cut options) or to keep what is common between the existing part and the completed loft using the Intersect option. By default, the Join operation is selected.

Closed Loop   Click to join the first and last sections of the loft feature to create a closed loop.

Merge Tangent Faces   Click to join the sections together with no edge between the tangent faces.

CONDITIONS TAB

The Conditions tab, shown in the following image, allows you to control the boundary, angle, and weight condition of the loft feature.

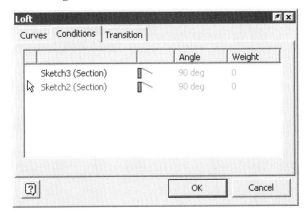

Figure 8-72

Conditions   The column on the left lists the sketches that are specified for the sections. To change a sketch's condition, click on its name and then select a condition option.

Condition Boundaries   There are three boundary condition buttons available for each section, as shown in the following image.

Figure 8-73

   Free(top button)–With this option there is no boundary condition, and the loft will blend between the sections in the most direct fashion.

Direction (middle button)–This option is only available when the curve is a 2D sketch. When selected, you can specify the angle at which the loft will intersect or transition from the section.

Tangent(bottom button)–When selected, the loft will be tangent to the adjacent section or face.

Angle   This option is only enabled for a section when the boundary condition is Tangent or Direction. Default is set to 90° and is measured relative to the profile plane. The option sets the value for an angle formed between the plane that the profile is on and the direction to the next cross-section of the loft feature. Valid entries range from 0.0000001° to 179.99999°.

Weight   Default is set to 0. The weight value controls how much the angle influences the tangency of the loft shape to the normal of the starting and ending profile. A small value will create an abrupt transition, and a large value creates a more gradual transition. High weight values could result in twisting the loft, and may cause a self-intersecting shape.

TRANSITION TAB

The Transition tab, shown in the following image, allows you to specify point sets. A point set is used to define section point relationships and control how segments blend from one section to the segments of the adjacent sections. Points are reoriented or added on two adjacent sections.

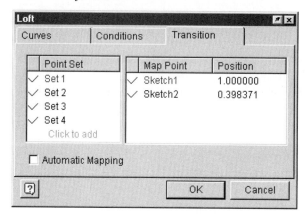

Figure 8-74

Point Set   The name of the point set appears here.

Map Point   The corresponding sketch for the selected point set appears here.

Position   The location of the selected map point appears here. You can modify the position by either entering a new value or dragging the point to a new location in the graphics window.

To modify the default point sets, or to add a point set, follow these steps:

1. Click on the Transition tab and clear the Automatic Mapping box. The dialog box will populate the automatic point data for each section. The list is sorted in the order in which the sections were specified on the Curve tab.

2. To modify a point's position, first select the point set in which the point is specified. When you select its name, it will become highlighted in the graphics window.

3. Click in the Position section of the map point that you want to modify, and either enter a new value or drag the point to a new location in the graphics window.

4. To add a point set, click Click to add in the point set area.

5. Click a point on the profile of two adjacent sections. As you move the cursor over a valid region of the active section, a green point appears. As the points are placed, they are previewed in the graphics window, as shown in the following image.

6. You can modify the new point set in a similar way to the default point set.

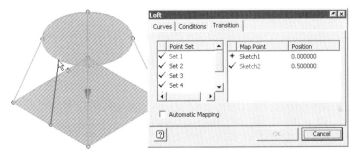

Figure 8-75

The following image shows a loft created from two sections and one rail. The image on the left shows two sections and a rail in their sketch stages shown in the top view. The image on the right shows the completed loft.

Figure 8-76

EXERCISE 8-4 Creating Loft Features

In this exercise, you use loft options and controls to define the shape of a razor handle.

1. Open *ESS_E08_04.ipt*.
2. In the Browser, double-click Sketch1 to edit the sketch, then use the Look At tool to set up a view normal to the sketch.
3. Click the Point, Hole Center tool.
4. Create a point coincident with the spline near the bottom of the curve (make sure the coincident glyph is displayed as you place the point).

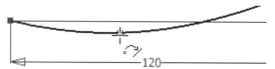

Figure 8-77

5. Draw construction lines coincident with the sketched point and the spline points on both sides of the Point that you just created.

Figure 8-78

6. Place an equal constraint between the construction lines to parametrically position the sketched point midway between the spline points.
7. Change the 120 mm dimension to **130 mm** and verify that the sketched point moves along the spline to maintain its position midway between the spline points.

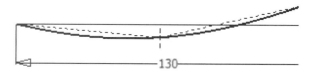

Figure 8-79

8. Click Return to finish editing the sketch.
9. Change to the Isometric view.
10. Click the Work Plane tool.
    - Create a work plane perpendicular to the spline click the sketched point you just created, then click the spline (at any point that is not close to the construction lines).

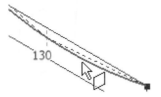

Figure 8-80

11. Next, you create a loft profile section on the new work plane. Create a new sketch on the new work plane.

12. Project the sketched Point onto the sketch.

13. Draw an ellipse with its center coincident with the projected point and the second point horizontally constrained to the Point and click a third point as shown in the following image.

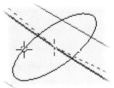

Figure 8-81

14. Place a **4 mm** and **9 mm** dimension to control the ellipse size as shown in the following image.

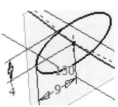

Figure 8-82

15. Click Return to finish the sketch.

16. Click the Work Plane tool.
    - Click the spline's right endpoint, then click the spline to create the work plane perpendicular to the spline curve.

Figure 8-83

17. Create a new sketch on the new work plane.
18. Project the spline end point onto the sketch.
19. Draw a circle coincident with the projected point.
20. Create a **7 mm** diameter dimension to control the diameter.

Figure 8-84

21. Click Return to finish the sketch.
22. Click the Loft tool.
    - For the first section, click the concave 3D face shown in the following image.

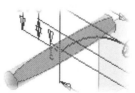

Figure 8-85

23. Click the other four profile sections in order (left to right) and click OK when the preview looks like the following image.

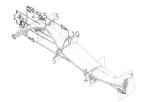

Figure 8-86

24. In the Browser, turn off the visibility of Sketch1 and all four work planes, then use the Rotate and other view tools to examine the loft.

Figure 8-87

25. Optional: Add fillets on the circular edge of the flat end and further refine the loft shape by adding new sections, adding rail curves, or inserting additional spline control points.

26. Close the file and do NOT save changes.

PART SPLIT AND FACE SPLIT

The Split tool allows you to split a part by removing one portion of the part, or to split individual or all faces. A typical application of this tool is to allow the application of face drafts to the split faces of a part. You can use the Split tool to:

- Split an entire part by sketching a parting line, or placing a work plane and then using it to cut material from the part in the direction you specify. The side that is removed is suppressed rather than deleted. To create a part with the other side removed, edit the split feature and redefine it to keep the other side, then save the other half of the part to its own file using the Save Copy As option, or create a derived part.

- Split individual faces by using a surface, sketching a parting line, or placing a work plane, and then selecting the faces to split. You can edit the split feature and modify it to to add or remove the desired part faces to be split.

The Split tool is located on the Part Features panel bar, as shown in the following image. Once selected, the Split dialog box will appear.

Figure 8-88

The Split dialog box contains the following sections:

METHOD

Split Part Click the left button in the Method section, as shown in the following image, to split an entire part using a selected work plane, surface, or sketched geometry to cut (remove) material. If you select this option, you are prompted to choose the direction of the material that you want to remove.

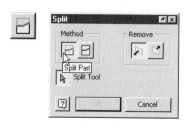

Figure 8-89

Split Face Click the right button in the Method section, as shown in the following image, to split individual faces of a part by selecting a work plane, surface, or sketched geometry, and then selecting the faces to split. The Split face method can split individual faces or all the faces on the part. When you select the Split Face method, the Remove area in the dialog box will be replaced with the Faces area as shown on the right in the figure.

Figure 8-90

Remove

The option to remove material is only available when you use the Split Part method (see the following image). After splitting a part, you can retrieve the cut material by editing the split feature and clicking to remove the opposite side, or by deleting the Split feature.

Figure 8-91

FACES

The Faces option is available only when you select the Split Face method.

All Click this button, as shown in the following image, to split all faces of the part that intersect the Split tool.

Figure 8-92

Select Click this button, as shown in the following image, to enable the selection of specific faces that you want to split. After clicking the Select button, the Faces to Split tool becomes active.

Figure 8-93

Faces to Split Click this button, as shown in the following image, and select the faces of the part that you want to split. Any faces selected will be split where they intersect the Split tool.

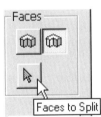

Figure 8-94

Split Tool Click this button, shown in the following image, and select a surface, work plane, or sketch that you want to use to split the part.

Figure 8-95

EXERCISE 8-5 Splitting a Part

In this exercise, you use split features to complete a model of a wireless phone handset. Three mating parts of the phone will be created from a single model.

1. Open ESS_E08_05.ipt.

2. Next, you use a work plane to split the handset into upper and lower halves. Click the Split tool.
   - Click Work Plane1 in the graphics window or Browser.
   - In the Split dialog box, click the Split Part method.
   - Click the left Side to Remove button so the dialog box and preview resemble the following image.

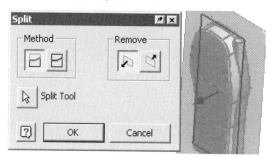

Figure 8-96

3. Click OK to remove the top half.

4. Right-click Work Plane1 in the graphics window or Browser and turn off its visibility.

5. Next, you use a sketch to split the part again to create the battery cover. From the Browser, turn on the visibility of the YZ plane from the Origin folder.

6. Make the YZ plane the new sketch.

7. Click the Look At tool and click the YZ plane.

8. Use the Common View tool to rotate the model until it is orientated like the following image.

9. In the graphics window right-click and select Slice Graphics.

10. Click the Project Cut Edges tool from the 2D Sketch Panel Bar. Notice that the edges of the sliced graphic are automatically projected and displayed in a different color.

11. Press L for the Line tool and draw two lines shown in the following image (the sketch lines should be parallel to the projected left vertical and bottom lines while sketching, or use the parallel constraint).

12. Click the General Dimension tool. Place **8 mm** and **90 mm** dimensions as shown in the following figure (select the lines to automatically align them).

Chapter 8 Advanced Part Modeling Techniques   595

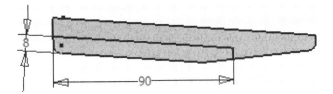

Figure 8-97

13. Click the Return tool on the Standard toolbar.

14. Click the Split tool.
    - Select the Split Part method.
    - Select the Split Tool arrow if it's not selected.
    - Select the profile you drew in the previous steps.
    - In the Remove area of the dialog box, click the right Side to Remove button to retain the battery cover and remove the rest of the part.
    - When your preview resembles the following figure, click OK.

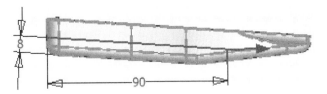

Figure 8-98

15. The part is split to create the battery cover. Press F4 and rotate the part to examine it. **Note**: This single file now contains model data for all three parts (upper cover, lower cover, and battery cover). In the next section, you edit features and save copies of this file to create separate files for each part.

16. In the Browser or graphics window turn off the visibility of the YZ Plane.

17. Next, you create mating part files from the original file. From the File menu, select Save Copy As. Save the file under the name *ESS_E08_05-BatCover.ipt* in the same directory as the original file. The battery cover part is now complete.

18. In the Browser, double-click the Split3 feature to edit it.
    - When the Split dialog box is displayed, verify the Split Part method is current.
    - Click the left Side to Remove button to reverse the remove direction.
    - Click OK to finish editing the split feature and create the mating part.

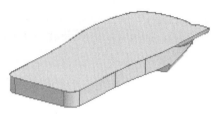

Figure 8-99

19. From the File menu, select Save Copy As. Save the file under the name ESS_E08_05-Lower_Cover.ipt in the same directory as the source file.

20. In the Browser, right-click Split 3 and select Delete.

21. When the Delete Features dialog box is displayed, click OK.

22. In the Browser, double-click the Split 2 feature.
    - When the Split dialog box is displayed, verify the Split Part method is current.
    - In the Remove area of the dialog box select the right Side to Remove button to reverse the remove direction.
    - Click OK to complete the edit.

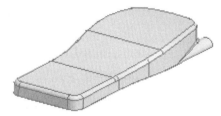

Figure 8-100

23. From the File menu, select Save Copy As. Save the file under the name ESS_E08_05-Upper_Cover.ipt in the same directory as the source file.

24. Note: In your files, you would complete the parts by changing properties in the new files. If you need to change the design, you can change either the battery cover or lower cover (each of these parts contains features for all three parts), and then use Save Copy As to create the mating part files.

25. Close the file and do NOT save changes.

COPYING FEATURES

To increase your productivity, you can copy features from one face of a part to another face on the same part or to a different part. If you copy a feature from one part to another, both files must either be open in Autodesk Inventor or be instances within the same assembly file. When making a copy of a feature, you have the option to designate whether you want the feature, its dependent feature(s), and its parameters to be dependent on or independent of the parent feature(s).

Dependent

The dimensions of the copied feature(s) will be equal to the dimensions of the parent feature(s). If the parent feature was defined with a value of d4=8 inches, for example, the copied feature would have a parameter similar to d14=d4. The Dependent option is only available when copying features within the same part.

Independent

The dimensions of the copied feature(s) will contain their own values. Initially, the value will equal the value of the parent feature. If the parent feature was defined with a value of d4=8 inches, for example, the copied feature will have a parameter similar to d14=8 inches. The value of d14 can be modified independently of the value of d4.

The following rules apply when creating an independent copy of a feature:

- You can only copy selected features by default–not children features.
- Upon pasting the feature, you can choose to copy dependent features.
- When pasting a copied feature, you can specify what plane and orientation the new feature will have.
- The newly pasted feature is completely independent and contains its own sketches and feature definitions within the Browser.

Follow these steps to copy and paste a feature within Autodesk Inventor:

1. Right-click on a feature's name (in the Browser), or change the Select option on the standard toolbar to Feature Priority (as shown in the following image), and then move the cursor over the feature to copy.

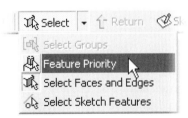

Figure 8-101

2. Select Copy from the menu to copy the feature to the clipboard.
3. Issue the Paste command by doing one of the following:
   1. Right-click and select Paste from the menu.
   2. Click Paste on the Edit menu.
   3. Press both the CNTRL and V keys on the keyboard at the same time.

4. The Paste Features dialog box will appear, as shown in the following image.

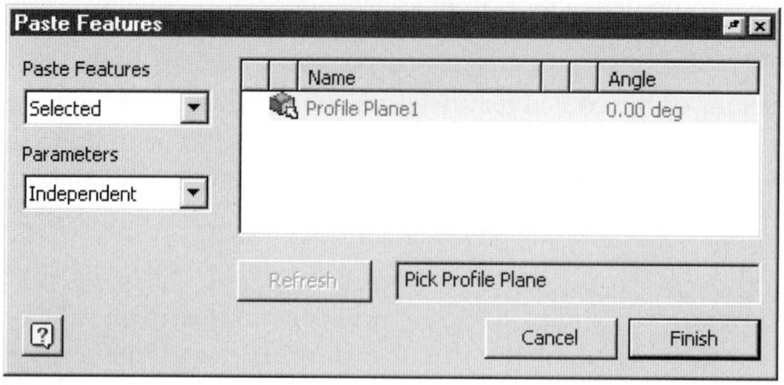

Figure 8-102

5. Click a plane on which to place the feature. The placement plane can be any planar face on the part or a work plane.

6. You can move or rotate the new feature roughly by using the compass in the preview image, as shown in the following image.

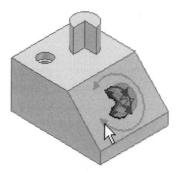

Figure 8-103

7. Click the four-headed arrows to move the profile, or click the arc around the arrows to rotate the profile. You can also enter the precise value for the angle of rotation of the profile in the dialog box.

8. To locate the pasted feature accurately, edit its sketch after placement and add constraints and dimensions as needed.

9. In the Paste Features dialog box, adjust the settings as needed.

The following are definitions of the various options in the Paste Features dialog box:

Paste Features Select these options to paste the selected feature(s) or the selected feature(s) and dependent feature(s) from the droplist, as shown in the following image.

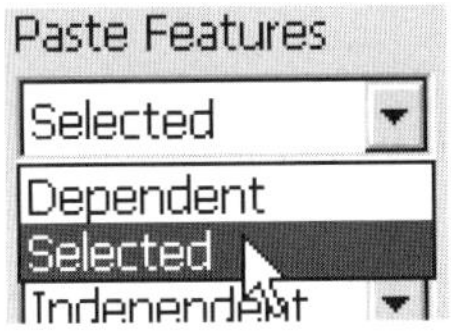

Figure 8-104

    Selected—This will copy only the features that were selected when initiating the Copy command.

    Dependent—This will copy the dependent feature(s) of the selected feature(s).

Parameters   Select these options to make the parameters of the newly copied feature dependent on the parent feature or independent of the parent feature, as shown in the following image.

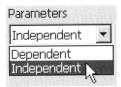

Figure 8-105

    Dependent—The dimensions of the copied feature(s) will be equal to the dimensions of the parent feature(s). If the parent feature was defined with a value of d4=8 inches, for example, the copied feature would have a parameter similar to d14=d4. The Dependent option is only available when copying features within the same part.

    Independent—The dimensions of the copied feature(s) will contain their own values. Initially the value will equal the value of the parent feature. If the parent feature was defined with a value of d4=8 inches, for example, the copied feature will have a parameter similar to d14=8 inches. You can modify the value of d14 independent from the value of d4.

Pick Profile Plane   Information about the pasted feature will appear in this area. You can select a new plane by clicking on its name in the Pick Profile Plane area of the Paste Features dialog box and then selecting a new plane in the graphics window. You can change the rotation angle of the pasted feature by clicking in the Angle section and entering a new value (see the following image).

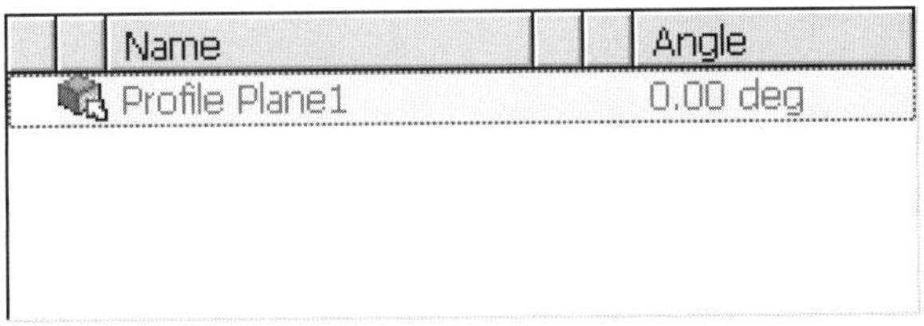

Figure 8-106

MIRRORING FEATURES

When creating a part that has features that are mirror images of one another, you can use the Mirror Feature tool to mirror a feature(s) about a planar face or plane instead of recreating each of the features from scratch. Before mirroring a feature, a plane or planar face must exist that will be used as the mirror plane. The plane must be located between the existing feature(s) and the new mirrored feature(s). It can be a planar face, a work plane on the part, or an origin plane that has the correct location and orientation. The mirrored feature(s) will be dependent on the parent feature; if the parent feature(s) change, the resulting mirror feature will also update to reflect the change. To mirror a feature or features, use the Mirror Feature tool on the Part Features panel bar, as shown on the left side of the following image. The Mirror Pattern dialog box will appear, as shown on the right. The following sections explain the options in the Mirror Pattern dialog box.

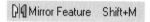

Figure 8-107

Mirror individual features   Click this option, shown in the following image, to mirror a feature or features.

Figure 8-108

Mirror the entire solid   Click this option, shown in the following image, to mirror an entire solid body.

Figure 8-109

Mirror Plane   Click and choose a planar face or work plane on which to mirror the feature(s).

Creation Method   Access the Creation Method tools by clicking the More (>>) button.

Optimized–Create mirrored occurrences which are direct copies of the original features (this option will compute the fastest but will not adjust to the model).

Identical–Create mirrored features that remain identical regardless of where they intersect other features on the part. Use this option when you want to mirror a large number of features that share a termination plane.

Adjust to Model–Use this option when you want to calculate an independent termination for each feature that is being mirrored. Use this option if the feature(s) being mirrored uses a model face as the termination option.

To mirror a feature or features, follow these steps:

1. Click the Mirror Feature tool on the Part Features panel bar. The Mirror Pattern dialog box will appear.

2. Click either the Mirror individual features or Mirror the entire solid option.

3. Select the feature(s) or solid body to mirror.

4. Click the Mirror Plane button and then select the plane on which the feature will be mirrored.

5. In the More (>>) area, determine if the mirrored feature(s) termination will be optimized, identical, or will adjust to the model.

6. Click the OK button to complete the operation.

The following image shows the front features on the cylinder being mirrored about a work plane.

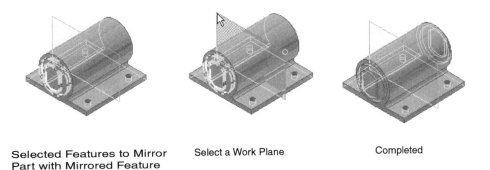

Selected Features to Mirror   Select a Work Plane   Completed
Part with Mirrored Feature

Figure 8-110

SUPPRESSING FEATURES

You can suppress a model's feature or features to temporarily turn off their display. Feature suppression can be used to simplify parts, which may increase system performance. This capability also shows the part in different states throughout the manufacturing process and can be used to access faces and edges that you would otherwise not be able to access. If you need to dimension to a theoretical intersection of an edge that was filleted, for example, you could suppress the fillet and add the dimension, then unsuppress the fillet feature. If the feature you suppress is a parent feature for other features (dependent), the child features will also be suppressed. Features that are suppressed appear gray in the Browser and have a line drawn through them, as shown in the following image. A suppressed feature will remain suppressed until it is unsuppressed, which will also return the Browser display to its normal state. The following image also shows the suppressed child features that are dependent on the suppressed extrusions.

Figure 8-111

To suppress a feature on the active part, follow one of these methods:

1. Right-click on the feature in the Browser and select Suppress Features from the menu, as shown in the following image.

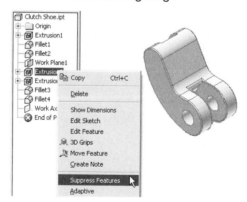

Figure 8-112

or

1. Click the Feature Priority option from the standard toolbar, as shown in the following image. This option allows you to select features on the parts in the graphics window.

2. Right-click on the feature in the graphics window and click the Suppress Features option from the menu.

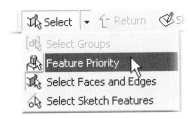

Figure 8-113

To unsuppress a feature, right-click on the suppressed feature's name in the Browser and select Unsuppress Features from the menu.

CONDITIONALLY SUPPRESS FEATURES

Another method to suppress features is to apply a parameter condition to a feature. For example, if a part has a feature that should be suppressed if the size of another feature is reduced, you can set the features properties to be suppressed when the parameter's value is changed. To conditionally suppress a feature, follow these steps:

1. Create part and rename parameters for important dimensions.
2. Right-click on the features to be suppressed if a set value of a parameter is reached and click Properties as shown on the left side of the following image.
3. In the Feature Properties dialog box from the droplist select the parameter that will determine suppression of the feature. The following image on the right shows that the Length parameter has been selected.
4. From the droplist select the mathematical condition.
5. Enter a value limit for the parameter that will suppress the feature. The right side of the following image shows the value of 20 mm.
6. Click OK.
7. To see the results of the supression, change the value of the Length parameter to reach or exceed the limit and update the part.

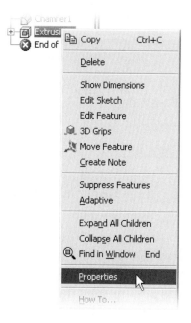 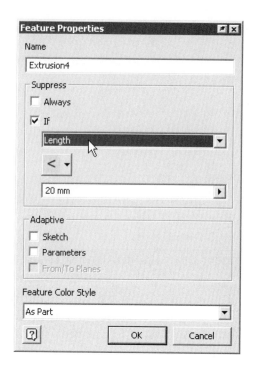

Figure 8-114

REORDERING FEATURES

You can reorder features in the Browser. If you created a fillet feature using the All Fillets or All Rounds option, and then created an additional extruded feature (like a boss), you can move the fillet feature below the boss and include the edges of the new feature in the selected edges of the fillet. To reorder features, follow these steps:

1. Click the features name or icon in the Browser and holding down the left mouse button (click-and-drag) move the feature to the desired location in the Browser. A horizontal line will appear in the Browser to show you the feature's relative location while reordering the feature. The left side of the following image shows a hole feature being reordered in the Browser.

2. Release the mouse button, and the model features will be recalculated in their new sequence. The right side of the image shows the Browser and reordered hole feature.

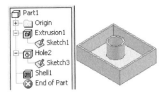
Browser and Part After Feature Reorder

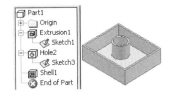
Browser and Part After Feature Reorder

Figure 8-115

If you cannot move the feature due to parent-child relationships with other features, Autodesk Inventor will not allow you to drag the feature to the new position. The cursor will change in the Browser to a No symbol instead of the horizontal line, as shown in the following image.

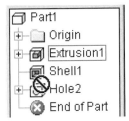

Figure 8-116

FEATURE ROLLBACK

While designing, you may not always place features in the order that your design later needs. Earlier in this chapter you learned how to reorder features, but reordering features will not always allow you to create the desired results. To solve this problem, Autodesk Inventor allows you to roll back the design to an earlier state and then place the additional new features. To roll back a design, drag the End of Part marker in the Browser to the location where the new feature will be placed. To move the End of Part marker, follow these steps:

1. Move the cursor over the End of Part marker in the Browser.

2. With the left mouse button pressed down, drag the End of Part marker to the new location in the Browser. While dragging the marker, a line will appear, as shown on the left in the following image.

3. Release the mouse button and the features below the End of Part marker are removed temporarily from calculation of the part. The End of Part marker will be moved to its new location in the Browser. The following middle image shows the End of Part marker in its new location.

4. Create new features as needed. The new features will appear *above* the End of Part marker in the Browser.

5. To return the part to its original state (including the new features), drag the End of Part marker below the last feature in the Browser.

6. If needed, you can delete all features below the End of Part marker by right clicking on the End of Part marker and click Delete all features below EOP as shown in the image on the right.

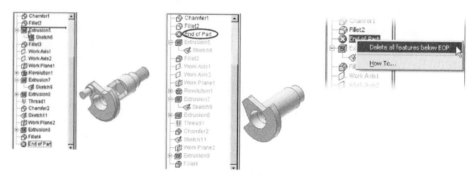

Figure 8-117

FILE PROPERTIES

You can specify properties for the files you create in Autodesk Inventor using the iProperties option on the File menu or by right-clicking on the files name in the Browser and selecting iProperties from the menu while in a part file or Properties while in an assembly file. After selecting the iProperties option, the Properties dialog box for the active file will appear. The following image shows the Properties dialog box with the Summary tab active.

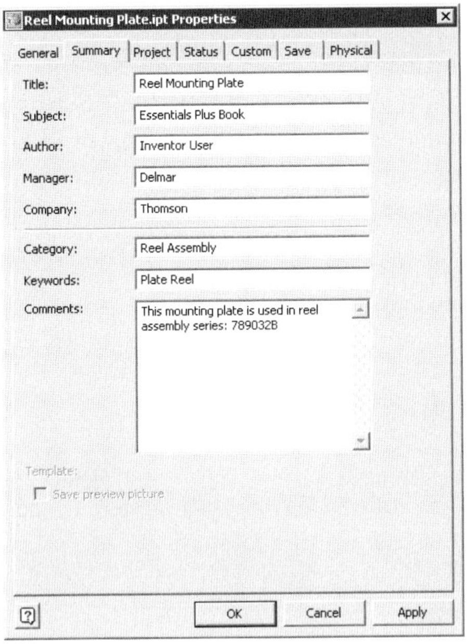

Figure 8-118

The Properties dialog boxes for part file contain the seven tabs described below. The Properties dialog box for a drawing file contains six tabs–drawing files do not include the Physical tab. The properties of your files can be used to classify and manage your Autodesk Inventor files. You can also use properties that have been populated with data for search criteria, for creating reports, and for updating your title blocks and parts lists in drawings and bills of material automatically. The following sections describe the tabs and their contents.

GENERAL

This tab lists the file name, location, size, date created, date modified, and last date the file was accessed. You cannot modify the information in this tab via the dialog box.

SUMMARY

The Summary tab contains fields in which you can enter data for the file: Title, Subject, Author, Manager, Company, Category, Keywords, and Comments. The previous image shows an example of data entered in the Summary tab.

PROJECT

The Project tab contains fields in which you can enter data for the file: Location, File Subtype, Part Number, Description, Revision Number, Project, Designer, Engineer, Authority, Cost Center, Estimated Cost, Creation Date, Vendor, and WEB Link.

STATUS

The Status tab contains fields in which you can enter data or change the status via droplists for the file: Part Number (not editable), Status, Design State, Checked By, Checked Date, Eng. Approved By, Eng. Approved Date, Mfg. Approved By, Mfg. Approved Date, File Status, Checked out By, Checked out, Checkout Workgroup, and Checkout Workspace.

CUSTOM

You can add custom properties to the active file using the Custom tab. The general Microsoft custom properties are also available and can be added to your Autodesk Inventor files.

SAVE

On the Save tab, you can specify whether you want to save a preview picture of your files for the preview image (and how the image is acquired), or you can specify an image file to use as the preview picture.

PHYSICAL

On the Physical tab you can calculate and display the physical and inertial properties for a part or assembly file, including Mass, Surface Area, and Volume. Physical properties of the model change as you create the structure of your files. To get accurate results for the mass, you will need to select a material for each part. If you are in an assembly, this will need to be done by making each part active and then adjusting the material. Clicking the Update button on this tab updates the physical properties based on changes to your models. If the material is changed, the part's color will also reflect the change.

CENTER OF GRAVITY

As you are designing, it may be helpful to know where the center of gravity is for the active part or assembly. To view a symbolic representation (X-, Y-, and Z-axis arrows) of the center of gravity for the active part or assembly, follow these steps:

1. Click Center of Gravity on the View menu.
2. If the mass properties of the active document are not up to date, Autodesk Inventor will prompt you to update them, as shown in the following image.

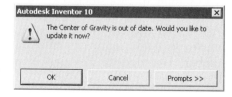

Figure 8-119

3. Click OK to update the mass properties, and the Center of Gravity symbol will appear at the location of the center of gravity.

4. To see the numeric value of the center of gravity, move the cursor over (or click) the sphere of the Center of Gravity symbol (you may need to use the Select Other tool) and the X-, Y-, and Z-axis location will appear, as shown on the left in the following image. These values are relative to the origin point for the active document.

5. If the part's physical size changes or parts in an assembly change location, the Center of Gravity symbol will appear colorless. This indicates the center of gravity is not up to date. The image on the right in the figure shows an example of an out-of-date center of gravity after a fillet was added and the holes' sizes were edited.

6. Either click Update Mass Properties on the Tools menu or click the Update tool on the Physical tab of the Properties dialog box to update the Center of Gravity to reflect these changes.

7. To turn off the Center of Gravity symbol, click Center of Gravity on the View menu.

Figure 8-120

PART MATERIALS

To change a part's physical properties and appearance to a specific material, click the iProperties option on the File menu or right-click on the part's name in the Browser and select a material from the material droplist on the Physical tab, as shown on the left side of the following image. As you learned in the File Properties section in this chapter, the properties will reflect the attributes of this material. Click the OK button to complete the operation, and the color of the part should change to reflect the material. If the material was not updated on the part, click the As Material option in the Color area of the standard toolbar, as shown below the image. If you selected a color from the Color droplist on the standard toolbar, it will only change the appearance of the part, and will not change the physical material or properties of the part.

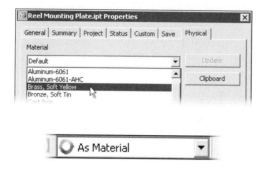

Figure 8-121

OVERRIDE MASS AND VOLUME PROPERTIES

While designing, you may not always draw parts that are 100¢\% complete; for example, you may model only the bounding area and critical features of a purchased part. You still want the mass and volume to be represented accurately in the Properties dialog box, however. You can override the mass and volume values of a part or assembly by following these steps:

1. Click iProperties on the File menu, or right-click on the part's name in the Browser and select iProperties from the menu.

2. The Properties dialog box will appear, as shown on the left in the following image. Click the Physical tab.

3. If a material has not been assigned, select a material from the Material droplist.

4. Click the Update button to update the properties if they are displaying N/A. Notice the calculator symbol next to the Mass and Volume values; the symbol shows that Autodesk Inventor has calculated these properties based on the material and size of the part.

5. To override a value for Mass or Volume, click in its text area and enter a new value. The symbol next to the overridden value will change to a hand to reflect that the value has been overridden, as shown below the image. Notice the * in the Inertial Properties and Center of Gravity areas state that the "Values do not reflect user-overridden mass or volume."

6. To change the overridden value back to the default value, delete the information in the text box area of Mass or Volume and click the Apply button.

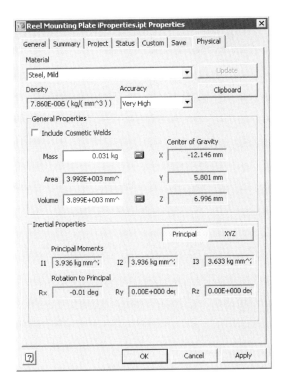

Default Properties

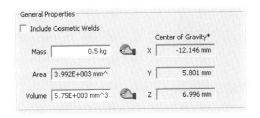

Overridden Mass and Volume

Figure 8-122

To copy the information on the Physical tab of the Properties dialog box into another application, such as Microsoft Word, click the Clipboard button on the Physical tab. Next, make the other application active and paste the information into that file. The following image shows the physical property information of a part file when copied into Microsoft Word.

```
Physical Properties for
General Properties:
    Material: {Steel, Mild}
    Density:    7.860E-006 ( kg/( mm^3 ) )
    *Volume:    2.5E+003 mm^3
        *Mass: 0.2 kg
    Area:   3.970E+003 mm^2
**Center of Gravity:
    X:   -12.146 mm
    Y:    5.801 mm
    Z:    7.129 mm
**Mass Moments of Inertia
    Ixx         3.816 kg mm^2
    Iyx Iyy     1.776E-015 kg mm^2      3.816 kg mm^2
    Izx Izy Izz 3.553E-016 kg mm^2      4.767E-005 kg mm^2      3.436
kg mm^2
**Principal Moments of Inertia
    I1:     3.816 kg mm^2
    I2:     3.816 kg mm^2
    I3:     3.436 kg mm^2
**Rotation from XYZ to Principal
    Rx:    -0.01 deg
    Ry:     0.00E+000 deg
    Rz:     0.00E+000 deg
*Calculations are based on user overwritten values
**Values do not reflect user-overridden mass or volume
```

Figure 8-123

IPARTS

iParts enable design intent and design knowledge to be shared and reused. They enable you to store multiple part parameters and properties and then calculate unique part file versions based on certain configurations of the stored parameters and properties. Using the Parameters dialog box, you can store one value for each parameter. In industry, configured parts have been referred to as tabulated parts, charted parts, or a family of parts. You can also use iParts to create part libraries that allow design data to be reused. An iPart is generated from a standard Autodesk Inventor part file (*.ipt*). When you activate the Create iPart tool, the standard part file is converted into an iPart. You can add individual members (or configurations) to the iPart; it is then referred to as an iPart factory.

There are two stages to the use of iParts: authoring the iPart and placing the iPart version. In the authoring or creation stage, you design the part and establish all possible versions of the design in a table. The rows of the table describe the members of the iPart. An iPart that has multiple members is called an iPart factory.

In the placement stage, you select a member from the iPart factory, and an iPart version is published and inserted into your assembly.

iParts allow you to suppress specific features, control iMates and thread properties, add or modify file properties to the different members that are contained within the iPart factory, and allow user-defined input for specific parameters. You can further enhance inputs to include an element of control. For instance, you can apply values within a predetermined range or from a list to a parameter. After creating an iPart, you can place it into an assembly, where you can select a specific member (or configuration). The following sections will explain how to create and then place an iPart into an assembly.

An example of an iPart is a simple bolt. The bolt has a number of different sizes that are associated with it. An iPart allows you to define these different sizes and configurations (material, part properties, and so on) and have them reside within a single file (factory). When placing the bolt in an assembly, you can select which size (version/member) of the bolt (iPart factory) you want to use. You can then constrain the placed member in the assembly like any other part file.

CREATING IPARTS

There are two types of iParts that can be created: standard and custom.

Standard iPart (Factories) You cannot modify standard iPart values; when placing a standard iPart into an assembly, you can only select the predefined members for placement. A standard iPart that you place in an assembly cannot have features added to it after placement.

Custom iPart (Factories) Custom iParts allow you to place a unique value for at least one variable. A custom iPart that you place in an assembly can have features added to it.

To create an iPart, follow these steps:

1. Create an Autodesk Inventor part or a sheet metal part.

2. Add the dimensions of the geometry of the design to be changed to the iPart Author table.

3. For easier creation of the member table, use descriptive names for the values of the parameters.

4. If you do not use parameter names, you will need to determine what each parameter (or d#) represents within the parts geometry.

TIP Any parameters that have a name other than d# are added automatically to the parameter table during the creation of the iPart.

5. Issue the Create iPart tool on the Tools menu (as shown in the following image) to add the members or configurations to the iPart Author table.

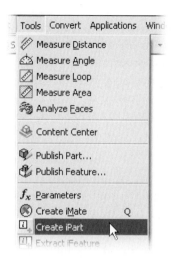

Figure 8-124

The iPart Author dialog box appears, as shown in the following image.

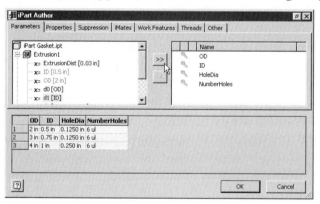

Figure 8-125

To create the iPart members, follow these steps:

1. Add to the right side of the dialog box (parameter list) all parameters and dimensions that will be configured. All named parameters are added automatically to the right side.

2. To add a d# dimension to the list on the right side, select it in the left column and then click the Add parameters (>>) button, as shown in the previous image. You can also double-click on the parameter to add it.

3. To remove a parameter from the right side of the dialog box, select its name and click the Remove parameters (<<) button.

4. After you have added the parameters, you need to define the keys. Keys identify the column whose values are used to define the iPart member when

the part is published (or placed) in an assembly. For example, setting a parameter as a primary key allows the designer placing the part to choose from all available values for that parameter in the selected items list.

5. To specify the key order, click an item in the key column of the selected parameters list to define it as a key, or right-click on the item and select the key sequence number, as shown in the following image. Selected keys are blue; items that are not selected as keys have dimmed key symbols. You can decide not to add primary keys or to add one or more additional keys as needed. You cannot specify custom table columns as keys.

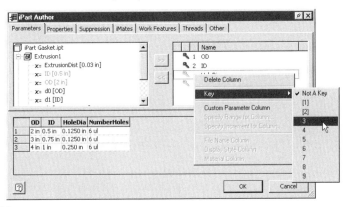

Figure 8-126

6. Click on the other tabs in the dialog box to perform other specific operations. These items are not required to define the iPart factory. The following sections describe them in more detail.

Properties Tab   The Properties tab lists all of the summary, project, and physical properties for the file. If you use the material property on this tab to control the material of your iParts, you have to use the Material Column option in the table. This option is described in the Special Table Elements section later in this chapter. You must also set the current color of the iPart to As Material prior to saving the iPart. For part properties to be used in drawings, Bill of Materials (BOMs), and other downstream purposes, you have to include the properties in the table even if their values do not vary between members.

Suppression Tab   The Suppression tab gives you the ability to suppress features of specific members of the iPart factory. If you enter Suppress in a cell for the feature, the feature will be suppressed when the version is calculated. When the cell contains Compute, the feature is not suppressed.

iMates Tab   On the iMates tab, select one or more iMates to include in the iPart. iMates are included in the iPart if their status is Compute. If their status is suppressed, they are not included in the iPart. Control the iMate properties for each member using the iMates, Parameters, and Suppression tabs. On the iMates tab, you can

- Define different offset values for different iPart members
- Suppress iMates or Composite iMates for different iPart members
- Change the matching name for different assembly configurations
- Change the sequence of the iMates

Since iMate names are not unique, each iMate is assigned an index tag in the iPart Author. Each iMate property (except the offset value) uses this tag to identify a specific iMate property in the table. Tags are not assigned to offset values because iMate offset values are parameters, which are unique by definition.

Work Features Tab   On the Work Features tab, select the user-defined work features to be included in the iPart. Origin work features are included automatically in an iPart factory and iPart members, but they are not listed. The work features are managed in the iPart if its table status is Include. If the status is Exclude, the work features are not managed in the iPart.

Threads Tab   You can control thread features in the iPart factory to create table-driven items for regular or tapered thread features. Use Family to identify the thread standard, Designation to control the size and pitch, and Class to define fit based upon the standard. You can assign the thread Direction as right or left in the table. In addition, options to modify the pipe diameter, class, and family are available.

Other Tab   You can use the Other tab to create custom table items. For example, you can add a column that represents the name of the iPart member. You can create a column named Version to identify the appropriate member when it is placed in an assembly. You can also create custom values such as Color if you want to control the display style of the iPart.

Special Table Elements   You can right-click on a cell of a member of the table or a column label to access some additional custom table capabilities. You can define the name of the iPart file when it is published using the File Name Column option. Control the color of the iPart upon publishing using the Display Style Column option. You can also specify material of the iPart using the Material Column option. These options are only available for part properties and the table elements created using the Other tab. They are not available on parameter items.

7. Next, add a member in the iPart table by right-clicking on a row number at the bottom of the dialog box and selecting Insert Row from the menu, as shown in the following image.

Chapter 8 Advanced Part Modeling Techniques 617

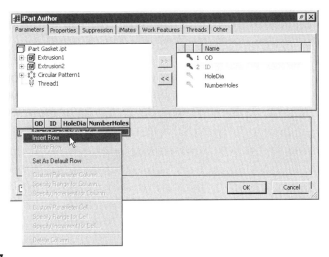

Figure 8-127

8. Edit the cell contents by clicking in the cell and typing new values or information as needed. To delete a row, right-click the row number and select Delete Row from the menu. Each row that you add in the bottom pane of the dialog box represents an additional member within the iPart factory. The table functions similarly to a spreadsheet. In addition to adding or deleting members of the iPart factory, you can also use the table to change members by modifying cell values.

To allow the designer placing the iPart to specify a custom value for a given column, right-click on the column name and select Custom Parameter Column from the menu, as shown in the following image. To make a specific cell custom, right-click in the individual cell and select Custom Parameter Cell from the menu.

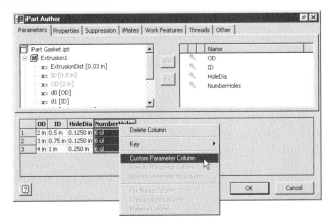

Figure 8-128

After designating a custom parameter column or cell, you can set a minimum and maximum range of values by right-clicking in the column heading or cell and selecting Specify Range for Column or Range for Cell from the menu. The Specify Range dialog box appears, as shown in the following image. Select the options as needed. Custom columns and cells are highlighted with a blue background in the table. You can also specify the increment that can be entered for a custom column or cell using the Specify Increment for Column option from the menu.

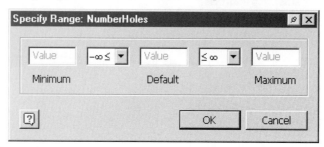

Figure 8-129

9. Set the member that will be the default by right-clicking on the row number and selecting Set As Default Row from the menu. You may select the other options as needed.

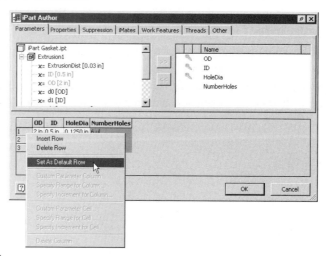

Figure 8-130

10. Click OK when you have finished defining the contents of the table. The iPart Author dialog box closes, and the part is converted to an iPart factory. The table is saved, and a table icon appears in the Browser. The part icon in the Browser also changes to display a factory, as shown in the following image.

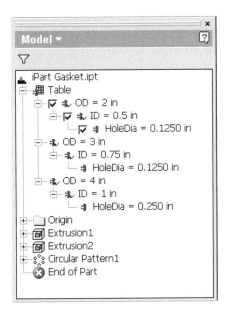

Figure 8-131

You can expand the Table icon in the Browser to view the iPart members based on the keys and values that you define. The active (calculated) version appears with a check mark.

EDITING IPARTS

There are a number of operations that you can perform on an iPart factory after you have created it. You can delete the table, modify the parameters or properties for individual members, add or delete additional members, and so on. Right-click the Table icon in the Browser to delete the table and convert the iPart factory back to a part, edit the table with the iPart Author dialog box using the Edit Table option, or edit the table with Microsoft Excel using the Edit via Spread Sheet option, as shown in the following image. Make changes as needed and then save the file.

 Note Changes made to iPart factories will not be updated automatically in members that you have previously placed in assemblies. To update the iParts in an assembly, open the assembly and click the Global Update tool on the standard toolbar.

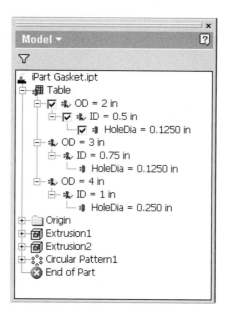

Figure 8-132

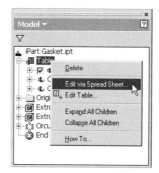

Figure 8-133

When editing the spreadsheet using Microsoft Excel, you can incorporate spreadsheet formulas, conditional statements, and multiple sheet data extractions, but you cannot modify spreadsheet formulas and conditional statements from the iPart Author dialog box. These types of cells are inactive and are highlighted in red, as shown in the following image (row 3/ column ID).

	OD	ID	HoleDia	NumberHoles
1	2 in	0.5 in	0.1250 in	6 ul
2	3 in	0.75 in	0.1250 in	6 ul
3	4 in	1	0.250 in	6 ul

Figure 8-134

IPART PLACEMENT

You place standard iParts in assemblies using the Place Component tool. When you select a standard iPart factory, an additional Place Standard iPart dialog appears. It allows you to use the Keys tab to identify an iPart member by selecting the key values, use the Tree tab to locate a member by expanding the key values, use the Table tab to identify an iPart member by selecting a row in the table, and place multiple instances of different iPart members.

To place a standard iPart into an assembly, follow these steps:

1. Start a new assembly or open an existing assembly in which to place the iPart.

2. Click the Place Component tool, and navigate to and select the iPart to place in the assembly in the Open dialog box.

3. Click the Open button and the Place Standard iPart dialog box will appear, as shown in the following image. If there is a custom cell(s) or column(s), the Place Custom iPart dialog box appears.

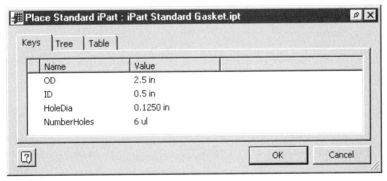

Figure 8-135

4. To select from the member list, select the Keys, Tree, or Table tab and then select the member that defines the part you want to place.

5. If you are placing a custom part, select the Keys, Tree, or Table tab and enter a value in the right side of the dialog box, as shown in the following image. The value must fall within the limits set in the iPart factory; otherwise, an alert appears.

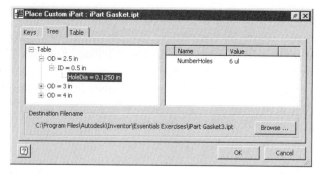

Figure 8-136

6. Place the part in the graphics window.

7. Continue placing instances of the iPart as needed; when you are finished, click the Dismiss button.

8. To change an iPart in an assembly to a different configuration, expand the part in the Browser, right-click on the table name, and select Change Component. Then select a new member from the Keys, Tree, or Table tab.

The Keys tab displays the primary and any secondary keys defined in the iPart factory. You can select the values of the keys to identify unique members. The Tree tab displays a hierarchical structure of the keys. If secondary keys exist, the values of the keys are filtered progressively as you choose to expand the key values. The Table tab displays the entire table rather than just the keys. To identify a unique iPart member, select a row and then click the OK button to place the iPart.

When you place the first version of a standard iPart into an assembly, a folder is created with the same name as the iPart factory. This folder is created in the same folder as the iPart factory file by default. As you place additional standard iParts into your assemblies, this folder is checked for existing iPart files prior to creating new iPart files.

STANDARD IPART LIBRARIES

In a collaborative design environment, the best way to manage iPart factories and the iParts published from those factories is through the use of libraries. Define a library within your project and place your iPart factories in this library's path. You can also specify that a standard iPart factory publish standard iParts to a different (proxy) folder. To do this, create another library entry using the same name as the iPart factory library prefixed with an underscore, as shown in the following image, or right-click on the library's name and click Add Proxy Path.

```
     Libraries
        iPart Factories - C:\Inventor\Factories
        _iPart Factories - C:\Inventor\Factory_Parts
```

Figure 8-137

If you have an iPart factory for bolts, for example, you could store the iPart factory in a library named Bolts, and then you could define an additional library where the individual members of the Bolt iPart factory will be stored. The library should be named _Bolts. Autodesk Inventor will automatically store all iParts published by the factory in the library directory specified by _Bolts. If you store an iPart factory in a workspace or a workgroup search path, the published iParts are stored in a subdirectory with the same name as the factory.

 NOTE It is recommended that you do not store iParts in the same folders as iPart factories.

CUSTOM IPARTS

Custom iParts are placed into assemblies in the same way as standard iParts. If you select an existing custom iPart member file, that member of the part is placed into your assembly with no option to define the value of the custom parameters. When you select a custom iPart factory, the Place Custom iPart dialog box appears. As with the Standard iPart Placement dialog box, you can use the Keys tab to identify an iPart member by selecting the key value, the Tree tab to expand key values and select a member, or the Table tab to identify an iPart member by selecting a row in the table. The Place Custom iPart dialog box provides additional options so you can enter values for custom parameters, define a destination and filename for the custom iPart by selecting Browse in the dialog box, and place multiple instances of different iPart members.

 NOTE Since each custom iPart is unique, you have to provide a different filename as you place different members.

After you have placed a custom iPart in an assembly, you can add more features to the iPart. The following image shows the Browser of a custom iPart named Gasket-B that has had a fillet and extrusion feature added to it. The capability to add features to an iPart makes the custom iPart behavior similar to that of a derived part.

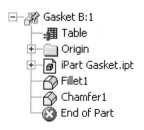

Figure 8-138

EXERCISE 8-6 Creating and Placing iParts

In this exercise you convert an existing part to a Standard iPart Factory. You then place Standard iParts from that factory into an assembly.

1. Open *ESS_E08_06.ipt*.

2. Click the Parameter tool and review the parameter names. In this exercise you will use the parameter HoleDia to drive the iPart factory.

3. Click Done at the bottom of the Parameters dialog box.

4. From the Tools menu click Create iPart.

5. In the iPart Author dialog box expand the Hole1 feature in the Parameters tab. Notice that the system parameter HoleDia has automatically been added to the list of table-driven items and a column has been added to the table. User parameters and renamed system parameters are automatically added to the list.

6. HoleDia will be used as the primary key for this iPart Factory. Click the key next to HoleDia.

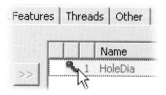

Figure 8-139

7. You also will control the material from the iPart Factory. To add the Material property to the table:
   - Click the Properties tab.
   - Collapse Summary.
   - Expand Physical.
   - Double-click Material. Notice that the material Aluminum (that was defined in the Properties dialog box in the Physical tab) is added to the table.

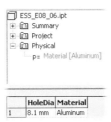

Figure 8-140

8. To identify a material when placing this Standard iPart, the Material property must also be defined as a key. To define Material as a secondary key, click the key next to Material. The number next to the key shows the key priority.

Figure 8-141

9. You can identify which column in the table is used to control the Material for each iPart version. In the table, right-click on the Material column label then click Material Column. There should be a checkmark next to the Material Column option. If one exists, do not deselect it.

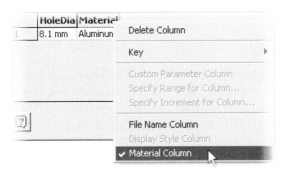

Figure 8-142

10. You will also control the name of the Standard iPart when it is published. To define a custom table item, click the Other tab.
    - Click the **Click here to add value** entry in the table-driven list.
    - Enter **PartName** then press Enter.

Figure 8-143

11. Similar to the Material Column, you can also identify which column in the table is used to control the file name for each iPart version. In the table, right-click on the PartName column label then click **File Name Column**.

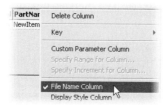

Figure 8-144

12. To define the file name for the first iPart version double-click on the first cell in the PartName column and enter **Large-Alum** then press Enter.

Figure 8-145

13. The iPart Factory will contain three different hole sizes and two different materials to choose from. You will create the first three versions using the table at the bottom of the iPart Author dialog box. The last three versions are defined in a later step using Microsoft Excel to edit the embedded spreadsheet. To create a new row in the table, right-click on the first row label then click **Insert Row**.

Figure 8-146

14. Repeat the last step to insert a third row.

15. Double-click on individual cells and enter the values as shown in the following image. The row highlighted in green is the default row. This defines

the default iPart version when you place an iPart from this factory into an assembly.

	HoleDia	Material	PartName
1	8.1 mm	Aluminum	Large-Alum
2	7.1 mm	Aluminum	Medium-Alum
3	6.1 mm	Aluminum	Small-Alum

Figure 8-147

16. To convert the part to a Standard iPart Factory using the data defined in the table click OK at the bottom of the iPart Author dialog box.

17. In the Browse,r expand the Table and expand each item under the table. The Browser represents your part as a Standard iPart Factory using a Table entry. Each version is shown under the table sorted by the primary and secondary key names and their values, and the active version has a check mark.

Figure 8-148

18. Notice that for each hole diameter in the table, only one material is defined—Aluminum. To add more versions to the table using Microsoft Excel right-click the Table icon in the Browser, then click **Edit via Spread Sheet**.
    - Copy cells **A2** through **C4** and paste them to cell **A5**.
    - Modify cells **B5** through **C7** to be consistent with the following image.

	A	B	C
1	HoleDia<k	Material [Pl	PartName<filenam
2	8.1 mm	Aluminum	Large-Alum
3	7.1 mm	Aluminum	Medium-Alum
4	6.1 mm	Aluminum	Small-Alum
5	8.1 mm	Copper	Large-Copper
6	7.1 mm	Copper	Medium-Copper
7	6.1 mm	Copper	Small-Copper

Figure 8-149

19. When you finish modifying the contents of the spread sheet in Excel, click Save.

20. In Excel, click Close & Return to *ESS_E08_06.ipt*.

21. You can check each version in the iPart Factory prior to publishing them for use in your assemblies by changing the active version. To test different versions right-click on different Material keys in the Browser then click **Compute Row**.

22. Make the **Aluminum** version of **HoleDia = 8.1 mm** the active version.
23. To save the Standard iPart Factory with a different name click File>Save Copy As and enter the file name *Bushing* and click to save it in the same location as the exercise files.
24. Close the original file. Do not save changes.
25. In the next portion of this exercise, you insert iParts from this iPart factory into an assembly.
26. Open *ESS_E08_06B.iam*.
27. From the Assembly Panel bar, click Place Component and click the *Bushing.ipt* file you just created, then click Open.
28. When you select a Standard iPart Factory, the Place Standard iPart dialog box allows you to place different versions of the part. The default values are shown on the Keys tab. Place two default versions of the Bushing by selecting the locations shown in the following image.

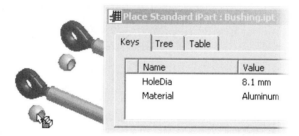

Figure 8-150

29. To place a version with a hole size of 6.1 mm and a material of copper click the value next to HoleDia then click **6.1 mm** from the list of available hole diameters as shown on the left side of the following image.
30. Click the value next to Material then click **Aluminium** from the list of available Materials as shown on the right side of the following image.

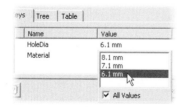 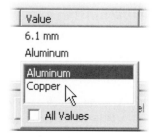

Figure 8-151

31. Click the placement location for the third iPart as shown in the following image. Notice that the color of the iPart reflects the material you specified.

Figure 8-152

32. To place a version with a hole size of 7.1 mm and a material of aluminum, click the Table tab and click the Second row in the table and click a location for the fourth iPart as shown in the following image.

Figure 8-153

33. Click Dismiss in the Place Standard iPart dialog box.

34. To change the version of the iPart you just inserted, in the Browser expand Medium-Alum:1, right-click the Table and click Change Component. Notice that the names of the iParts in the Browser reflect the names you specified when defining the PartName column in the iPart Factory.

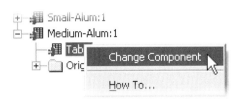

Figure 8-154

35. In the Place Standard iPart dialog box change the HoleDia to 6.1 mm and Material to Copper.

Figure 8-155

36. Click OK in the Place Standard iPart dialog box and verify that the Bushing was replaced in the graphics window.

Figure 8-156

37. Close the file. Do not save changes.

IFEATURES

iFeatures give you the ability to reuse single or multiple features from a part file in other Autodesk Inventor files. iFeatures capture the design intent built into a feature(s) that is going to be reused, such as the name or the size and position parameters. You can also embed or attach a file to the iFeature to be used as Placement Help when you place the iFeature into another design.

You can include reference edges as position geometry in your iFeatures. Reference edges allow you to capture additional design intent, but require that the iFeature be positioned in the same way it was designed originally in every part in which it is placed. iFeatures are saved in their own type of file that has an *.ide* extensionand a unique icon, as shown in the following image.

Hole-Pattern.ide

Figure 8-157

iFeatures are stored in a catalog. The catalog is a directory on your computer or on a server that is set up to store all of the iFeatures that you create. You can browse the iFeature catalog at any time by clicking the View Catalog tool on the Part Features panel bar (or toolbar), as shown in the following image.

Figure 8-158

When the View Catalog tool is selected, Windows Explorer opens to the directory specified in the iFeature Root setting on the iFeature tab of the Application Options dialog box. In the iFeature Root folder you can view, copy, edit, or delete iFeatures from your catalog.

CREATE IFEATURES

You create iFeatures using the Extract iFeature tool that is available on the Tools menu, as shown in the following image.

Figure 8-159

Once selected, the Create iFeature dialog box appears, as shown in the following image.

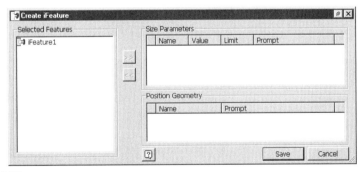

Figure 8-160

The Create iFeature dialog box contains the following sections.

Selected Features

This area of the dialog box, shown in the following image, lists the features selected to be included in the iFeature. Dependent features of a selected feature are included by default, but you can remove them from the Selected Features list. You can rename features in the list to have more descriptive names to assist in working with the iFeature at a later time.

Use the Add parameters (>>) button and the Remove parameters (<<) button to move parameters from the highlighted features in the Selected Features list to the Size Parameters table.

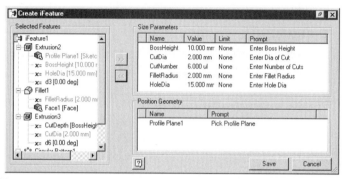

Figure 8-161

Size Parameters

The Size Parameters table lists all of the parameters that will be used for interface when the Feature is placed into another file. You can select the parameters by expanding the features listed in the Selected Features list or directly in the graphics window.

 NOTE Any parameters that have been given a name in the Parameters dialog box appear automatically in the Size Parameters pane of the Create iFeature dialog box upon iFeature creation. Renaming parameters that you want to include in an iFeature can speed up the process of creating them.

Name Specify a descriptive name for the parameter. Use names that describe the purpose of the parameter.

Value Place a value to be the default for the parameter when inserting your iFeature into a file. The value is restricted by settings in the Limit column.

Limit Place restrictions on the values that are available for the parameter by using one of three options: None, Range, or List, as shown in the following image.

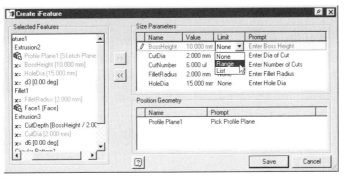

Figure 8-162

None specifies that no restrictions be placed on the Value field. Range gives you the ability to specify a minimum and maximum value, including less than, equal to, and infinity (shown on the left in the image). You can specify the default value to use. The List option, as shown on the right in the following image, allows you to predefine a list of values. Upon placement, the user can choose from these values for the size parameters.

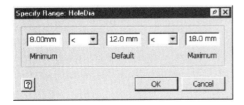
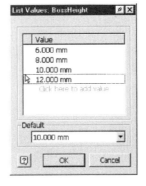

Figure 8-163

Prompt

Enter descriptive instructions to explain further why the parameter is used. The text entered in the prompt field appears in a dialog box during the iFeature placement.

Position Geometry

Specify the geometry of the iFeature that is necessary to position it on a part. You can add or remove geometry to or from the Position Geometry list in the Selected Features tree by right-clicking the geometry and selecting Expose Geometry. You remove geometry from the Position Geometry list by right-clicking it and selecting Remove Geometry.

Name   This describes the position geometry.

Prompt

Enter descriptive instructions to prompt a user for position geometry. The text entered in the prompt field appears in a dialog box during the iFeature positioning.

You can customize the position geometry by right-clicking it and selecting one of the two additional options available on the menu.

Make Independent

Select this option to separate entries for geometry shared by more than one feature.

Combine Geometry

Select this option to combine listings of geometry shared by more than one feature in a single entry.

INSERT IFEATURES

Insert iFeatures using the Insert iFeature tool that is available on the Part Features panel bar, as shown in the following image.

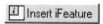

Figure 8-164

Once selected, the Insert iFeature dialog box appears, as shown in the following image. You can select tasks from the tree structure in the left pane of the dialog box or move forward and backward using the Next and Back buttons of the dialog box. Click the Browse button to navigate the iFeatures folder structure and select an iFeature to place.

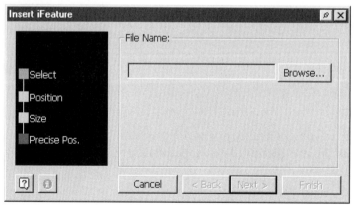

Figure 8-165

The Insert iFeature dialog contains the following sections.

Select   Choose the iFeature that you want to insert using the Browse button.

Position   This lists the names of the interface geometries specified during the creation process of the iFeature. After selecting a planar face, work plane, or series of face edges, click the arrowhead on the positioning symbol to either move or rotate the symbol, as shown in the following image.

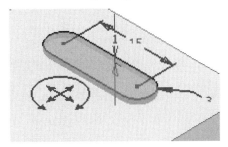

Figure 8-166

You can also specify a precise rotation value directly in the angle field of the dialog box. After the requirement has been satisfied, a check mark will be placed in the left column, as shown in the following image.

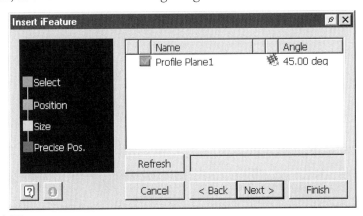

Figure 8-167

Name

This section lists the named interface geometry.

Angle

This section shows the default of the placement geometry on the iFeature.

Move Coordinate System

This section defines horizontal or vertical axes when the iFeature has horizontal or vertical dimensions or constraints included.

Size

This shows the names and default values specified for the iFeature, as shown in the following image. Click in the row to edit the values and then click Refresh to preview the changes.

Name

This section lists the name of the parameter.

Value

This section lists the value of the parameter.

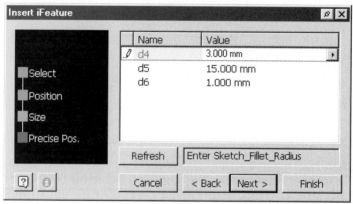

Figure 8-168

Precise Position

This further refines the position of the iFeature, using either dimensions or constraints after you have placed it.

Activate Sketch Edit Immediately

Click this option to activate the sketch of the iFeature and the 2D Sketch panel bar. You can then apply additional dimensions and/or constraints to position the iFeature on the part.

Do not Activate Sketch Edit

Click this option, as shown in the following image, to position the iFeature without applying additional constraints or dimensions.

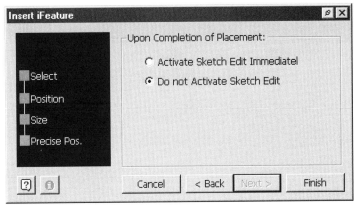

Figure 8-169

EDITING IFEATURES

You can edit an iFeature by opening the *.ide* file in Autodesk Inventor. There are three tools available in the iFeature environment when the file is opened: Edit iFeature, View Catalog, and iFeature Author Table.

The Edit iFeature tool, shown in the following image, opens the Create iFeature dialog box. You cannot change which parameters are used to define the iFeature after it has been created, but you can modify the size parameters and position geometry by editing the properties for the name, value, limit, and prompt.

Figure 8-170

The iFeature Author Table tool, shown in the following image, allows you to create iFeatures that are driven by a table.

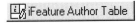

Figure 8-171

Once selected, the iFeature Author dialog box appears, as shown in the following image. The table in the iFeature Author tool functions the same as it does when working with iParts. You can insert, delete, and specify the default row. You can also define custom parameter columns or cells and include a range for the value. The main difference is that there is no option to set a column as a file name, display style, or material. If you are converting an iFeature to a table-driven iFeature that contains a parameter with a list of possible values, the table is populated automatically with the different sizes from the list as rows of the table. You can also edit the table via a spreadsheet.

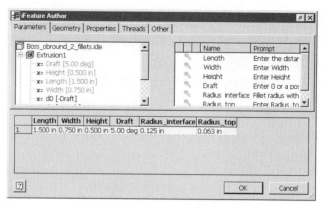

Figure 8-172

Parameters Tab

The Parameters tab is the default tab that appears in the iFeature Author dialog box. This tab lists all of the parameters used to create the iFeature. You cannot change which parameters make up the iFeature after it has been created. The Size Parameters pane on the right of the dialog displays all of the parameters used in the definition and their names and prompts. You can define keys for the parameters that function the same as when working with iParts.

Geometry Tab

The Geometry tab, shown in the following image, allows you to add or remove interface geometry for any iFeatures created after the table-driven iFeature is saved. Previously placed iFeatures are not affected. You can also modify the labels during editing.

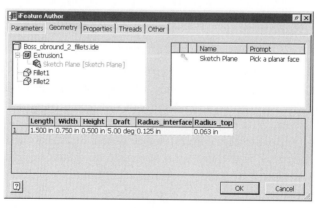

Figure 8-173

Properties Tab

The Properties tab, shown in the following image, lists all of the Summary and Project properties for the file. This tab functions the same as it does when working with iParts, with the exception that you cannot declare a column to be a File Name, Display Style, or a Material column, and that Physical Properties are not available.

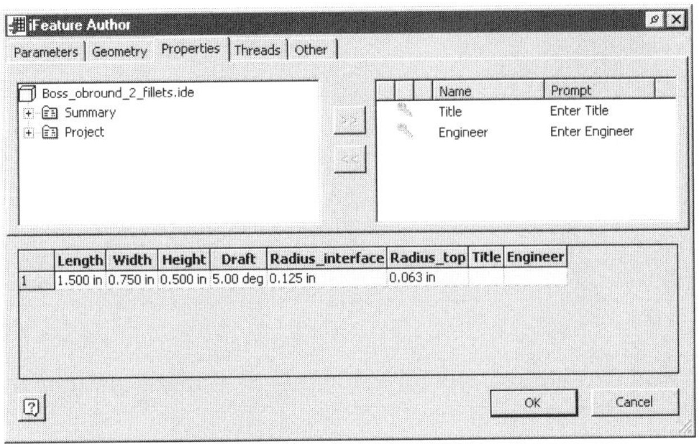

Figure 8-174

Threads Tab

You can control thread features in the iFeature similar to the way you do when working with iParts. Use Family to identify the thread standard, Designation to control the size and pitch, and Class to define fit based upon the standard. You can assign the thread Direction as right or left in the table. In addition, options to modify the pipe diameter, class, and family are available.

Other Tab

You can use the Other tab to create custom table items. This tab functions the same as it does when working with iParts, with the exception that you cannot declare a column to be a File Name, Display Style, or Material column.

INSERTING TABLE-DRIVEN IFEATURES

You insert table-driven iFeatures the same way that you would insert a typical iFeature. The wizard displays the key parameters as a droplist while defining the Size parameters, and any custom parameters willl have the edit fieldavailable for entry. In the example shown in the following image, the parameters that are key values display a droplist from which you can select the appropriate value. The Draft parameter is a custom parameter, which means you can enter the draft angle directly in the dialog box.

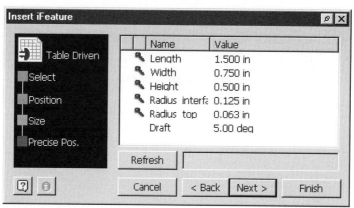

Figure 8-175

The Browser displays table-driven iFeatures in a similar manner to that used with iParts.

When you place a table-driven iFeature in a part file, a new *.ide* file is not created. The placement is similar to that of an iFeature that is not table-driven. The file in which the iFeature is placed contains an independent copy of the iFeature that is not associative to the original *.ide* file. The spreadsheet that contains the table information is not visible in the Browser, and you cannot edit it once you place it in the part file.

EXERCISE 8-7 Creating and Placing iFeatures

In this exercise, you will create a hex drive iFeature from an existing component. You then insert the iFeature into another part file.

1. Open *ESS_E08_07.ipt*.

2. Click the Parameters tool, review the list of Model Parameters. The renamed diameter *(Hex_Dia)* and depth *(Hex_Depth)* parameters of the hex drive extrusion are the variables that will be extracted with the iFeature. The geometry of the raised ring around the hex drive is linked to the diameter of the hex drive.

3. In the Parameters dialog box, click Done.

4. From the Tools menu click Extract iFeature.

5. Click feature named *Hex* in the Browser. All dependent features are selected and shown in the Create iFeature dialog box.

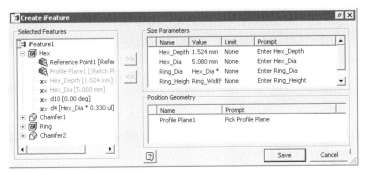

Figure 8-176

6. Click the *Pick Profile Plane* prompt under Position Geometry and enter **Select Plane to Position the Hex Drive** as the new prompt.

7. Expand *Hex* under Selected Features and right-click *Reference Point1* and click **Expose Geometry** from the menu.

Figure 8-177

8. The point is added to the Position Geometry list. This point constrained the hex drive extrusion concentric to the circular face on the end of the cylinder. The point referenced will allow you to select model geometry to position the iFeature during placement in a new part. Click the *Pick Reference Point* prompt under Position Geometry.
   - Highlight the text and replace it with **Select Point to Position Hex Drive Centerline** as the new prompt.
   - Click and drag the Reference Point1 position geometry below Profile Plane1 as shown in the following image.

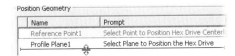

Figure 8-178

- The following image shows the completed operation.

Name	Prompt
Profile Plane1	Select Plane to Position the Hex Drive
Reference Point1	Select Point to Position Hex Drive Centerli

Figure 8-179

9. There are three parameters; *Ring_Height, Ring_Width* and *Ring_Dia* that are linked to the hex drive diameter and are not required in the iFeature. To remove these parameters from the list do the following
   - In the Size Parameters area, click *Ring_Height and* click the Remove parameters button (<<).
   - In the Size Parameters area, click *Ring_ Width and* click the Remove parameters button (<<).
   - In the Size Parameters area, click *Ring_ Dia and* click the Remove parameters button (<<).

Figure 8-180

10. Only two parameters should exist in the SizeParameter areal Hex_Depth and Hex_Dia.

11. Click Save and enter **Hex Drive** in the File name edit box and place the file in the same location as the exercise files.

12. Close the file. Do not save changes.

13. Open *ESS_E08-07-Lock Post.ipt*.

14. Click the Insert iFeature tool.
    - Click the Browse button in the Insert iFeature dialog box.
    - Navigate to the directory where the exercises were installed.
    - Select *Hex Drive.ide*.
    - Click Open.

15. To locate Profile Plane1 click the front face of the Lock Post as shown in the following image.

Chapter 8 Advanced Part Modeling Techniques 643

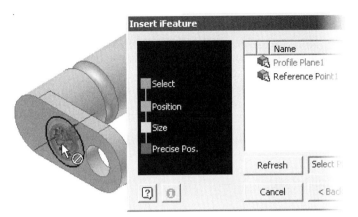

Figure 8-181

16. Click Reference Point1 in the Insert iFeature dialog box.
    - Click the outer circular edge of the lock post.
    - Click the point that is displayed on the part as shown in the following image.

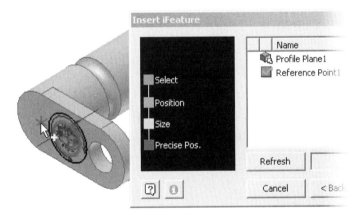

Figure 8-182

17. Click Next.

18. Leave the *Hex_Depth* and *Hex_Dia* variables as their default values.

19. Click Next.

20. Click Do Not Activate Sketch Edit.

21. Click Finish.

22. The hex drive iFeature is placed with its center point coincident with the selected center point of the arc.

Figure 8-183

23. Close the file. Do not save changes.

PUBLISHING PARTS AND FEATURES

Another method that allows you to reuse existing parts in other assemblies or reuse existing features in other parts is to publish parts and features to the content library. Once the parts and features are published they can be reused in another assembly or a part. To publish a part or feature, follow these steps.

PUBLISHING PARTS

To publish a part, follow these steps

1. Open the part file that represents the part, that will be published.
2. From the Tools menu click Publish Part.
3. The Publish dialog box appears. Expand the category that the part will be published to or create a new category, by right-clicking in the Category View area either on an existing category to create a subcategory or click in a blank area to create a new category and then click Add Category as shown in the following image.

Figure 8-184

4. Enter a name in the Family Name area near the bottom of the dialog box. The following image shows the name Gasket with Slot entered for the family name.

Figure 8-185

5. Click the Publish button.

 Note For more information about publishing parts see the help system.

PUBLISHING FEATURES

To publish a feature follow these steps

1. Open the part file that contains the feature that will be published.

2. From the Tools menu click Publish Feature.

3. You will be prompted to select a feature to publish, click the feature in the Browser or in the graphics window then click Next.

 Note Only one feature can be selected, dependent features won't be included.

4. The Author dialog box will appear (the name will be different depending upon the feature that was selected) and the Parameters tab will be active. The selected feature's parameters populate the Available Parameters area. Select which parameters will be published by either double clicking on their name or click on their name and click the Add parameters (>>) button.

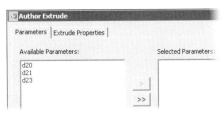

Figure 8-186

5. Click on the Properties tab (the name will be different depending upon the feature that was selected). In the Drop area you determine what type of location the sketch will be located; aligned (flat) or tangent (circular face).

Figure 8-187

6. The Publish dialog box will appear. Expand the unit type you want to publish the feature to. Then click the category that the part will be published to or create a new category right-clicking in the category area and click Add Category as shown in the following image. You can also add subcategory to an existing category.

Figure 8-188

7. Enter a name in the Family Name area near the bottom of the dialog box. The following image shows the name Rectangular Slot entered for the family name.

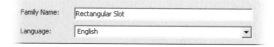

Figure 8-189

8. Click the Publish button.

 Note For more information about publishing features see the help system.

PLACING PUBLISHED PARTS AND FEATURES

Once parts or features have been published you use the Content Center tool to place them into an assembly or part. The Content Center also contains numerous predefined common parts and features that can be utilized.

PLACING A PUBLISHED PART

To place published part, follow these steps.

Chapter 8 *Advanced Part Modeling Techniques* 647

1. Open an assembly that the part will be placed into.
2. Click Content Center from the Assembly Panel bar.
3. Click the Parts tab as shown in the following image.

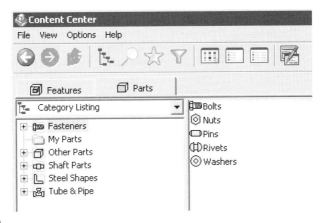

Figure 8-190

4. Click the desired category in the Category Lisiting area. If subcategories exist expand the main category.
5. In the area on the right, double click on the name of the part.

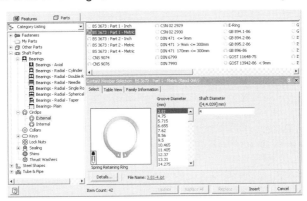

Figure 8-191

6. Select from the available parameters (if any) to define the details of the part.
7. Click in the preview area and with the mouse button depressed drag the part in to the assembly and click to place the part in the desired location.
8. If needed save the file with a new part name.

Note After a part has been saved to the Content Center it cannot be edited.

PLACING A PUBLISHED FEATURE

To place published feature follow these steps.

1. Open the part that the feature will be inserted into.
2. Click Content Center from the Part Features Panel bar.
3. Click the Features tab if it is not current.
4. Click the desired category in the Category Listing area. If subcategories exist expand the main category.
5. In the area on the right, double-click on the desired feature.

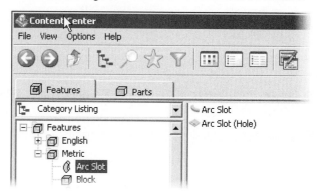

Figure 8-192

6. Select from the available parameters (if any) to define the details of the feature.
7. Click in the preview area and with the mouse button depressed drag the feature in to the part.

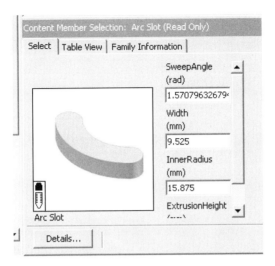

Figure 8-193

8. Click on the face to place the feature, right-click to get options with which to cancel or complete the operation, or move the feature.

9. Edit the feature as any other feature.

DESIGN ACCELERATOR

Autodesk Inventor's Design Accelerator enables you to quickly create complex parts and features based on engineering data such as ratio, torque, power, and material properties. The Design Accelerator consists of engineer's handbook, mechanical calculators and component generators.

- Engineer's handbook contains engineering theory, formulas and algorithms used in machine design.
- Mechanical calculators use standard mathematical formulas and theories to create parts. The following calculators are available, weld and solder joints, plates, bearings, brakes, clamping joints and fit and tolerance.
- Component and feature generators consist of mechanical connections, shafts, hubs, gears, belt and chain drives, power screws, springs, and o-rings

To create parts and features with the Design Accelerator follow these steps.

1. Open the assembly in which to place the part, or the part in which to place the feature.

2. Click Design Accelerator from the Tools menu,

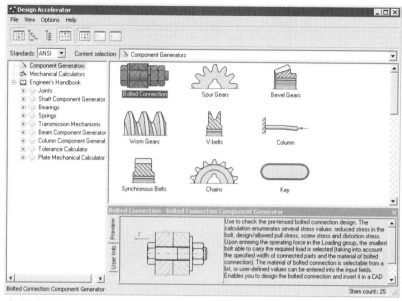

Figure 8-194

3. From the Design Accelerator dialog box select the desired operation you need to perform, expand the category as needed.

4. A wizard will walk you through the steps of the operation.

 Note For more information about the Design Accelerator see the Help system.

CHAPTER SUMMARY

To	Do This	Tool
To extrude an Open Profile	Click the Extrude tool, click the Open Profile, and click on the side of the part that will be filled in.	Extrude +E
To create a rib or web	Click the Rib tool and select an open profile.	Rib
To change the extrusion termination	Click the Extrude tool, click on the More tab, and then click Minimum Solution.	Extrude +E
Place text on a sketch	Click the Create Text tool on the 2D Sketch Panel Bar.	Text
Emboss text	Click the Emboss tool on the Part Features Panel Bar.	Emboss

To	Do This	Tool
Create a sweep feature	Click the Sweep tool on the Part Features Panel Bar.	Sweep Shift+S
Create a coil feature	Click the Coil tool on the Part Features Panel Bar.	Coil
Create a loft feature	Click the Loft tool on the Part Features Panel Bar.	Loft Shift+L
Split a face or part	Click the Split tool on the Part Features Panel Bar.	Split
Copy a feature	Right-click on a feature's name in the Browser, click Copy from the menu, and then Paste the feature.	
Reorder a feature	Click on the feature's name in the Browser and, with the left mouse button depressed, drag the feature to the desired location.	
Mirror a feature	Click the Mirror Feature tool from the Part Features Panel Bar.	Mirror Feature Shift+M
Suppress a feature	Right-click on the feature's name in the Browser and select Suppress Features on the menu.	
Rollback features	Click the End of Part marker in the Browser and, with the left mouse button depressed, drag the End of Part marker to the new location.	
Adjust a file's properties	Click the iProperties option on the File menu.	
View the center of gravity of a part or assembly	Click Center of Gravity on the View menu.	
Create an iPart	Click Create iPart on the Tools menu.	
Create a custom iPart	Right-click a column in the iPart Author dialog and select Custom Parameter Column.	
Place an iPart	Click Place Component tool from the Assembly Panel Bar	
Create an iFeature	Click Extract iFeature on the Tools menu.	
Insert an iFeature	Click Insert iFeature on the Part Features Panel Bar or toolbar.	
Create a table-driven iFeature	Click the iFeature Author Table tool on the iFeature Panel Bar or toolbar.	

APPLYING YOUR SKILLS

SKILL EXERCISE 8-1

1. In this exercise, you use complex part modeling techniques to create a housing for an electronic device. Using the knowledge you gained through this course, you will utilize the lofting, splitting and embossing tools to create a joystick handle.

2. Open ESS_E08_08.ipt.

3. Click the Loft tool.
   - In the graphics window, click the three elliptical sections in order from top to bottom.
   - In the Loft dialog box, select *Click to add* under Rails.
   - In the graphics window, click the two splines. Your screen should resemble the following image.

Figure 8-195

4. In the Loft dialog box, select OK.

5. Next, you remove the top portion of the part using a surface and create a fillet. In the Browser, turn on the visibility of ExtrusionSrf1.

6. Click the Split tool.
   - In the graphics window, click the surface as the split tool.
   - In the Split dialog box, click Split Part, click the Direction button to remove the top portion of the part, then click OK.

Figure 8-196

7. Turn off the visibility of ExtrusionSrf1.

8. Use the Fillet tool to create a 3 mm fillet around the top edge.

Figure 8-197

9. Next, you create a sketch and use the Emboss tool to create a button. Click the 2D Sketch tool. In the Browser, click Button_Plane to define the sketch plane.

10. Use the Look At tool and click Button_Plane in the Browser.

11. Create a circle as shown below and constrain it by doing the following
    - Project the inside ellipse.
    - Add a **12 mm** diameter and a **16 mm** horizontal dimension.
    - Add a horizontal constraint to the center point of the circle and left quadrant of the projected ellipse.

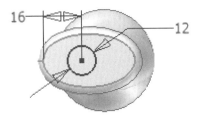

Figure 8-198

12. Click the Return tool to finish the sketch.

13. Change to an isometric view.

14. Click the Emboss tool.
    - Click inside the circle.
    - Enter a Depth of **2 mm**.
    - Click the Top Face Color button and select any color.
    - Verify that the Emboss from Face option is selected.
    - Click the Wrap to Face option, then in the graphics window click the top face of the part.
    - Click OK.

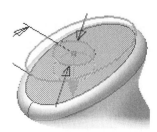

Figure 8-199

15. To finish the design, create a **1 mm** fillet around the top edge of the button.

Figure 8-200

16. Close the file and do NOT save changes.

CHECKING YOUR SKILLS

Use these questions to test your knowledge of the material covered in this chapter.

1. True__ False__   When creating a single rib or a web feature, you can only select a closed profile as the profile.

2. True__ False__   Both the Extrude and Revolve tool can use the minimum or maximum extrusion solutions.

3. True__ False__   You can only place embossed text on a planar face.

4. True__ False__   A sweep feature requires three unconsumed sketches.

5. True__ False__   You can create a 3D curve with a combination of both 2D and 3D curves.

6. Explain how to create a 3D path using geometry that intersects with a part.

   _____

   _____

7. True__ False__   The easiest way to create a helical feature is to create a 3D path and then sweep a profile along this path.

   _____

   _____

8. True__ False__   You can control the twisting of profiles in a loft by defining point sets.

9. Explain how to save both halves of a part after splitting it.

   _____

   _____

10. True__ False__   You can copy features between parts using the Copy Feature tool on the Part Features Panel Bar or toolbar.

11. Explain the difference between suppressing and deleting a feature.

    _____

    _____

12. True__ False__   After mirroring a feature, the mirrored feature is independent from the parent feature. If the parent feature changes, the mirrored feature will not reflect this change.

13. Explain why you would want to override a part's mass and volume properties.

    _____

    _____

14. True__ False__   After changing a part's physical material properties, the part's color in the graphics window will change to match the material.

15. True__ False__   Multiple versions of an iPart can be placed in an assembly.

16. True__ False__   When you make changes to an iPart factory, the changes are updated automatically in iParts that have been placed in assemblies.

17. True__ False__   You can add features to standard iParts after they have been placed in an assembly.

18. True__ False__   Named parameters are added automatically as Size parameters during iFeature creation and cannot be removed.

19. What happens when you create a table-driven iFeature and one of the original parameters contains a list of values for the parameter?

_____

_____

CHAPTER 9

Sheet Metal Design

In this chapter you will learn how to create sheet metal parts in Autodesk Inventor. Assemblies often require components that are manufactured by bending flat metal stock to form brackets or enclosures. Cutouts, holes, and notches are cut or punched from the flat sheet, and 3D deformations such as dimples or louvers are often formed into the flat sheet. The punched sheet is then bent at specific locations using a press brake or other forming tools to create a finished part. The sheet metal environment in Autodesk Inventor presents specialized tools for creating models in both the folded and unfolded states.

CHAPTER OBJECTIVES

After completing this chapter, you will be able to

- Start the Autodesk Inventor sheet metal environment
- Modify settings for sheet metal design
- Create sheet metal parts
- Modify sheet metal parts to match design requirements
- Create sheet metal flat patterns
- Create drawing views of a sheet metal part

INTRODUCTION TO SHEET METAL DESIGN

A metal blank folded into a finished shape is a common component in a mechanical assembly. Examples include enclosures, guards, simple-to-complex bracketry, and structural members. Although the term *sheet metal* is often associated with these components, heavy plates (1"+) can also be formed using similar methods.

Common sheet metal components include electronic and consumer product chassis and enclosures, lighting fixtures, support brackets, frame components, and drive guards.

Sheet metal fabrication can include a number of processes:

- Drawing
- Stamping
- Punching and cutting

- Braking
- Rolling
- Other, more complex operations

The tools in the Autodesk Inventor sheet metal environment enable you to create press brake-formed models. You can also create rolled shapes including cones and cylinders. In addition, sheet metal parts can include formed features such as nail holes, lances, and louvers. Using lofts and surfacing tools, you can create parts that are fabricated through deformation processes such as drawing or stamping, but Autodesk Inventor cannot unfold these parts to determine the shape of the flat blank.

SHEET METAL FABRICATION

Brake-formed parts must be described in a minimum of two states: the flat blank shape prior to bending and the finished part in its folded form. Manufacturing processes are often performed on the flat sheet before folding the part into a finished shape. Holes and other openings are punched into the flat stock, and deformations such as dimples are stamped into the blank with forming tools and dies. Preparing the flat stock can be done manually, or more commonly with CNC punching machines. High-volume sheet metal parts are often created on progressive die lines where a continuous feed of flat stock (from a coil) is passed through a series of forming dies that remove material or form the shape of the component. For lower-volume components, manual or CNC press brakes are the primary tools used to bend the finished blank into its bend up form.

In most sheet metal designs, the folded shape of the part is known. You can create the folded model using various techniques, the most basic being a process of adding individual faces, flanges, and other sheet metal features to build the final state of the folded part. A key face is the first feature added to the part; additional "as bent" features are added and joined to open edges of the part. During feature creation, a bend is automatically created at the intersecting edge of the new feature and the existing part. You can also add bends between disjointed faces, a technique that is very helpful when designing a sheet metal component in the context of an assembly. When the folded design is complete, Autodesk Inventor can create a flattened version of the model, commonly called a flat pattern. The flat pattern locates the position of bend centerlines used to form the part in a press brake or other sheet metal forming tools. The flat pattern also contains any holes, cutouts, and other features placed on faces of the folded part. Settings in the current sheet metal style determine the size of the flat pattern.

When a metal blank or sheet is bent in a press brake, the metal in the bend area deforms. Material on the inside of the bend compresses, and material on the outside of the bend is stretched. This deformation must be taken into account when calculating the flattened state of the folded model. The amount of deformation is dependent on the material, the radius and angle of the bend, and the process and equipment used to create the bend. In the sheet, there is a plane where the material neither compresses nor stretches. If the location of this plane is known, the length of the flat sheet can be calculated. For many materials and processes, the location of this neutral plane is known and can be expressed as a percentage of the thickness of the part, as measured from the surface on the inside of a bend. This offset value is referred to as a kFactor. The following image shows the neutral plane of a sheet metal bend. The location of the neutral plane is kFactor * Thickness, where 0 < kFactor < 1. The default sheet metal style uses a kFactor of 0.44. This value is appropriate for many common materials and processes.

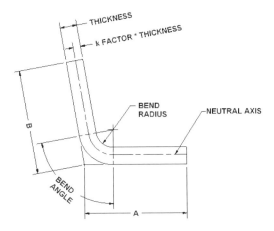

Figure 9-1

You can edit the default sheet metal style or add additional styles. You store sheet metal styles in a template and apply the appropriate style for the design. The current sheet metal style applies to all sheet metal features in the part. Although you cannot apply separate styles to different sheet metal features in a part, you can override default values during feature creation or editing. For example, some materials have different bending properties dependent on the orientation of the bend relative to the grain of the material. As the sheet is manufactured, the material microstructure aligns with the rolling direction. This results in a sheet with bend properties related to its orientation in the press brake. Stainless steel sheets often display this anisotropic behavior.

The use of a constant kFactor is not always appropriate. For some materials and processes, when formed parts require precise tolerances, you can use a bend table in place of the kFactor. A bend table uses empirically derived information to apply length adjustments to bends. With overrides, you can apply a separate kFactor or bend table to each bend on a sheet metal part.

Bend Tables

In the calculation of the unfolded length of a bend, a bend table replaces the kFactor with a set of known adjustments to the unfolded length. A bend table can provide greater precision of unfolded dimensions because the values are derived from your measurements of bend length of a particular material on a specific machine. When you use a bend table to calculate an unfolded length, the following formula applies:

L = A + B - x

Referring to the previous equation, the variables in the formula are:
L Unfolded length
A Length of folded face 1
B Length of folded face 2
x Adjustment from bend table

The value of x is dependent on the sheet thickness, the angle of the bend, and the inner radius of the bend.

Note that the measurements of A and B are to the projected intersection of the extended outer faces on either side of the bend. This intersrection is used when the angle of the bend is less than or equal to 90°. The following image applies when the bend angle is greater than 90°. The measurements are parallel to the face and tangent to the outer surface of the bend. The same formula is used to determine the unfolded length of the part.

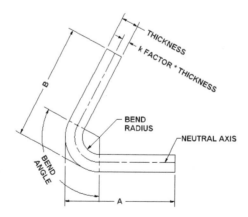

Figure 9-2

Autodesk Inventor can use a bend table that is a plain text file (*.txt* extension), but you can create or edit a spreadsheet version of the bend table using Microsoft Excel and then save the table in text format.

Autodesk Inventor includes both metric (mm) and English unit (in) bend tables that you can modify. Each bend table is supplied in both *.txt* and *.xls* formats. The following image illustrates the angle information required in a bend table. A flat sheet bent at an angle of 110° will have an opening angle of 70°. The opening angles are entered in the bend table.

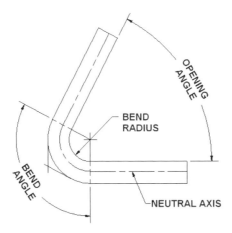

Figure 9-3

SHEET METAL PARTS

Sheet metal parts can be designed separately or in the context of an assembly. Since sheet metal parts are often used as supports or enclosures for other components, designing in the assembly environment can be advantageous.

SHEET METAL DESIGN METHODS

Sheet metal parts are most often created in the folded state. The model is then unfolded into a flat sheet, as shown in the following image, using the Flat Pattern tool. To create the folded model, extrude the first sketch the thickness of the sheet to create a face. Add additional faces or flanges to open edges of the part and add bends automatically between the features. Add cuts and other special sheet metal features to the part as required.

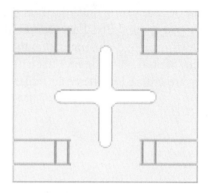

Figure 9-4

Autodesk Inventor's ability to create models with disjointed solids (unconnected features) is very helpful when building sheet metal parts in an assembly (see the following image). You can build separate faces of a sheet metal part quickly by referencing faces on other parts in the assembly. Additional sheet metal features can then join the distinct faces.

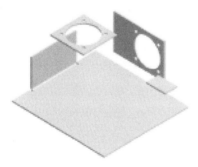

Figure 9-5

You can also create sheet metal parts from standard parts that have been shelled. All walls of the shelled part must be the same thickness and match the thickness of the flat sheet as defined by the style. The solid corners of the shelled part are ripped open to enable the model to unfold, and appropriate bends are added along the edges between faces.

CREATING A SHEET METAL PART

The first step in creating a sheet metal part is either to click the New icon in the What To Do section or to click New on the File menu on the standard toolbar. You can then click the *Sheet Metal.ipt* icon on the Default tab, as shown in the following image.

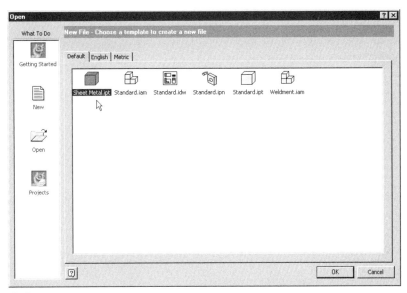

Figure 9-6

You can switch between the standard modeling environment and the sheet metal environment at any time by clicking Sheet Metal on the Convert menu, as shown in the following image. You can use Autodesk Inventor's general modeling tools to add features to a sheet metal part, even when the Sheet Metal application is active.

Figure 9-7

The creation of a sheet metal part usually begins by specifying the sheet metal style or settings, such as sheet thickness, material, and bend radius. You then use sheet-metal-specific tools to create sheet metal parts by either adding faces at existing edges or connecting disjointed faces with bends. Use the optimized sheet metal tools to add, cut, and clean up features of the part. Sheet metal faces and parts can be adaptive and their size adjusted to meet design rules specified by assembly constraints.

SHEET METAL TOOLS

After creating the new sheet metal document from a sheet metal template file, Autodesk Inventor's Panel Bar tools change to reflect the sheet metal environment. As with standard parts, a sketch is created and becomes active. The sketch tools are common to both sheet metal and standard parts. The first sketch of a sheet metal part must be either a closed profile that is extruded the sheet thickness to create a sheet metal face or an open profile that is thickened and extruded as a contour flange. Use additional tools to add sheet metal features to the base feature. The following is a description of the tools used to create sheet metal features. You can also add standard part features to a sheet metal part, but these features may not unfold when a flat pattern is generated from the folded model.

Use sheet metal tools to:

- Build a sheet metal part by adding faces along edges of existing faces.
- Create individual key faces of the sheet metal part and then connect these disjointed faces by adding bends between them.
- Extend faces automatically to create corner seams where face or flange edges meet.
- Cut shapes from faces with tools enhanced for sheet metal design.
- Add standard features, such as chamfers and fillets, using tools that are optimized for working with thin sheets.
- Create a flat pattern model of the part with a single button click. This flat pattern model is updated automatically as features are added, removed, or edited.
- Create drawing views of the folded part and flat pattern to document your design for manufacturing.

SHEET METAL STYLES

The sheet metal-specific parameters of a part are stored in a sheet metal style. These include the thickness and material of the sheet metal stock, a bend allowance factor to account for metal stretching during the creation of bends, and various parameters dealing with sheet metal bends and corners. You can create additional sheet metal styles to account for various materials and manufacturing processes or material types. If you create sheet metal styles in a template file, they are available in all sheet metal parts based on that template.

Figure 9-8

The thickness of the sheet metal stock is the key parameter in a sheet metal part. Sheet metal tools such as Face, Flange, and Cut automatically use the Thickness parameter to ensure that all features are the same wall thickness, a requirement for unfolding a model. In the default sheet metal style, all other sheet metal parameters, such as Bend Radius, are based on the Thickness value.

To edit or create a sheet metal style, follow these steps:

1. Click the Sheet Metal Styles tool from the Sheet Metal Features Panel Bar. The Sheet Metal Styles dialog box appears, as shown in the following image.

2. To create a new sheet metal style, click the New button in the Sheet Metal Styles dialog box. This creates a copy of the style that is currently selected in the Style List section of the dialog box.

3. Edit the values on the Sheet, Bend, and Corner tabs to define the default feature properties of parts created with this sheet metal style.

4. Rename the style and click the Save button.

5. When multiple sheet metal styles exist in a part, set the active style by selecting it in the Active Style list.

Changing to a different sheet metal style or making changes to the active style updates the sheet metal part to match the new settings.

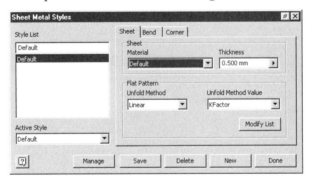

Figure 9-9

The following is a description of the settings in a sheet metal style. The parameters with numeric values, such as Thickness and Bend Radius, are saved as model parameters. The parameter values are updated when you modify a sheet metal style or activate a new sheet metal style. Other model parameters and user-defined parameters can reference these parameters in equations.

Sheet Tab

The Sheet tab contains settings for the sheet metal material and the unfolding method. You can create multiple styles that contain different materials and thicknesses to be applied to a sheet metal model. The previous image shows the Sheet tab.

Material Specify a material from the list of defined materials in the document. The material color setting is applied to the part. You can define new materials using the Styles Editor.

Thickness Specify the thickness of the flat stock that will be used to create the sheet metal part.

Unfold Method Specify the method used to calculate bend allowance. Bend allowance accounts for material stretching during bending. Options are Linear or Bend Table for more complex or precise requirements.

Unfold Method Value Specify the default value used for calculating bend allowances. A kFactor is a value between 0 and 1 that indicates the relative distance from the inside of the bend to the neutral axis of the bend. A kFactor of 0.5 specifies, the neutral axis lies at the center of the material thickness. The Unfold Method Value is combined with the Unfold Method.

Modify List Click to see a list of kFactors and named Bend Tables included in the sheet metal style. Each bend can have the default bend allowance overridden by a value in this list.

Bend Tab

The Bend tab contains settings for sheet metal bends, as shown in the following image. Typically, most bend settings are defined as a function of the sheet thickness. Bend Radius is the inside radius of the completed bend. This setting is the default for all bends, but you can override it during the creation of any bend.

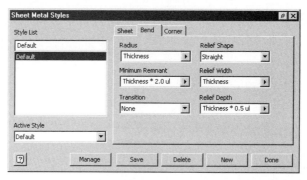

Figure 9-10

You can specify bend reliefs when a bend zone (the area deformed during a bend) does not extend completely beyond a face (see the following image).

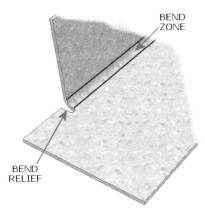

Figure 9-11

If bend reliefs are used, they are incorporated in the flat blank prior to folding. You can generate bend reliefs by punching or with laser, water-jet, or other cutting methods. The creation of bend reliefs may increase production costs. It is common practice with thin, deformable materials, such as mild steel, to add bends without bend reliefs. The material is allowed to tear or deform where the bend zone intersects the adjacent face, as shown in the following image.

Figure 9-12

Descriptions of the settings on the Bend tab follow.

Radius Define the inside bend radius value between adjacent, connected faces.

Minimum Remnant Specify the distance from the edge at which, if a bend relief cut is made, the small tab of remaining material is also removed.

Transition Control the intersection of edges across a bend in the flattened sheet. For bends without bend relief, the unfolded shape is a complex surface. Transition settings simplify the results, creating straight lines or arcs, which can be cut in the flat sheet before bending. The following image shows four of the five transition types. Trim to Bend, which is not shown, does not create a transition.

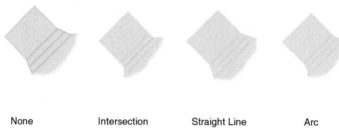

None   Intersection   Straight Line   Arc

Figure 9-13

Relief Shape If a bend does not extend the full width of an edge, a small notch is cut next to the end of the edge to keep the metal from tearing at the edge of the bend. Select from Straight, Round, or None for the shape of the relief. The following image shows the Straight and Round shapes.

Relief Width Specify the width of the bend relief.

Relief Depth Specify the distance a relief is set back from an edge. Round relief shapes require this be at least one half of the Relief Width value.

Figure 9-14

Corner Tab

A corner occurs where three faces meet. The corner seam feature controls the gap between the open faces and the relief shape at the intersection. As with bend reliefs, corner reliefs are added to the flat sheet prior to bending. You can choose from five corner relief options, four of which are shown in the following image. The Linear Weld option (not shown) creates a narrow gap for corners that are to be welded. You can view the shape of the corner relief in either the flat pattern or folded view of the model when a corner seam feature has been applied.

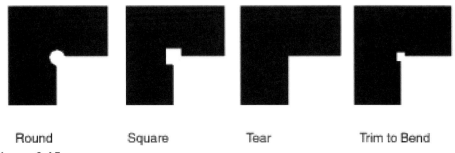

Round   Square   Tear   Trim to Bend

Figure 9-15

The Corner tab, shown in the following image, is where you set how corner reliefs will be applied to the model. You can designate corner relief size and shape.

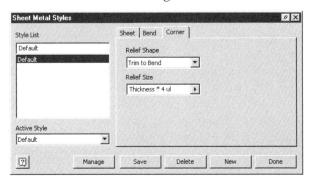

Figure 9-16

Relief Shape   Specify the shape of the corner relief. The five shapes available are Round, Square, Tear, Trim to Bend, and Linear Weld.

Relief Size   Set the size of the corner relief.

FACE

The Face tool, shown in the following image, extrudes a closed profile for a distance equal to the sheet metal thickness. If the face is the first feature in a sheet metal part, you can only flip the direction of the extrusion. When you create a face later in the design, you can connect an adjacent face to the new face with a bend. If the sketch shares an edge with an existing feature, the bend is added automatically. As an option, you can select a parallel edge on a disjointed face. This will extend or trim the attached face to meet the new face, with a bend created between the two faces.

Figure 9-17

To create a sheet metal face, follow these steps:

1. Create a sketch with a single closed profile or a closed profile containing islands. The sketch is most often on a work plane created at either a specific orientation to other part features or by selecting a face on another part in an assembly.

   NOTE You can create a single face from multiple closed profiles.

2. Click the Face tool on the Sheet Metal Features Panel Bar. The Face dialog box appears, as shown in the following image.

If a single closed profile is available, it is selected automatically. If multiple closed profiles are available, you must select which one defines the desired face area.

3. If required, flip the thickness direction to extrude the profile.

4. If the face is not the first feature and the sketch does not share an edge with an existing feature, click the Edges button and select a parallel edge to which to connect the face. The two faces are extended or trimmed as required, meeting at a bend. The bend is listed as a child of the new face in the Browser. If the face attached to the selected edge is parallel but not coplanar with the new face, a set of double bend options are presented. An additional face is added to connect the two parallel faces. The orientation and shape of this face is determined by the selected double bend options.

The Unfold Options and Relief Options tabs contain settings for overriding the default values in the sheet metal style. Overrides are applied to individual features.

5. Click OK.

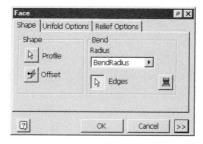

Figure 9-18

Shape Tab

The Shape tab is where you select the profile and direction of the face. You can also override the bend radius specified in the sheet metal style and select edges of other existing faces or edges that exist in the sheet metal part to apply a bend feature upon creation of the face. The following options are available on the Shape tab.

	Profile	Select a profile(s) to extrude a distance equal to the Thickness parameter.
	Offset	Toggle the direction for the creation of the feature.
	Radius	This displays the default bend radius. If the face will be attached to another edge of the part, you can specify the value for the bend radius to be used. You can modify the default value on a *per feature* basis.
	Edges	Select an edge to include in the bend. When selected, the edge is extended to match the edge of the face. A bend feature is created between the two faces.
	Extend Bend Aligned to Side Faces	Click to extend material along the faces of the edges connected by the bend instead of perpendicularly to the bend axis. An image of a part before this setting is applied is shown on the left side in the following image, and an example of the part when this option is active is shown on the right side.

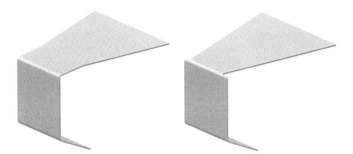

Figure 9-19

There are options for creating a double bend or full-radius bend that you can access by clicking the More (>>) button. These options are discussed in the section on bends later in the chapter.

Unfold Options Tab

The Unfold Options tab is available so you can override the values set in the Sheet Metal Style dialog box, if necessary. This tab also exists when working with other sheet metal tools (i.e., Flange, Contour Flange, Hem, Bend, and so on). The following image exhibits the Unfold Options tab; the following sections describe its components.

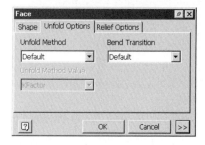

Figure 9-20

Unfold Method   Set the unfold method. Available methods are Default (reverts to the setting in the Sheet Metal Style), Linear, and Bend Table. Linear and Bend Table give you the ability to choose a different Unfold Method if one has been defined in the Sheet Metal Style.

Unfold Method Value   Specify a value for the Unfold Method. The list provides access to the values defined in the active Sheet Metal Style.

Bend Transition   Set the transition type for the sides of a bend. The following six types are available:

    Default–Creates a bend using the value in the active style.
    None–Uses a spline to join the tangencies.
    Intersection–Joins adjacent edges in the bend zone.
    Straight Line–Uses a straight line to join the tangencies.
    Arc–Uses an arc to join the tangencies.
    Trim to Bend–Does not create a transition.

Relief Options Tab

The Relief Options tab, shown in the following image, is available so you can override the values set in the Sheet Metal Style, if necessary. This tab also exists when working with a number of the sheet metal tools (e.g., Flange, Contour Flange, Hem, Bend, and so on). These settings allow you to override how bends and bend reliefs are created on a per feature basis.

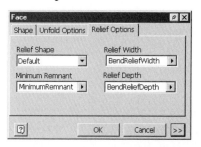

Figure 9-21

Relief Shape   Override the relief shape specified by the active sheet metal style. Available shapes are:

    Default–Uses the setting in the active sheet metal style.
    None–Does not create a bend.
    Round–Creates a relief with full radius corners.
    Straight–Creates a relief with square corners.

Minimum Remnant   Specify a value for the minimum remnant. Any value entered will override the value specified in the active sheet metal style.

Relief Width   Specify a value for the relief width. Any value entered will override the value specified in the active sheet metal style.

Relief Depth   Specifies a value for the relief depth. Any value entered will override the value specified in the active sheet metal style.

CONTOUR FLANGE

You create a contour flange (see the following image) from an open sketch profile. The sketch is extruded perpendicular to the sketch plane and the open profile is thickened to match the sheet metal thickness. The profile does not require a sketched radius between line segments–bends are added at sharp intersections. Arc or spline segments are offset by the sheet metal thickness. A contour flange can be the first feature in a sheet metal part, or you can add it to existing features.

Figure 9-22

To create a contour flange, follow these steps:

1. Create a sketch with a single open profile. The sketch can contain line, arc, and spline segments, and it can be constrained to projected reference geometry to define a common edge between the contour flange and an existing face.

2. Click the Contour Flange tool on the Sheet Metal Features Panel Bar. The Contour Flange dialog box appears, as shown in the following image.

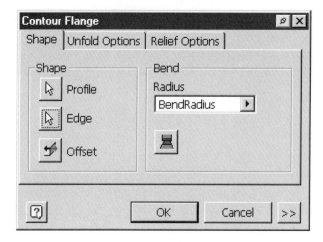

Figure 9-23

3. Click the open sketch profile to define the shape of the contour flange.
4. If required, flip the side to offset the profile.
5. If the contour flange is the first feature in the part, enter an extrusion distance and direction. If the contour flange is not the first feature, you can select an edge perpendicular to the sketch plane to define the extrusion extents. If the sketch is attached to an existing edge, select that edge. If the sketch is not attached to any projected geometry, the contour flange is extended or trimmed to meet the selected edge. The selected edge is typically the closest edge to the end of the sketch.
6. Apply any unfold and bend overrides on the Unfold Options and Bend Relief Options tabs.
7. If you select an edge to define the extrusion extents, you can select from three options to further refine the length of the contour flange.
8. Click OK.

Shape Tab

The Shape tab is where you select the profile, edge, and offset direction of the contour flange. You can also override the bend radius that is specified in the sheet metal style. The following options are available on the Shape tab.

	Profile	Select an open profile to extrude for the shape of the contour flange.
	Edge	Specify an existing edge for termination of the feature.
	Offset	Toggle the direction for the creation of the feature.

Radius This displays the default bend radius. Bends are added to sharp edges of the profile.

 Extend Bend Aligned to Side Faces Click to extend material along the faces on the side of the edges connected by the contour flange.

More Button (>>)

The More button provides you the option to specify the type of extents for the feature, as shown in the following image.

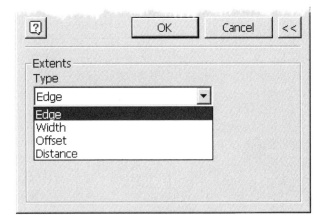

Figure 9-24

These extent types are also available for flange and hem features. The flange and hem features do not have the Distance option because the distance is specified in the main body of the dialog box. Descriptions of the four options follow.

Edge Select so that the contour flange extends the full length of the selected edge.

Width Select so that a point defines one extent of the contour flange. The flange can be offset from this point and extends a fixed distance. You define the starting point by selecting an endpoint on the selected edge, a work point on a line defined by the edge, or a work plane perpendicular to the selected edge.

Offset Select so that, similar to the Width option, two selected points define both extents of the flange. You can specify an offset distance from each point.

Distance Select so that you can enter a distance over which you want the contour flange to be extruded and can specify a direction, as shown in the following image.

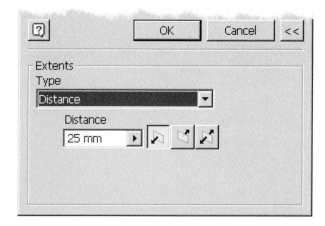

Figure 9-25

The following image shows samples of the Edge, Offset, and Width extent types.

Figure 9-26

FLANGE

A sheet metal flange is a simple rectangular face created from an existing face edge. A sketch is not required when creating a flange. The flange can extend the full length of the selected edge. The Flange tool, shown in the following image, adds a new sheet metal face and bend to an existing face. You set the flange length and the angle relative to the adjacent face in the Flange dialog box. The selected edge defines the bend location between the two faces.

Figure 9-27

 NOTE A minimum of one sheet metal face must exist before creating a flange.

To create a flange, follow these steps:

1. Click the Flange tool on the Sheet Metal Features Panel Bar. The Flange dialog box appears, as shown in the following image.
2. Select an edge of an existing face.
3. Enter the distance and angle for the flange in the Flange dialog box. The flange preview updates to match the current values.
4. If required, flip the direction for the flange and offset for the thickness.
5. Expand the dialog box and select the appropriate extent type. See the Contour Flange section in this chapter for additional information on extents.
6. Click Apply to continue creating flanges, or click OK to apply the flange and exit the dialog box.

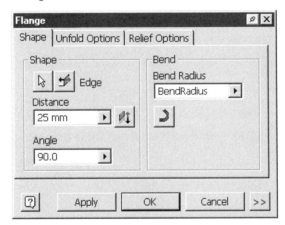

Figure 9-28

Shape Tab

The Shape tab is where you select the edge, edge offset, flange distance, and direction and bend angle for the flange. You can also override the bend radius specified in the sheet metal style. The following options are available on the Shape tab.

 Select Edge Select an edge along which to create the flange.

 Flip Offset Toggle the flange creation to be on the inside or outside of the face.

Distance Enter a distance for the depth of the flange.

Flip Direction Toggle the side of the face used to create the flange.

Angle Enter an angle for the flange. The value must be less than 180°.

	Bend Radius	Override the default bend radius set by the active sheet metal style.
	Bend Tangent to Side Face	Click to create a flange that is tangent to the side face.

The options available on the Unfold Options and Relief Options tabs were discussed in the Face section. The More button provides access to the Edge, Width, and Offset extent types, which are discussed in the Contour Flange section.

EXERCISE 9-1 Creating Sheet Metal Parts

In this exercise, you create sheet metal faces, flanges, and seams to build a small enclosure.

1. Open *ESS_E09_01.ipt*.
2. Click the Sheet Metal Styles tool.
3. On the Sheet tab:
    a. Select Aluminum-6061 from the Material drop-down list.
    b. Enter a value of **1.6** in the Thickness field.
    c. Select the Bend tab.
    d. Select Round from the Relief Shape drop-down list.
    e. Select the Corner tab
    f. Select Round from the Relief Shape drop-down list.
    g. Click Save.
    h. Click Done.

The existing sheet metal face is updated to reflect the revised sheet metal style settings.

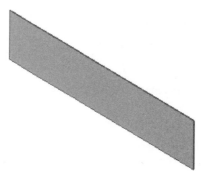

Figure 9-29

4. Create a new sketch on the long, thin face of the existing feature.

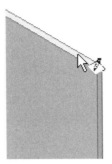

Figure 9-30

5. Click the Look At tool, and click Sketch2 in the Browser.
6. Click the Two Point Rectangle tool.
7. Click the upper horizontal edge when the coincident icon is displayed.
8. Create a rectangle (approximately) as shown.
9. Add a **127** mm dimension to the vertical edge of the sketch rectangle.

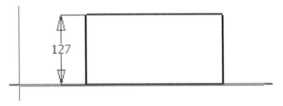

Figure 9-31

10. Click the Return button to exit the sketch environment.
11. Click the Face tool.
12. Select the rectangle profile.
13. Click OK to accept the default settings.
14. Right-click the graphics window.
15. Select Isometric View from the menu.
16. Zoom in on a corner of the bend and examine the bend relief notch in the original part face.
17. Use the Zoom, Rotate and Pan tools to reorient the view as shown.

Figure 9-32

    18. Click the Flange tool

    19. In the graphics window, click the edge on the horizontal surface as shown.

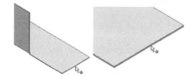

Figure 9-33

    20. In the Flange dialog box:

        a. Click the arrow next to the Distance field.

        b. Select Show Dimensions.

        c. Click the first face you created.

        d. Highlight all of the text in the Distance field.

        e. Click the displayed 127 mm dimension.

The dimension parameter name replaces the text in the Distance field.

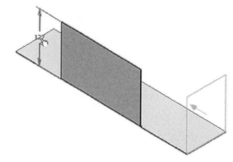

Figure 9-34

21. Click OK to create the flange.
22. Orient your view as shown in the following image.

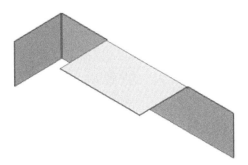

Figure 9-35

23. Click the Flange tool.
24. Click the right-side (nearest) edge of the base feature as shown.

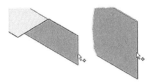

Figure 9-36

25. In the Flange dialog box:
    a. Click the arrow next to the Distance field and select Show Dimensions.
    b. Click the first face you created.
    c. Highlight all of the text in the Distance field and click the displayed 127 mm dimension.
    d. Complete the equation in the field by inserting **/sin(60)** after the dimension parameter name.
    e. Enter a value of **60** for the Angle.

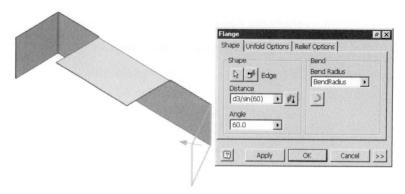

Figure 9-37

26. Click OK to complete the flange.

 NOTE The expressions entered for the two flange distances allow you to control the length of the face and the two flanges by editing the dimension of the first face you created.

27. Click the Rotate tool.

    a. Press the SPACEBAR to change to Common View.

    b. Click the upper-left isometric direction arrow as shown in the following image.

    c. Right-click in the graphics window and select Done.

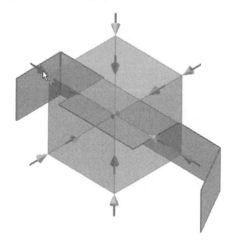

Figure 9-38

28. Click the Corner Seam tool.

    a. Select the two edges as shown.

NOTE The order of selection and whether the top or bottom model edge is selected, is not important.

Figure 9-39

29. Enter Thickness in the Gap field.
30. Click each of the three Seam buttons and examine the resulting seam preview.
31. Click the No Overlap button.
32. Click OK to create the corner seam.
33. Create a second corner seam between the opposite end of the face and the sloped flange.
34. Use the same settings as the first corner seam.
35. Click OK to create the corner seam.
36. Click the Flange tool.
    a. Select the bottom edge on the outer surface of the first face you created.

Figure 9-40

37. In the Flange dialog box:

    a. Enter a value of **20** in the Distance field.

    b. Enter a value of **90** in the Angle field.

    c. Zoom in on the lower right-hand corner of the flange.

    d. Click the Flip Offset and Flip Direction buttons and observe the effect on the flange preview.

    e. Return the Flip Offset and Flip Direction buttons to their original (raised) state.

    f. Click OK.

38. Click the Wireframe Display tool.

39. Click the Flange tool.

    a. Select the outside edge of the sloped flange, as shown in the following image.

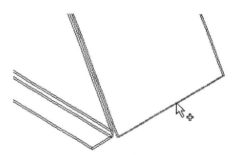

Figure 9-41

NOTE This edge is selected to ensure that the bottom surfaces of the mounting flanges will be coplanar.

40. Enter a value of **20** in the Distance field.
41. Enter a value of **60** in the Angle field.
42. Click OK to complete the flange.
43. Click the Shaded Display tool.
44. Click the Zoom All tool.
45. Reorient the view and create a 20 mm, 90° flange on the opposite end of the enclosure to complete the part as shown.

Figure 9-42

46. Close the file. Do not save changes. End of exercise.

HEM

Hems eliminate sharp edges or strengthen an open edge of a face. Material is folded back over the face with a small gap between the face and the hem. A hem does not change the length of the sheet metal part; the face is trimmed so the hem is tangent to the original length of the face. Create hems using the Hem tool (see the following image).

Figure 9-43

 NOTE A minimum of one sheet metal face must exist before creating a hem.

To create a hem, follow these steps:

1. Click the Hem tool on the Sheet Metal Features Panel Bar. The Hem dialog box appears, as shown in the following image.

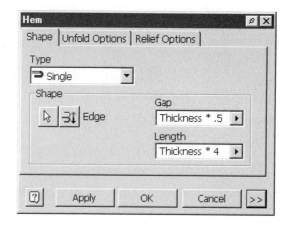

Figure 9-44

2. Select an open edge on a sheet metal face.

3. Select the hem type (examples shown in the following image):

- Single–A 180° flange
- Double–Single hem folded 180° resulting in a double-thickness hem
- Teardrop–A single hem in a teardrop shape
- Rolled–A cylindrical hem

4. Enter values for the hem. Teardrop and rolled hems require radius and angle values, while single and double hems require gap and length values. The hem preview changes to match the current values.

5. Expand the dialog box by clicking the More (>>) button and selecting Edge, Width, or Offset for the hem extents.

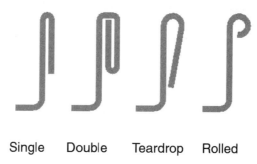

Single   Double   Teardrop   Rolled

Figure 9-45

6. Click Apply to continue creating hems, or click OK to complete the hem and exit the dialog box.

Shape Tab

The Shape tab is where you select the edge and type of the hem to create. Based on the type of hem that you want to create, different options will become active in the Hem dialog box. The following options are available on the Shape tab.

	Type	Select one of the four hem types.
	Select Edge	Select the edge along which the hem will be created.
	Flip Direction	Toggle the direction that the hem will be created.
	Gap	Specify the distance between the inside faces of the hem. This is available when you select the single or double hem type.
	Length	Specify the length of the hem. This is available when you select the single or double hem type.
	Radius	Specify the bend radius to apply at the bend. This is available when you select the rolled or teardrop hem type.
	Angle	Specify the angle applied to the hem. This is available when you select the rolled or teardrop hem type.

The options available on the Unfold Options and Relief Options tabs were covered in the Face section. The More (>>) button provides access to the Edge, Width, and Offset extent types, which are discussed in the Contour Flange section.

EXERCISE 9-2 Hems

In this exercise, you create two hem features.

1. Open ESS_E09_02.ipt.
2. Click the Hem tool.
    a. Click the edge shown in the following image.

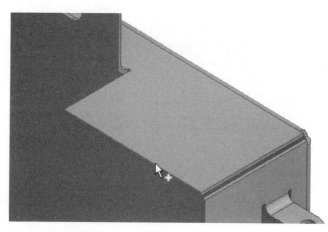

Figure 9-46

3. In the Hem dialog box:

    a. Enter **15**mm in the Length edit box.

    b. If the hem previews to the inside of the box, click the Flip Direction button.

    c. Enter **0.5**mm in the Gap edit box.

    d. Click OK to close the Hem dialog box.

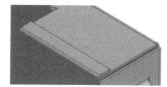

Figure 9-47

4. Use the Zoom and Rotate tools to reorient your view to match the following image.

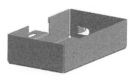

Figure 9-48

5. In the browser, right-click the Hem Extents WorkPlane.

6. Select Visibility from the menu.

7. Click the Hem tool.

8. Click the edge shown in the following image.

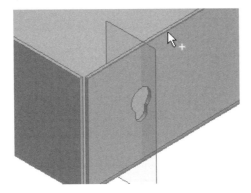

Figure 9-49

9. In the Hem dialog box:

    a. Select Rolled from the Type list.

    b. Enter **2**mm in the Radius edit box.

    c. Enter **265**deg in the Angle edit box.

    d. Click the Flip Direction button if the hem previews to the inside of the box.

    e. Click the More button.

    f. Select Offset from the Extents Type list.

    g. Click the arrow next to Offset1.

    h. In the graphics window, click the hem extents workplane.

    i. Click the far end point of the selected edge.

    j. Enter **0**mm in the Offset1 edit box.

    k. Enter **10**mm in the Offset2 edit box.

    l. Click OK to close the Hem dialog box.

Figure 9-50

10. Experiment with the other Hem types and edges as desired.

11. Close the file. Do not save changes. End of exercise.

FOLD

An alternate method for creating sheet metal features is to start with a known flat pattern shape and then add folds to sketched lines on a face. The Fold tool, shown in the following image, can create sheet metal shapes that are difficult to create using Face or Flange tools.

Figure 9-51

To create a fold, follow these steps:

1. Create a sketch on an existing face. Sketch a line between two open edges on the face.

 NOTE The sketched line endpoints must be coincident to the face edge.

2. Click the Fold tool on the Sheet Metal Features Panel Bar. The Fold dialog box appears, as shown in the following image.

3. Select the sketch or bend line. The fold direction and angle are previewed in the graphics window. The fold arrows extend from the face that will remain fixed. The face on the other side of the bend line will fold around the bend line.

4. If required, flip the fold direction and side.

5. Enter the angle of the fold.

6. Select the positioning of the fold with respect to the sketched line. The line can define the start, centerline, or end of the bend. The fold preview updates to match the current settings.

7. Make any changes to Unfold or Bend Relief options.

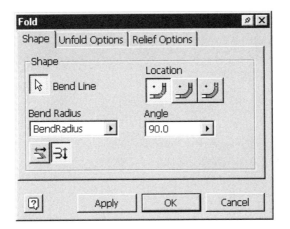

Figure 9-52

Shape Tab

The Shape tab is where you select the bend line for the fold to be created. You can set the location of the selected bend line relative to the fold feature that determines the folded shape. The angle for the fold and direction are also specified on the Shape tab. You can also override the bend radius that is specified in the sheet metal style. The following options are available on the Shape tab:

	Bend Line	Select a sketch line to use as the fold line.
	Bend Radius	Override the active sheet metal style used by default.
	Flip Side	Flip the side used for the angle of the fold.
	Flip Direction	Toggle the direction that the fold will be created.
	Centerline of Bend	Determine the centerline of the fold from the selected sketch line.
	Start of Bend	Determine the start of the fold from the selected sketch line.
	End of Bend	Determine the end of the fold from the selected sketch line.
	Angle	Specify the angle to apply to the fold.

The options available on the Unfold Options and Relief Options tabs are covered in the previous section on creating faces.

BEND

You create bends with the Bend tool (see the following image) as child objects of other features when the feature connects two faces. You can also create bends as independent objects between disjointed faces.

Figure 9-53

A preview of a bend is shown in the following image, and the faces being joined are either trimmed or extended to connect the faces.

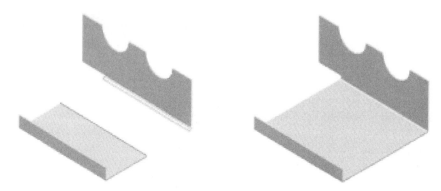

Figure 9-54

When you select parallel faces for the bend feature, a joggle or z-bend will be created, depending on the edge location. Joggles are often used to allow overlapping material. The following image shows a sample of a joggle feature (on the left). You can also create double bends when the two faces are parallel and the selected edges face the same direction.

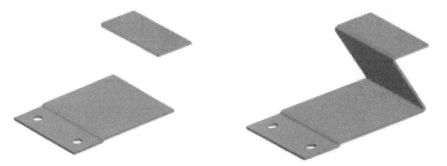

Figure 9-55

You can create the bend to be two 90° bends with a tangent face between the bends or a single full-radius bend between the two faces, as shown in the following image.

Chapter 9 Sheet Metal Design   693

Figure 9-56

 NOTE To apply a bend, the sheet metal part must have two disjointed or (sharp-cornered) intersecting faces. An example of intersecting faces is a shelled box that is being changed into a sheet metal part. The intersection of the box base and a wall is an edge that can be changed to a bend.

To create a bend, follow these steps:

1. Click the Bend tool on the Sheet Metal Features Panel Bar. The Bend dialog box appears, as shown in the following image.

2. Select the common edge of two intersecting faces, or select two parallel edges on disjointed, non-coplanar faces. If the two faces are parallel, a set of Double Bend options is presented. Depending on the position of the two faces, the Double Bend options will be either 45 Degree and Fix Edges or 90 Degree and Full Radius.

3. Make any changes to Unfold or Bend Relief options.

4. Click Apply to continue creating bends, or click OK to complete the bend and exit the dialog box.

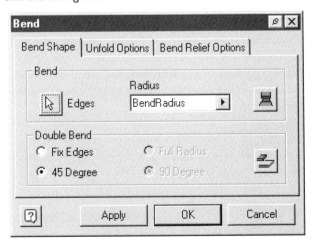

Figure 9-57

Shape Tab

The Shape tab is where you select the edges between which to create the bend and specify the type of bend that you want to create. You can also override the bend radius that is specified in the sheet metal style. The following options are available on the Shape tab.

 Edges — Select the edges where the bend will be applied. The selected edges will be trimmed or extended as needed to create the bend feature.

Radius — Override the default bend radius set by the active sheet metal style.

Extend Bend Aligned to Side Faces — Click to extend material along the faces of the edges connected by the bend instead of perpendicularly to the bend axis.

Fix Edges — Click to create equal bends to the selected sheet metal edges.

45 Degree — Click to create 45° bends on the selected edges.

Full Radius — Click to create a single semicircular bend between the selected edges.

90 Degree — Click to create 90° bends between the selected edges.

 Flip Fixed Edge — Click to reverse the order of the edges being fixed. Normally, the first edge selected is fixed by default, and the second edge will be trimmed or extended. This button reverses the order.

The options available on the Unfold Options and Bend Relief Options tabs were covered in the Faces section.

You can edit a bend that is not listed under a face in the Browser at any time, allowing you to reselect the edges that define the bend. You can even edit the bend to connect two faces that were not joined at the bend initially.

EXERCISE 9-3 Modifying Sheet Metal Parts

In this exercise, you complete the design of a sheet metal bracket within the context of an assembly and then modify the sheet metal part to eliminate component interference. You add bends to connect disjointed sheet metal faces and modify one bend to eliminate interference with a component in the assembly. You then add a hinge feature to the sheet metal part using the general modeling tools.

1. Open *ESS_E09_03.iam*.

2. Apply the "Start" design view representation. The assembly should appear as shown in the following image.

Figure 9-58

 NOTE The sheet metal bracket consists of several unconnected faces. Three faces act as mounting surfaces for assembly components; the two connected faces form the base of the bracket. The unique Autodesk Inventor method of creating sheet metal bends between disjointed faces is used to complete the sheet metal bracket.

3. In the Browser, right-click *ESS_E09_03-Bracket.ipt:1*
4. Select Edit from the menu.
5. Reorient the assembly as shown.

Figure 9-59

6. Click the Bend tool.
   a. Click the two edges as shown in the image below.
   b. Leave the bend radius at the default Sheet Metal setting, BendRadius.

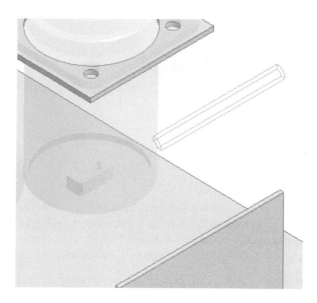

Figure 9-60

    7. Click OK to place the bend.

    8. Click the Bend tool.

        a. Click the two edges as shown in the image below.

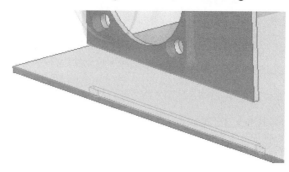

Figure 9-61

    9. Click OK to place the bend.

   10. Reorient the view of the assembly as shown in the following image.

Figure 9-62

11. Click the Bend tool
    a. Click the back edge of the small face shown as Edge #1 in the figure below.
    b. Click Edge #2 as shown in the image below.

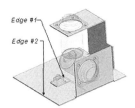

Figure 9-63

12. In the Bend dialog box:
    a. Click the Fix Edges option.
    b. Click OK to place the bend.

The bend angles required to create the Z-bend may be difficult to obtain. You can easily edit the bend to return to the default 45-degree double bends.

13. In the Browser:
    a. Right-click the last bend listed.
    b. Select Edit Feature from the menu.

14. In the Bend dialog box:
    a. Click the 45 Degree option in the Double Bend area.
    b. Click OK to modify the bend

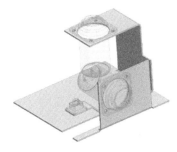

Figure 9-64

 NOTE The small tab next to the vertical mounting face is not required.

15. Click the large base face.
16. Press the S key to create a new sketch.
17. Use the Zoom and Pan tools to enlarge the view of the small tab.
18. Click the Line tool.
19. Draw a line connecting the corner of the bend relief and the edge of the cutout for the 45-degree jog bend.

 NOTE Ensure that the line is constrained to the corner point of the bend relief, and is perpendicular to the cutout edge

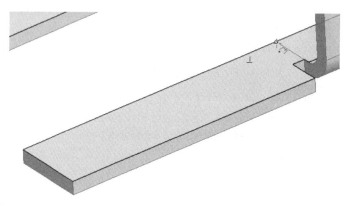

Figure 9-65

20. Click the Cut tool.
21. Select the small tab as the profile.
22. Set the Extents to Distance and a value of Thickness.
23. Click OK to complete the cut.
24. Click the Zoom All tool.
25. In the Browser:
    a. Double-click ESS_E09_03.iam to return to the assembly environment.
    b. Right-click ESS_E09_03-Control_Box.ipt:1.
    c. Select Visibility from the menu.

 NOTE The Control Box interferes with the sheet metal part

Chapter 9 Sheet Metal Design   699

Figure 9-66

To eliminate the interference, you can edit the bend causing the conflict. In the graphics window:

26. Double-click *ESS_E09_03-Bracket:1* to activate it for editing.

27. Use the Rotate and Zoom tools on the Standard toolbar to orient the assembly as shown below.

Figure 9-67

28. In the Browser:

    a. Move the cursor over the first bend under *ESS_E09_03-Bracket.ipt*. The bend adjacent to the interference should be highlighted in the graphics window.

    b. Right-click this bend and select Edit Feature from the menu.

29. In the Bend dialog box:

    a. Click the Edges button.

    b. Hold down the CTRL key and deselect the two edges that define the bend.

    c. Select the top edge of the face supporting the horizontal cylinder, and the adjacent parallel edge on the face supporting the vertical cylinder.

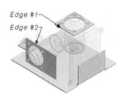

Figure 9-68

30. Click OK to create the modified bend.

The control box no longer interferes with the sheet metal bracket. The vertical face adjacent to the control box can be modified to provide support for the control box. That modification is not covered in this exercise.

31. Reorient the view of the assembly as shown in the following image.

Figure 9-69

32. In the Browser:

    a. Right-click Sketch_Hinge.

    b. Select Visibility from the menu

33. Click the Extrude tool.

Verify the selected profile matches the profile shown in the following image.

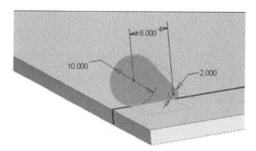

Figure 9-70

34. In the Extrude dialog box:

a. Enter a value of **200** as the Extents Distance.

b. Click the middle Extents Direction button to flip the extrusion direction.

c. Click OK to complete the weld-on hinge feature

35. Click the Zoom All tool.

36. In the Browser:

37. Double-click the top level assembly, ESS_E09_03.iam, to activate the assembly environment.

Figure 9-71

38. Close the file. Do not save changes. End of exercise.

CUT

The Cut tool, shown in the following image, is a sheet-metal-specific implementation of the standard Extrude tool. The Cut tool always performs an extrude cut—join and intersection are not available. The default extents are a blind cut at a distance equal to the sheet metal Thickness parameter. This ensures that the cut only extends through the face containing the sketch and not through other faces that may be folded under the sketch face.

Figure 9-72

Most cuts are manufactured on the flat sheet stock before the sheet is bent to form the folded part. The cuts often cross bend lines. Because the part is modeled in the folded state, representing cuts that cross bend boundaries requires the bend be unfolded to represent the flat sheet. When the bend is refolded, the cut deforms around the bend to ensure that the extrusion remains perpendicular to the sheet metal faces on both sides of the bend and deforms throughout the bend if required. When sketching a cut profile that will cross a bend, the Project Flat Pattern tool projects the unfolded flat pattern geometry onto the sketch face, as shown in following two images.

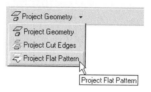

Figure 9-73

Figure 9-74

To create a cut feature, follow these steps:

1. Create a sketch on a sheet metal face that includes one or more closed profiles representing the area(s) to cut. If required, use the Project Flat Pattern tool to project the unfolded geometry of connected faces onto the sketch.

2. Click the Cut tool on the Sheet Metal Features Panel Bar. The Cut dialog box appears, as shown in the following image.

3. Select the profiles to cut.

4. If a profile crosses a bend, click Cut Across Bend.

5. Click OK to complete the cut and exit the dialog box.

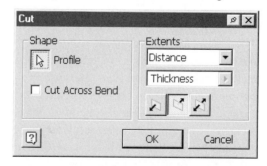

Figure 9-75

The following items are available in the Cut dialog box:

Chapter 9 Sheet Metal Design 703

 Profile   Select the profile(s) to be cut into the sheet metal part.

Cut   Click to project the profile across faces that are bent in the sheet metal
Across   part.
Bend

Extents   Choose one of the typical extrusion options: Distance, To Next, To,
           From To, All.

Direction Specify the direction for the cut to be created in the part.

EXERCISE 9-4 Cut Across Bend

In this exercise, you complete a sheet metal bracket using a fold and double bend. You then use the Project Flat Pattern tool to create a cut across bends.

1. Open *ESS_E09_04.ipt.*

2. Use the Zoom and Rotate tools to examine the part. Reorient your view to match the following image.

Figure 9-76

3. In the Browser:

    a. Right-click Face_Sketch and select Visibility from the menu.

The sketch includes projected geometry from the adjacent face.

4. Click the Face tool

5. Click OK in the Face dialog box.

Figure 9-77

6. Reorient your view to match the following image.

Figure 9-78

You will create a double bend between the two parallel faces.

    7. Click the Bend tool

    8. Click the two edges as shown in the following image.

Figure 9-79

    9. In the Bend dialog box:

        a. Click Full Radius.

        b. Click 90 degree.

        c. Click OK

   10. Create a new sketch on the face highlighted in the following image.

Figure 9-80

   11. Click the Line tool.

   12. Create a line as shown in the following image.

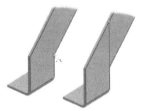

Figure 9-81

13. Right-click in the graphics window and select Done.
14. Right-click in the graphics window and select Finish Sketch.
15. Click the Fold tool and select the sketched line.
16. In the Fold dialog box:
    a. If required, click the Flip direction button to match the following image.
    b. Click the End of Bend button.

Figure 9-82

   c. Click OK

You will use the Project Flat Pattern tool to create a sketch. You then use this sketch to create a cut across bends.

17. Create a new sketch on the face highlighted in the following image.

Figure 9-83

18. Click the Project Flat Pattern tool.
19. Select the face highlighted in the following image.

Figure 9-84

20. Click the Look At tool.
21. Click one of the projected flat pattern edges.
22. Create the sketch shown in the following image.

 NOTE The sketch is centered on the face by two equal length construction lines.

Each horizontal construction line connects the projected edge midpoint and the sketch line midpoint.

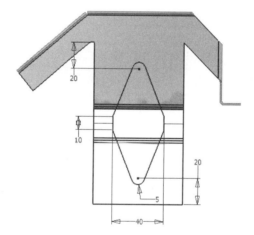

Figure 9-85

23. Right-click the graphics window and select Finish Sketch.
24. Click the Cut tool.
25. Click inside the sketch if it is not already selected.
26. Ensure that Cut across bend is checked in the Cut dialog box.
27. Click OK.

Chapter 9 Sheet Metal Design 707

Figure 9-86

28. Close the file. Do not save changes. End of exercise.

CORNER SEAM

When three faces meet in a sheet metal part, a gap is required between two of the faces to enable unfolding. Using a box as an example, the walls are connected to the floor, and gaps between the walls enable the box to be unfolded. The gap between adjacent faces is a corner seam, and you create it using the Corner Seam tool, as shown in the following image.

Figure 9-87

You can also use corner seams to create mitered gaps between coplanar faces, as shown in the following image. You can also create a corner seam by ripping open a corner at an intersection between faces.

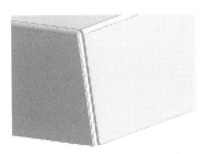 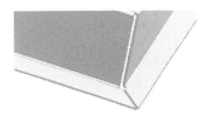

Figure 9-88

To create a corner seam, follow these steps:

1. Click the Corner Seam tool on the Sheet Metal Features Panel Bar. The Corner Seam dialog box appears, as shown in the following image.

2. Select face edges that meet one of the following criteria:

- Two open edges on faces that share a common connected edge (two walls that share a connection to a floor).

- Two nonparallel edges on coplanar faces (creates a mitered joint with a gap between the extended faces).
- A single edge at the intersection of two faces (the wall intersections of a shelled box).

3. Select the Seam option in the Corner Seam dialog box.
4. Enter a value for the corner seam gap.
5. Make any changes to the corner options.
6. Click Apply to continue creating corner seams, or click OK to complete the corner seam and exit the dialog box.

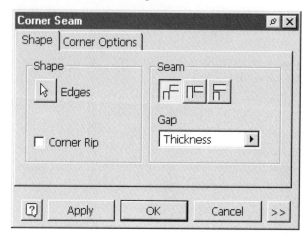

Figure 9-89

Shape Tab

The Shape tab is where you select the edges where the corner seam will be created. You can also specify the orientation of the seam and whether to rip a corner. The following options are available on the Shape tab.

Edges	Select the edges of the model on which you will create a corner seam.
Corner Rip	Click to rip open a square corner of a part.
No Overlap	Click to extend edges and create no overlaps between the edges.
Overlap	Click to extend edges and cause the first selected edge to overlap the second edge.
Reverse Overlap	Toggle the second selected edge so that it is extended to overlap the first edge.

Gap   Enter a gap for the clearance distance between the edges. The default value is Thickness.

Corner Options

The Corner Options tab allows you to override the settings for relief and bend transition on a per feature basis, as shown in the following image. By default, the settings specified in the sheet metal style are used. The following options are available:

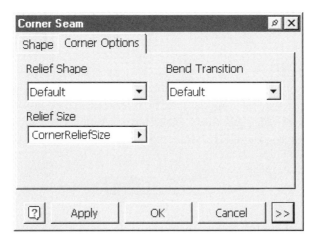

Figure 9-90

Relief Shape   Specify the shape of the relief generated by the corner seam. There are six available shapes:
  Default–Uses the setting in the active sheet metal style.
  Round–Creates a circular corner relief.
  Square–Creates a square corner relief.
  Tear–Creates a torn relief.
  Trim to Bend–Creates no transition.
  Linear Weld–Creates a narrow gap for corners that are to be welded.

Relief Size   Override the Corner Relief Size set by the active sheet metal style.

Bend Transition   Specify the Transition type for the bend. Five types are available:
  Default–Creates bends using the value in the active style.
  None–Uses a spline to join the tangencies.
  Intersection–Joins adjacent edges in the bend zone.
  Straight Line–Uses a straight line to join the tangencies.
  Arc–Uses an arc to join the tangencies.

More Button

There are additional options available when you click the More (>>) button, as shown in the following image. The options are described below.

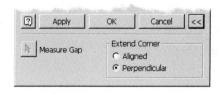

Figure 9-91

Measure Gap Select two edges to measure the distance that exists between them.

Aligned Click to make the first face align to the second face you select.

Perpendicular Click to make the first face perpendicular to the second face you select.

CORNER ROUND

The Corner Round tool, shown in the following image, is available when no sketch is active. It is a sheet-metal-specific fillet tool. All edges other than the ones at open corners of faces (all these edges have a Length = Thickness) are filtered out. This enables you to select these small edges easily without zooming in on the part.

Figure 9-92

To create a corner round, follow these steps:

1. Click the Corner Round tool on the Sheet Metal Features Panel Bar. The Corner Round dialog box appears, as shown in the following image.

2. Enter a radius for the corner round.

3. Select the Corner or Feature Select Mode.

4. Select the corners or features to include.

5. Add additional corner rounds with different radii, and select corners or features for the additional corner rounds.

6. Click OK to complete the corner round and exit the dialog box.

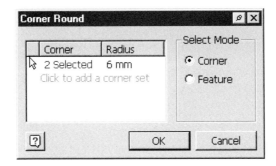

Figure 9-93

Select Mode

Selecting edges to which you apply a corner round is similar to selecting edges for the fillet feature. There are two select modes that enhance the feature for use in the sheet metal environment: Corner and Feature.

Corner Click to select individual thickness edges to be rounded or filleted.

Feature Click to select all thickness edges of a feature automatically.

CORNER CHAMFER

The Corner Chamfer tool, shown in the following image, is a sheet-metal-specific chamfer tool. As with the Corner Round tool, all edges other than the ones at open corners of faces are filtered out. Because the edges are always discontinuous, the Edge Chain and Setback settings available with the standard Chamfer tool are not available.

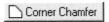

Figure 9-94

To create a corner chamfer, follow these steps:

1. Click the Corner Chamfer tool on the Sheet Metal Features Panel Bar. The Corner Chamfer dialog appears, as shown in the following image.

2. Select the chamfer style: One Distance, Distance and Angle, or Two Distances (see the following list for explanations of these styles).

3. Select the corners (or edge and corner for Distance and Angle) you wish to include.

4. Enter chamfer values.

5. Click OK to complete the corner chamfer and exit the dialog box.

Figure 9-95

|  | Distance | Creates a 45° chamfer on the selected edge. You determine the size of the chamfer by typing a distance in the dialog box. The value is offset from the two common faces. |

|  | Distance and Angle | Creates a chamfer offset from a selected edge on a specified face at an angle from the number of degrees specified. Enter an angle and distance for the chamfer in the dialog box, then click the face on which the angle is based and specify the face edge to be chamfered. |

|  | Two Distances | Creates a chamfer offset from two faces, each the amount that you specify. First, click a corner and enter a value for Distance1 and Distance2. A preview image of the chamfer appears. To reverse the direction of the distances, click the Flip button. |

|  | Corners | Click to select individual corners to chamfer. |

|  | Edges | Click to select an edge for chamfers when using the Distance and Angle option. |

|  | Flip | Click to flip the direction for the chamfer distance when using the Two Distances option. |

| | Distance | Enter a distance to be used for the chamfer feature. |

| | Angle | Enter an angle for the chamfer feature when using the Distance and Angle option. |

PUNCHTOOL

Cuts and 3D deformation features, such as dimples and louvers, are usually added to the flat sheet metal stock in a turret punch before the sheet is bent into the folded part. Turret punch tools are positioned on the flat sheet by a center point corresponding to the tool center at an angle to a fixed coordinate system. The PunchTool tool, shown in the following image, places specially designed iFeatures on sketch points that define the center point of the tool. A streamlined interface simplifies the selection and placement of the iFeatures.

Figure 9-96

A default Punch folder is installed under the top-level Catalog folder when you install Autodesk Inventor. A selection of punch tools is included in the folder. PunchTool lists all iFeatures in the designated punch folder. You can set the default Punch folder on the iFeature tab of the Application Options dialog box, as shown in the following image.

Figure 9-97

To qualify as a PunchTool, a saved iFeature must have a single hole center point in the sketch of the first feature included in the iFeature. The point corresponds to the center of the tool and will be used as the point of placement when you apply a PunchTool iFeature to a part face. Asymmetrical shapes must have the center point position controlled by geometric relationships or equations to ensure that it remains centered when the iFeature parameters are changed. Create the iFeature in a sheet metal part to ensure that sheet metal parameters such as Thickness are saved with the iFeature.

To place a PunchTool, follow these steps:

1. Create a sketch on a sheet metal face. Place at least one sketch point in the sketch. Hole Center points are selected automatically as punch centers during PunchTool placement. You can add sketch points, line or curve endpoints, and arc centers manually as additional punch centers.

2. Click PunchTool on the Sheet Metal Features Panel Bar. The PunchTool dialog box appears, as shown in the following image.

3. Select the desired PunchTool from the list. Click Next.

4. All hole centers are selected automatically. Hold down the SHIFT or CTRL key and select hole centers to exclude them, and select other center points as required.

5. Select a rotation angle for all occurrences of the punch. At 0° rotation, the X-axis of the first feature in the saved iFeature is aligned with the X-axis of the current sketch. Click Next.

6. Enter values for the PunchTool parameters.

7. Click OK to complete the PunchTool and exit the dialog box.

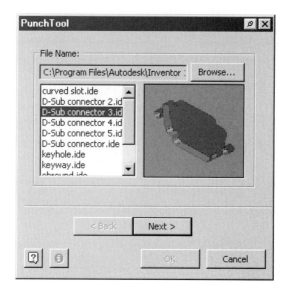

Figure 9-98

EXERCISE 9-5 Punch Tool

In this exercise, you create an iFeature that can be used as a punch and then use the PunchTool tool in another sheet metal part.

    1. Open *ESS_E09_05.ipt*.

You will complete the sketch and save the cutout as an iFeature. The visible sketch contains all geometry required for the cut. You will add construction geometry and a hole center to define the center of the punch.

    2. Click the Look At tool.

    3. Click the face containing the sketch.

    4. Right-click Sketch2 in the browser, and then select Edit Sketch.

    5. Click the Line tool.

    6. Select the Construction line style.

    7. Sketch the two lines shown in the illustration. Endpoints are on the arc centers. Line 2 connects to the midpoint of Line 1.

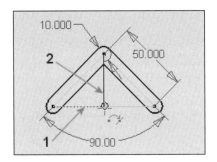

Figure 9-99

This iFeature is symmetrical in both X and Y. The midpoint of the vertical construction line is the center of the cut. For nonsymmetric shapes, you must use equations to ensure the hole center remains at the center of the iFeature.

    8. Click the Point, Hole Center tool.

    9. Place a hole center at the midpoint of the line shown in the following image.

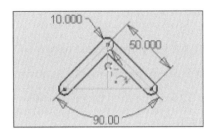

Figure 9-100

    10. Press S to finish the sketch.

    11. Click the Cut tool.

    12. Click OK.

You rename a parameter prior to saving the iFeature. Other required parameters have previously been renamed.

    13. Click the Parameters tool.

    14. Click the model parameter name cell containing d1.

    15. Enter Angle as the new parameter name.

    16. Click Done.

    17. Click the Extract iFeature tool.

NOTE Access to the Extract iFeature tool is located on the Tools pull-down menu.

18. Click Cut2 in the Browser.

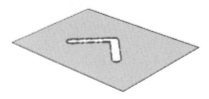

Figure 9-101

19. In the Create iFeature dialog box:

    a. Under Size Parameters, click the Limit cell in the Angle row.

    b. Select Range from the in-cell list.

20. In the Specify Range:Angle dialog box:

    a. Select < from the limit list between Minimum and Default.

    b. Select < from the limit list between Default and Maximum.

    c. Enter **5** deg in the Minimum edit box.

    d. Enter **175** deg in the Maximum edit box.

    e. Click OK.

21. Click Save in the Create iFeature dialog box.

22. Browse to the Catalog\Punches folder.

23. Enter **V Slot Punch** in the File name edit box.

24. Click Save.

Next, you place the punch feature into an existing part.

25. Open *ESS_E09_05a.ipt*.

You will create a sketch and use the V Slot punch tool.

26. Click the top face of the part.

27. Press S to create a sketch.

28. Click the Look At tool and select the sketch face.

29. Click the Point, Hole Center tool.

30. Place a hole center as shown in the following image.

Figure 9-102

Next, you create a rectangular pattern of points.

    31. Click the Rectangular Pattern tool.

    32. Click the Hole Center.

    33. Click the selection tool under Direction 1 in the Rectangular Pattern dialog box.

    34. Click the edge shown in the following image

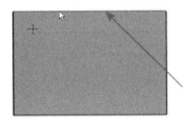

Figure 9-103

    35. Enter **5** in the Dirction 1 Count edit box.

    36. Enter **70** in the Direction 1 Spacing edit box.

    37. Click the selection tool under Direction 2.

    38. Click the edge shown in the following image.

Figure 9-104

    39. Enter **4** in the Direction 2 Count edit box.

40. Enter **50** in the Direction 2 Spacing edit box.

41. Click OK.

You sketch a line to enable manual placement of punch tool centers.

42. Click the Line tool.

43. Sketch the line as shown in the following image.

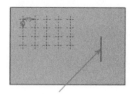

Figure 9-105

44. Press S to complete the sketch.

45. Click the PunchTool tool.

46. In the PunchTool dialog box:

    a. Click V Slot Punch.ide in the File Name list.

    b. Click Next.

The PunchTool shape is previewed at each Hole Center in the sketch. You add the line endpoints as tool centers.

47. Click the two line endpoints shown in the following image.

Figure 9-106

 NOTE The sketch plane is only reference for this punch. Any additional geometry references listed here must be satisfied before proceeding.

48. Click Next.

49. In the PunchTool dialog box:

    a. Enter **60** in the Angle Value cell.

    b. Enter **30** in the Arm_Length Value cell.

c. Enter **7** in the Slot_Width Value cell.

d. Click Refresh to update the preview with the new values.

e. Click OK.

Figure 9-107

50. Close all open files. Do not save changes. End of exercise.

FLAT PATTERN

The final step is the creation of a sheet metal flat pattern that unfolds all of the sheet metal features. The flat pattern represents the starting point for the manufacturing of the sheet metal part. The flat pattern can appear in a 2D drawing view, complete with lines indicating bend centerlines. The Flat Pattern tool, shown in the following image, creates a 3D model of the unfolded part.

Figure 9-108

You can also export the flat pattern directly as 3D (SAT file) or 2D (*.dxf* or *.dwg*) formats to be used to create machine tool programming for flat pattern punching or cutting. The following image shows a sheet metal part in its formed state on the left and the same part in its flat state on the right.

Figure 9-109

Manually modeling a flat pattern of a sheet metal part is straightforward, but it can take considerable effort for a complicated shape. Autodesk Inventor can create a 3D unfolded model of your sheet metal part quickly and update this model automatically as you modify the features that make up the part. You can create the flat pattern model at any time during the sheet metal modeling process.

The flat pattern model and icon are shown under the part name in the Browser, as shown in the following image. The flat pattern model is viewed in a separate window. Both the model and flat pattern windows can be open and visible at the same time. Select Arrange All from the Window menu to display both windows. Close this window or select the model window from the Window menu to return to the window displaying the folded part. The flat pattern model is updated automatically as you modify sheet metal features. Right-click the flat pattern in the Browser and select Open Window to view the current flat pattern shape.

If the flat pattern window has been opened already but is not the active window in Autodesk Inventor, the option appears as Show Window.

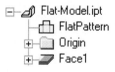

Figure 9-110

 NOTE If you delete the flat pattern model (from the Browser), you will not be able to create a drawing view of the flat pattern. If you delete the flat pattern model after the creation of a flat pattern drawing view, the drawing view will be deleted.

Right-click the flat pattern in the Browser and select Extents from the menu to display the dimensions of the smallest rectangle that encloses the flat pattern, as shown in the following image. Use this information for stock selection and multiple flat pattern layouts on large sheets.

 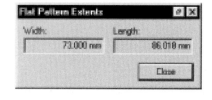

Figure 9-111

To create a flat pattern, follow these steps:

1. Select a face that you expect will not be removed in future edits. The selected face will remain fixed, and all other faces will unfold from this face.

NOTE This is not a strict requirement, but selecting a persistent face before creating a flat pattern is good practice.

2. Click the Flat Pattern tool on the Sheet Metal Features Panel Bar. The flat pattern model appears in a new window. The flat pattern model is saved in the same document as the folded model. A Flat Pattern node is also added to the Browser.

3. Right-click Flat Pattern at the top of the Browser. Menu items are available for saving the flat pattern, aligning the flat pattern to a model edge, and displaying the overall dimensions (extents) of the flat pattern.

4. To document a flat pattern, start a new drawing. Create a drawing view and select Flat Pattern from the Sheet Metal View list in the Drawing View dialog box, as shown in the following image.

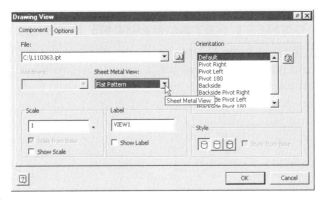

Figure 9-112

COMMON TOOLS

A number of tools on the Sheet Metal Features Panel Bar are common to sheet metal and standard parts. Following are short descriptions of how to use these tools in the sheet metal environment.

Work Features   Work planes are often used to define sketch planes for disjointed faces.

Holes   The hole feature is common to both the part modeling and sheet metal environments. You can enter sheet metal parameters such as Thickness in numeric fields such as hole depth.

Catalog Tools   You can save and place iFeatures in the sheet metal environment. See the PunchTool section for information on special iFeatures for sheet metal.

Mirror and Feature Patterns   You can mirror and pattern sheet metal features like any other modeled feature. You can also pattern or mirror flanges, holes, cuts, iFeatures, and features created with the standard part feature tools.

Promote You can import files which describe surface models (such as IGES files) in the sheet metal environment and turn them into solid bodies.

Derived Component You can reference another part or assembly (sheet metal or not). You can also reference and use parameters, sketches, work features, surfaces, and iMates from other files in a sheet metal part.

Parameters You can access model parameters and equations via the Parameters dialog box. You can add user parameters and link to an Excel spreadsheet to reference exported parameters from other models. There are also sheet-metal-specific parameters that are added to the Parameters dialog box when the sheet metal environment is activated: Thickness, BendRadius, BendReliefWidth, BendReliefDepth, CornerReliefSize, MinimumRemnant, and TransitionRadius.

Create iMate You can apply iMates to sheet metal components prior to placement in an assembly.

DETAILING SHEET METAL DESIGNS

You can create drawing views of both the 3D model and flat pattern when detailing a sheet metal part. The flat pattern drawing view enables all bend locations, bend and corner reliefs, and cutouts to be located and sized with dimensions. The 3D model views describe the folded shape of the part and allow the forming tool operator to validate the folded part.

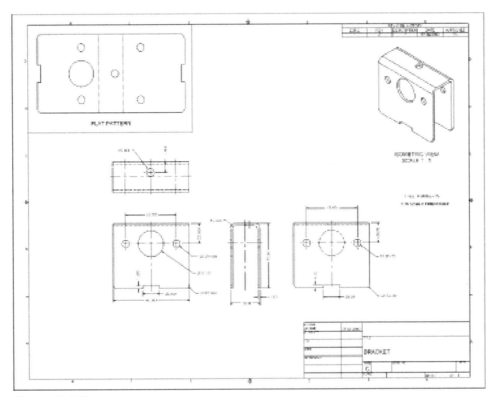

Figure 9-113

EXERCISE 9-6 Documenting Sheet Metal Designs

In this exercise, you create a flat pattern model of a sheet metal part and observe the live nature of the flat pattern as you add or modify features. You obtain the flat pattern extents and create model and flat pattern drawing views of the sheet metal part.

1. Open *ESS_E09_06.ipt*.
2. Click the Flat Pattern tool.

A new window displays the flat pattern model of the part.

Figure 9-114

3. Close the flat pattern window to return to the part window.

NOTE You can also display both windows at the same time. From the Menu Bar, click Windows > Arrange All. The part and flat pattern windows are both visible.

4. Click the Corner Chamfer tool.
5. Move the cursor over the top-right corner of the large face.
6. When the edge of the top right face is highlighted, click to select it.
7. Select all four corners of the large face.
8. In the Corner Chamfer dialog box, enter a value of 4mm in the Distance field.
9. Click OK to place the chamfers.

Figure 9-115

10. In the Browser, right-click Flat Pattern.
11. Select Open Window from the menu.

The flat pattern now includes the chamfers.

Figure 9-116

12. Activate the part window.

13. Click the Corner Round tool.

14. In the Corner Round dialog box:

    a. Enter a value of **1.5mm** for the radius.

    b. Click the Feature button under Select Mode.

15. Move the cursor over one of the four tabs.

16. Click when the tab is highlighted.

The small corners at each end of the tab are highlighted as the tab is selected.

17. Select all four tabs.

18. Click OK to place the corner rounds.

Tip: If you have both the part and flat patterns visible on the screen, note that both views are updated to show the corner round feature.

19. Click the Rotate tool.

20. Press the SPACE BAR if the common view box is not visible.

21. Select the upper-left isometric arrow as shown in the image below.

22. Right-click in the graphics window and select Done.

Figure 9-117

23. Click the Wireframe Display tool.
24. Click the Insert iFeature tool.
25. Browse to the exercises folder and select Power_Plug.ide.
26. Click Open to place the iFeature.
27. Click the large face on the part to place the iFeature.

The placement plane is the only input for this iFeature, the plug dimensions are fixed.

28. Click the crossed arrows centered on the iFeature and drag the shape to roughly place the iFeature as shown in the image below. Click to locate the feature.
29. Click Finish in the Place iFeature dialog box.

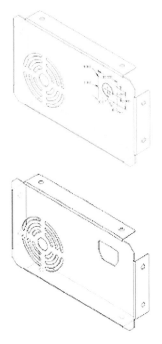

Figure 9-118

    30. Click the Shaded Display tool.

    31. In the Browser, right-click Flat Pattern and select Extents.

Dimensions for the flat pattern sheet size are shown in the Flat Pattern Extents dialog box.

    32. Click Close to close the Flat Pattern Extents dialog box.

    33. Close the Flat Pattern window.

    34. From the Menu Bar, select File > Save Copy As.

    35. Enter **End_Plate.ipt** as the file name.

    36. Click Save.

    37. Close the *ESS_09_06.ipt* file without saving it.

    38. Open *End_Plate.ipt*.

    39. Click the New tool

    40. Click the Metric tab.

    41. Double-click the DIN.idw template.

    42. Click the Base View tool.

NOTE If you have more than one part or assembly file open, select *End_Plate.ipt* from the File drop-down list.

The Sheet Metal View drop-down list displays Folded Model as the default view to be created.

43. Click on the drawing sheet to place a front view of the folded sheet metal part.

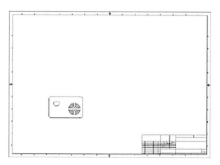

Figure 9-119

44. Click the Projected View tool.

45. Create a top, right, and isometric view as shown in the following image.

Figure 9-120

46. Click the Base View tool.

 NOTE If you have more than one part or assembly file open, select *End_Plate.ipt* from the File drop-down list.

47. In the Drawing View dialog box, select Flat Pattern from the Sheet Metal View drop-down list.

48. Click on the drawing sheet to place the flat pattern view as shown in the following image.

Chapter 9 Sheet Metal Design 729

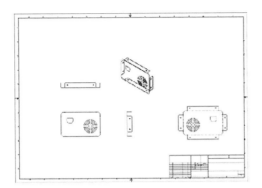

Figure 9-121

49. Click the General Dimension tool.

50. Dimension the tab length in the flat pattern view as shown in the following image.

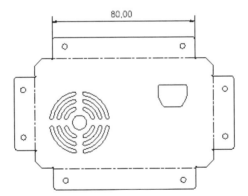

Figure 9-122

51. Close all open files. Do not save changes. End of exercise.

CHAPTER SUMMARY

To	Do This	Tool
Create a sheet metal part	Create a new part file using a *Sheet Metal.ipt* template file or click Sheet Metal on the Convert menu.	
Edit sheet metal styles	Click the Styles tool on the Panel Bar or on the Sheet Metal Features toolbar.	
Create a sheet metal face	Click the Face tool on the Panel Bar or on the Sheet Metal Features toolbar.	

To	Do This	Tool
Create a sheet metal flange	Click the Flange tool on the Panel Bar or on the Sheet Metal Features toolbar.	
Create a sheet metal bend	Click the Bend tool on the Panel Bar or on the Sheet Metal Features toolbar.	
Create a sheet metal corner seam	Click the Corner Seam tool on the Panel Bar or on the Sheet Metal Features toolbar.	
Create a sheet metal flat pattern model	Click the Flat Pattern tool on the Sheet Metal Features Panel Bar or on the Sheet Metal Features toolbar.	
Create a flat pattern drawing view	Click the Base View tool on the Panel Bar or on the Drawing Views toolbar and choose Flat Pattern from the Sheet Metal View: droplist.	
Create a PunchTool	Click Extract iFeature on the Tools menu.	

APPLYING YOUR SKILLS

SKILL EXERCISE 09-1

Using the knowledge you gained through this chapter, create the sheet metal part shown in the following image. As part of the exercise, create a drawing that details both the flat and folded states of the part.

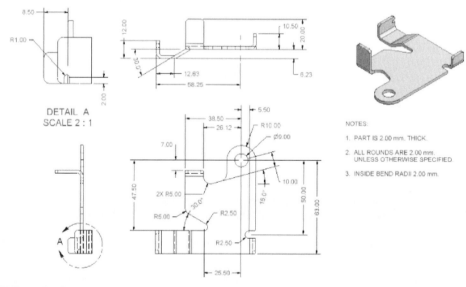

Figure 9-123

CHECKING YOUR SKILLS

Use these questions to test your knowledge of the material covered in this chapter.

1. The base feature of a sheet metal part is most often a:

    a. Revolve

    b. Face

    c. Extrude

    d. Flange

2. What is the procedure to change the edges connected by a bend feature?

    a. Suppress the existing bend and add a new bend.

    b. Delete the existing bend and add a new bend.

    c. Edit the bend and select the new edges.

    d. Create a corner seam between the desired faces.

3. Which tool would you use to create a full-length rectangular face off of an existing face edge?

    a. Flange

    b. Extrude

    c. Face

    d. Bend

4. What is required to update a flat pattern model?

    a. Right-click on the flat pattern in the Browser and select Update.

    b. Erase the existing flat pattern and recreate it.

    c. Create a new flat pattern drawing view.

    d. The flat pattern is updated automatically.

5. True__False__ Sheet metal parts can contain features created with Autodesk Inventor modeling tools.

6. True__False__ Sheet Metal Style settings cannot be overridden; a new Style must be created for different settings.

7. True__False__ During the creation of a sheet metal face, it can extend to meet another face and connect to it with a bend.

INDEX

Symbols
. (period) hot key, 181
/ (forward slash) hot key, 181
; (semicolon) hot key, 182
] (end bracket) hot key, 186
2D Sketch Panel, 364
2D sketch tools, 50–55
2D splines, 503–508
2D vs. 3D fillets, 143
3D features, overview, 96
3D Grips tool, 119–121
3D Intersection tool, 569–570
3D Move/Rotate tool, 104, 182
3D Sketch option, 42
3D sketches, 566–575
3D solids, importing, 83
45 Degree option, bends, 694
90 Degree option, bends, 694

A
Activate Contact Solver option, 325
Activate Sketch Edit Immediately option, 636
active components, 331
Active Only option, component opacity, 325
active sketch, 47
active sketch plane, 128–133
adaptivity
  definition, iv
  exercise, 384–391
  options, 375–376
  overview, 374–375
  parts at assembly level, 383
  properties for features, 378–381
  subassemblies, 382–383
  techniques, 430–432
  unconstrained features, 376–378
  work planes, 380–381, 520
Add to Accumulate option, 59

Adjust option, rectangular patterns, 195
Adjust to Model option
  circular patterns, 196
  mirroring features, 601
Align to Base option, 234
aligned sections, 246
aligning
  balloons, 474–476
  corner seams, 710
  drawing views, 279–280
  text, 302–303
All Fillets option, 146
All Hidden control, design views, 421
All Model Dimensions option, 233
All option
  Assembly options, 325
  extrusions, 106
  splitting, 593
All Rounds option, fillets, 146
All Visible control, design views, 421
ALT-drag constraining, 347–348
Alternate Solution option, extrusions, 108
Amount of Value option, 411
analysis tools
  Analyze Interference, 404–405
  assembly exercise, 405–409
  Center of Gravity, 403
  Enable Constraint Redundancy Analysis option, 325
  Measure tools, 56–59
Angle option
  assembly constraint, 342
  chamfers, 152, 712
  constraint offset, 339
  flanges, 677
  folds, 691
  hems, 687
  iFeatures, 635
  lofts, 586
  revolving, 112

angles
    aligned section views, 246
    constraints, 342, 432–434
    dimensioning, 74
    face drafts, 175–180
    measurement tool, 57
    options, 150, 196, 581–582
angular general dimensions, 291
Animation tool, 449, 453–455
annotations *See also* centerlines; lists of parts; text
    balloons, 464–477
    bill of materials, 327, 479–480
    drawing, 211–212, 296–304
    exercise, 307–318
    hole and thread notes, 304–306
    welding symbols, 234
ANSI (American National Standards Institute), 160, 209
Application Options
    Assembly, 22, 324–327
    Colors, 22
    Display, 22
    Drawing, 22, 207
    File, 21
    General, 21
    Hardware, 22
    iFeature, 22, 631
    Notebook, 22
    overview, 21–22
    Part, 22, 43–44
    Prompts, 22
    Save, 21
    Sketch, 22, 40–43
Apply Design View option, 428
Apply To options, centerlines, 300
arcs
    creating, 50
    dimensioning, 74–75
    Radius option, 42
    sketching with, 61–64
    spline curvature, 504
    tangent, 50
area measurement tool, 59
Around Placement mode, 469
arrowheads, 183, 293
assemblies *See also* components
    adaptivity, 374–391
    analyzing, 403–409
    balloons, 464–477
    bottom-up approach, xxviii, 328
    browser, 327
    constraints, 336–364
    design view representations, 415–429
    drawing views, 484–493
    driving constraints, 409–415
    enabling and disabling components, 391
    flexible, 430–432
    In-Place Activation tool, 372–373
    in-place component design method, 364–372
    iParts in, 612, 619, 621–630
    lists of parts, 478–483
    multiple part placement, 333
    options, 324–325
    overview, 323
    patterning components, 391–403
    positional representations, 432–448
    presentation files for, 448–463
    restructuring and reordering components, 335–336
    slice graphics tools, 520
    style standards, 220
    subassemblies, 334–335
    tools, 325–327
    top-down approach, xxviii, 329–332
assembly (.iam) file type, 19
Assembly Browser, 327, 372–374
Assembly options
    adaptivity, 375–376
    Application Options dialog box, 22
    global, 324–325
    presentation views, 450
Assembly Panel Bar, 325–327
associative component patterns, 392
associative design views, 426–427
Associative option, Pattern tools, 515
Attach Custom Balloon option, 476–477
Auto Balloon tool, 466–469
Auto Dimension tool, 51, 78–79, 533–538
Auto Explode method, 451
Auto-Bend Radius, 42
Auto-Hide In-Line Features option, 43
AutoCAD files, importing/opening, 82–92
AutoCAD splines, 506
Autodesk Inventor
    Design Accelerator, 649–650
    Design Support System, 59
    Getting Started screen, 1–4
    predefined styles, 220
    shortcut keys, 26
    XML file usage, 217
Autodesk Vault, 11, 14
Automated Centerlines, 298–300
automatic constraints, 54
Automatic Edge Chain option, 145, 147
Automatic explosion option, 450
Autoproject edges options, 42
Auxiliary View tool, 210
auxiliary views, 230, 241–242
AVI files, 410

axes
  default, 47
  work, 181–185
Axis option
  coils, 579
  revolving, 111

B

B hot key, 464
Balloon tool, 212, 464–466
balloons, 464–477, 492–493
base features, 97
Base View tool, 210
base views, 229, 234–235
Baseline Dimension Set tool, 211
Baseline Dimension tool, 211
Bend Extents option, 234
Bend Line option, 691
Bend Radius option
  Auto-Bend Radius, 42
  sheet metal tools, 678, 691
Bend tab parameters, sheet metal, 666–669
bend tables, 660–661
Bend Tangent to Side Face option, 678
Bend Transition option, sheet metal tools, 672, 709
bends, 666, 671–673, 675, 691–694, 701–707
Bill of Materials *See* BOM (Bill of Materials)
body pattern type, 191
Bolted Connection tool, 326
BOM (Bill of Materials) *See also* balloons; lists of parts
  auto ballooning settings, 466
  definition, 464
  editing, 479–480
  tool, 327
  views, 465
borders, creating, 214, 224–229
bottom-up assembly approach, ixxviii, 323
boundary conditions, loft features, 585
Boundary Patch tool, 103
Bowtie option, splines, 503–505
brake-formed parts, 658–659
Break Out View dialog box, 253–263
Break Out View tool, 210
break-out views, 230, 253–263
Broken View tool, 210
broken views, 230, 250–253
Browser, Autodesk Inventor, 24–25, 97–100, 327, 372–374
BSI (British Standards Institute), 160, 209

C

C hot key, 338
camera views, 31, 269
Catalog tools, 721
Caterpillar tool, 212
Center dimension text on creation option, Drawing options, 207
Center Line Bisector tool, 297
Center Mark tool, 211
center marks, 296–297, 300
center of gravity, 403, 608–609
center point (origin) *See* origin (center point)
Center Point Arc tool, 51
Center Point Circle tool, 50
Centered Pattern tool, 298
Centerline of Bend option, 691
centerlines
  automated, 298–300
  diametric dimensions, 113–114
  in annotations, 296–300
centers of holes, creating, 161
Centers option, Hole tool, 158
Chamfer tool, 51, 103, 149, 327
chamfers, 149–155, 711–712
charted parts (iParts), 612–630
child-parent relationships, 99
circles, dimensioning, 74–75
circular patterns, 183
Circular Pattern dialog box, 195
Circular Pattern tool, 51, 103, 327
circular patterns, 183–185, 191–192, 195–197, 199–200, 514–515 *See also* holes
Class option
  Hole tool, 160
  Thread tool, 168
Clear Accumulate option, 59
Clearance Hole option, 161
Close Spline option, 507
Closed Loop option, lofts, 585
Coil Ends options, 581–582
coil features, 578–582
Coil Shape options, 578–580
Coil Size options, 580–581
Coil tool, 102, 578
coincident constraints, 65, 67, 574–575
Coincident tool, 65
collaborative environments, 7, 14
collinear constraints, 66
Collision Detection option, 411
color
  constrained objects, 64
  faces, 558
  features, 123–124
  materials display, 609, 615, 665

restoring, 421
Colors options, 22
Column Chooser function, parts lists, 481
Column Count option, rectangular patterns, 193
Column Spacing option, rectangular patterns, 194
Combine Geometry parameter, iFeatures, 634
command bar, 25
command entry, 25–29, 230
Commands Tab and customizing shortcuts, 28
comments in parameters list, entering, 528
Common View (Glass Box) option, 33
Component Generators, 649
Component Opacity option, 325
Component options, drawing views, 231–232
components
  animating, 453–455
  assembly exercise, 354–364
  creating, 329–330
  derived, 104
  designing in-place, 364–372
  editing, 331–332
  enabling/disabling, 391
  exposing internal, 253
  interference between, 404–405
  moving, 348–349
  opacity, 325
  patterning, 324, 391–403
  placing in assemblies, 328, 331, 621–630
  project search order, 9
  reordering, 336
  restructuring, 335
  sheet metal, 657
  tweaking, 450–453
  visibility of, 374
Compute options, rectangular patterns, 195
concatenating text boxes, 301
Concentric constraint tool, 66
Concentric options, holes, 158
conditional suppression of features, 603
Conditions options, lofts, 585–586
configured parts (iParts), 612–630
conflicting constraints, 351
Constant options, fillets, 145–147
constraining
  ALT-drag, 347–348
  mirror sketched, 517–518
  sketches, 64–72
  splines, 508
Constraint Placement Priority option, 41
Constraint Sketch Plane to Selected Face option, 365

Constraint tool(s)
  2D sketch, 52
  assemblies, 326, 338
  Find Other Half tool, 352
  listing of, 65
constraints *See also* adaptivity; dimensions
  2D sketch tools, 52
  adding/applying, 67, 72
  angle, 432–434
  assembly, 329, 336–354
  automating of, 54, 78–79
  coincident, 574–575
  driving, 409–415
  hiding/showing, 68–69
  overconstrained dimensions/sketches, 42
  scrubbing, 54
  symmetric/symmetry, 517–518
  types of, 65
  underconstrained adaptive features, 376–378
constructed paths, 570–572
construction geometry, 500–501
consumed vs. unconsumed sketches, 97, 157
Contact Solver, 325
Content Center, 103, 326
Content Center Files location, 8
Contour Flange tool, 673–676
Coordinate System Indicator option, 41
Copy Component tool, 326
copying
  components (occurrences), 328, 330
  features, 596–599
Corner Chamfer tool, 711–712
Corner Rip option, 708
Corner Round tool, sheet metal, 710–711
Corner Seam tool, sheet metal, 707–710
Corner(s) options, sheet metal tools, 668-669, 709, 711–712
Counterbore Depth option, 381
Counterbore Diameter option, 381
counterbore holes, 158
countersink holes, 158
Create Component tool, 326, 329
Create Constraint tool, 67
Create Dimension tool, 73
Create iFeature dialog box, 631–634, 637
Create iMate tool, 104, 327, 722
Create In-Place Component dialog box, 329, 364
Create iPart tool, 613
Create Point tool, 182
Create Text tool, 52
Create Trails option, 450
Create Tweak options, 452
Create View tool, 449

Creation Method options
  circular patterns, 196–197
  mirroring, 600
Cross Part Geometry Projection options, 376
Curvature option, splines, 504
Curve Length option, rectangular patterns, 194
Curves options, lofts, 584–585
custom iParts, 613, 623
Custom tab, iProperties dialog box, 608
customizing
  Autodesk Inventor interface, 24–26
  custom parts in parts lists, 482
  iFeature Author tool, 639
  iPart Author tool, 616–619
  shortcuts, 27–29
  title blocks, 216–217
  toolbars, 25
Cut Across Bend option, 703
Cut option
  coils, 580
  extrusions, 105
  revolving, 112
  sweep features, 564
Cut tool, sheet metal, 701–707
cycling faces, 129–130

D

D hot key, 73
Datum Identifier Symbol tool, 212
Datum Target tool, 212
Decal tool, 103
default planes, 47, 364
Default project, 4–5
Defer Update option, 324
Definition in Base View option, drawing views, 234
deformation features, 658–659, 712
degrees of freedom (DOF), 336–337, 343
Delete Component Pattern Source(s) option, 324
Delete Face tool, 103
deleting
  component patterns, 394
  constraints, 68, 351
  drawing views, 278
  features, 124
  objects in sketches, 56
  sketches, 127
Delimiter, BOM, 466
Demote option, assembly components, 335
dependent features, 597, 631
Depth option, embossing, 558
Derived Component tool, 104, 722
Design Accelerator, 649–650

Design Doctor, 23, 351
Design Support System, 22–24, 61
Design View (.idv) file type, 20, 424
Design View/Presentation View option, 231
design views, assemblies, 415–429
Designation option for thread pitch size, 160, 168
Detail View tool, 210
detail views, 230, 248–250
detailing sheet metal designs, 722–729
diameter general dimensions, 291
Diameter option, holes, 160
diametric dimensions, 74, 113–114
Dimension Type Preferences option, 207
dimensioning
  drawing views, 283–293
  ellipses, 502
  general, 72–82
  relationships and parameters exercise, 533–538
  splines, 508
dimensions See also constraints
  adding, 307–318
  automatic, 79
  diametric, 114–116
  display options, 525
  drawing annotation tools, 211
  editing, 119
  entering and editing, 76–78, 287–288, 292–293, 526
  exercises, 82
  general, 72–75, 288–291
  holes, 159
  model, 207–208
  overconstrained, 42
  retrieving, 284–286
  units, 45
  viewing options, 233
  visibility of, 287
DIN (Deutsche Industrie Norm-The German Institute for Standardization), 160, 209
direct model edge referencing, 134
Direction option
  extrusions, 107
  Face Draft tool, 176
  Hole tool, 158
  rectangular patterns, 193–195
  ribs, 550
  sheet metal tools, 703
  Shell tool, 172
  Tweak tool, 452
disabled vs. enabled components, 391
disjointed faces, 662, 669, 693
Display Accumulate option, 59
Display Line Weights option, 208

Display option(s) *See also* visibility
  dimensions, 525
  Drawing options, 208
  Drawing View dialog box, 233–234
  overview, 22
  sketches, 41
  spline curvature, 507
  threads, 167
  trails, 452
  View Tools, 33
Distance and Angle option, chamfers, 150, 712
distance measurement tool, 57
Distance option(s)
  chamfers, 150, 152, 712
  exploding parts, 450
  extrusions, 107
  flanges, 675, 677
  rectangular patterns, 194
Do not Activate Sketch Edit option, 636
documents and style management, 219
DOF (degrees of freedom), 336–337, 343
double bends, 692
Draft Angle option, 176
Draft Type option, Face Draft tool, 176
Draft View tool, 210
draft views, 266–267
drafting standards, 3
dragging sketches, 67
drawing (.idw) file type, 20
Drawing Annotation Panel Bar, 210–212
Drawing Annotation tools, 296–304
Drawing options, 22, 207–208
Drawing Resources, 213–217
drawing sheets, 212–217, 224–229
Drawing View dialog box, 230–236, 426
Drawing Views Panel Bar, 210
drawing views
  annotations, 296–306
  auxiliary, 241–242
  base, 234–235
  break-out, 253–263
  broken, 250–253
  commands, 230
  detail, 248–250
  detailing sheet metal designs, 722
  dimensioning, 283–293
  draft, 266–267
  editing, 276–283
  exercises, 224–229, 236–241, 270–275, 280–283, 307–318, 484–493
  from design views, 427–429
  from positional representations, 438–439
  introduction, 207
  perspective, 268–270
  presentation files, 463

  projected, 235–236
  section, 242–247, 263–265
  selecting objects, 294–295
  sheet preparation, 212–217
  starting drawings, 208–209
  styles, 217–224, 305–306, 307–310
  templates, 229
  tools overview, 210–212, 229–230
Drawings style standards, 221
Drill Point option, 159
drilled holes, 158
Drive Adaptivity option, 411
Drive Constraint option, 351
Drive Constraints tool, 409–415
driven dimensions, 42, 77–78
DWG files, importing/opening, 83–89
DXF files, importing, 90
Dynamic Rotate tool, 32

E

E hot key, 105
Edge Chain option, 152
Edge mode, constant fillets, 145
Edges option(s)
  bends, 671, 694
  chamfers, 151, 712
  corner tools, 708
  flanges, 674–675
edges
  chamfering, 149–152
  Face Draft tool, 176
  fillets, 142–149
Edit Coordinate System tool, 53
Edit Dimension dialog box, 76
Edit dimensions when created option, 42
Edit Existing Trail button, 453
Edit Feature tool, 118
Edit iFeature tool, 637
Edit Parts List dialog box, 481
Edit Property Fields dialog box, 215–216
Edit Section Depth dialog box, 265
editing
  balloons, 470–477
  Bill of Materials, 479–480
  component patterns, 393–394
  components, 331–332
  constraints, 350–352
  dimensions, 75–78, 287–288, 292–293, 526
  drawing views, 276–283
  features, 117–127
  holes, 157, 306
  iFeatures, 637–639
  iParts, 619–620
  linked parameters, 532–533
  lists of parts, 480–482

patterns, 515
shell features, 172
sketches, 125–127
style libraries, 221–223
title blocks, 215–216
tweaks, 450–453, 457
Ellipse tool, 50, 502
ellipses, 502
Emboss from Face option, 558
Emboss tool, 103, 557–559
Emboss/Engrave from Plane option, 558
embossing, 556–563
Enable Constraint Redundancy Analysis option, 325
enabled vs. disabled components, 391
End of Bend option, 691
End of Part marker, 605–606
End option, coils, 581
Engineer's Handbook, 649
Engrave from Face option, 558
engraving, 556, 558
environments
  2D vs. 3D fillets, 143
  3D sketch, 566
  collaborative, 7, 14
  multi-user, 4, 6
  multiple document, 20
  network, 7, 217
  sheet metal, 663
  switching, 100–101
equal constraint, 66
Equation parameter, 528
equations, editing dimensions, 526–527
Excel, Microsoft
  bend tables, 660
  iPart editing, 619–620
  linked parameters, 527, 530–533
  thread data, 167
Expert mode, 25, 50
Explosion Method option, 450
Export function, parts lists, 482
Export Parameters Column, 528
Extend Bend Aligned to Side Faces option, 671, 675, 694
Extend Profile option, ribs, 551
Extend tool, 51
extension lines, 292
Extents option
  extrusions, 106
  revolving, 112
  ribs, 550–551
  sheet metal tools, 675, 703
Extract iFeatures tool, 631–634
Extrude dialog box, 105–110
Extrude tool, 102, 326
extruded features, adaptive, 379–380
extruding, 104–110, 547–548, 701

F

F4 Dynamic Rotation shortcut, 33
Face Draft tool, 103, 176–180
Face option(s)
  chamfers, 151
  embossing, 558
  Hole tool, 158
  Part Features tools, 103
  sheet metal tools, 694
  splitting, 593
  Thread tool, 167
Face tool, sheet metal, 669–673
Faces option, Face Draft tool, 176
Faces to Split option, 593
faces
  altering, 121–122
  colors, 558
  cycling, 129–133
  disjointed, 662, 669, 693
  drafts, 175–180
  options, 172–175, 584, 675, 678
  shelling, 171
  sketch options, 520–524
  splitting, 591–593
  tools, 327
  vs. work planes, 186
factories, iPart, 612, 615–620, 622
failed features, 124
family of parts (iParts), 612–630
fasteners, hole options, 160–161
Feature Control Frame tool, 211
Feature Identifier Symbol tool, 211
Feature option
  constant fillets, 145
  Corner Round tool, 711
feature pattern type, 190
Feature Priority option, suppressing features, 602
Features option
  circular patterns, 196
  rectangular patterns, 193
features *See also* iFeatures; sketches; *individual features*
  adaptive, 375–381
  Auto-Hide In-Line Features option, 43
  Browser tool, 98–99
  coil, 578–582
  copying, 596–599
  deformation, 658–659, 712
  Design Accelerator, 649–650
  editing, 116–127
  embossing and engraving, 556–563
  extruding, 104–110, 547–548

loft, 582–591
mirroring, 327, 600–601
overview, 96
placed, 142
projecting edges, 133–136
publishing, 645–646, 648–649
reordering, 604–605
revolving, 110–116
ribs and webs, 549–556
rolling back, 605–606
sketched, 128
suppressing, 601–603
sweep, 563–578
switching environments, 100–102
tools, 102–104
transformation from sketches to, 501
types of, 300
work, 616, 721
file extensions, 19–20
File Open Options dialog box, 450
File options
  Application Options dialog box, 21
  drawing views, 231
  presentation views, 450
File Properties (iProperties), 216, 606–608,
    See also BOM (Bill of Materials)
file-reservation mechanism, 4
files
  drawing view management, 229
  included, 6
  new, 3
  opening, 3, 20
  resolving links, 10
  saving, 20–21
  search options, 5, 9
  style management, 219
  template, 45–46
  types, 19–20
Fillet tool, 51, 102, 142–149, 527
fillets, 143–149, 155
Find Other Half tool, 352
Finite option, ribs, 551
Fit Method option, splines, 505–506
Fit option, hole clearance, 161
Fitted option, patterns, 197, 515
fix constraints, 65–66
Fix Edges option, bends, 694
Fixed Edge draft, Face Draft tool, 176
Fixed Plane draft, Face Draft tool, 176
Flange tool, sheet metal, 676–678
Flat Angle option, coils, 581
Flat option, splines, 505
Flat Pattern tool, sheet metal, 661
flat patterns
  projecting, 52, 135, 701
  sheet metal fabrication, 658, 661, 719–729

flexible assemblies, 430–432
Flip Direction option
  embossing, 558
  sheet metal tools, 677, 687, 691
Flip Fixed Edge option, bends, 694
Flip Offset option, flanges, 677
Flip option
  chamfers, 151, 712
  circular patterns, 196
  coils, 579
  Face Draft tool, 176
  rectangular patterns, 193–194
  ribs, 550
  threads, 167
Flip Side option
  folds, 691
  Hole tool, 158
Fold tool, sheet metal, 690–691
Folder Options, 8
Format Text dialog box, 301
From Point option, break-out view, 254–255
From Sketch option, holes, 158
From To option, extrusions, 106
From/To Planes option, 380
Full Depth option, tapped holes, 160
Full Length option, Thread tool, 167
Full Radius option, bends, 694
Full rotation option, 112

G

Gap option, sheet metal tools, 687, 709
GB (Guojia Biaozhun-The Chinese National
    Standard), 209
General Dimension tool, 51, 73, 211
general dimensions, 288–291
geometric constraints, 329
Geometry tab, iFeature Author tool, 638
geometry, construction, 500–501
Getting Started page, 1–4
Glass Box (Common View) option, 33
Glass View option, 418
graphics drivers, 22
graphics window, 25
gravity, center of, 403, 608–609
grid lines, display, 41
grips, 3D, 119–121
Ground Shadow option, 31
grounded components, 333
Grounded Work Points tool, 104, 182–183
Group Settings function, parts lists, 482

H

H hot key, 156

half sections, 246
Handle option, splines, 504
Hardware options, 22
hatch patterns, modifying, 247
Hatching option, drawing views, 234
Healing option, 103
Height option, coils, 580
helical shapes (coil features), 578–582
Help system, 22–24
Hem tool, 685–690
Hidden Edge Display tool, 33
Hidden Line Calculation option, drawing views, 233
highlighting presentations, 456
Hole Center tool, 162
Hole Depth option, 381
Hole Notes tool, 304–306
Hole Table tool, 212
Hole tool, 102, 155-166, 327, 721
Hole/Thread Note tool, 211
Holes dialog box, 156–157
holes
 creating, 156–166
 features, 381
 threads, 166–167, 170
 To Hole option, 254, 258–260
horizontal constraints, 41
Horizontal Placement mode, 468
hot keys
 . (period), 181
 / (forward slash), 181
 ; (semicolon), 182
 ] (end bracket) hot key, 186
 B, 464
 C, 338
 D, 73
 E, 104
 H, 156
 N, 329
 P, 328
 SHIFT+F, 143
 SHIFT+K, 149
 SHIFT+O, 192
 SHIFT+R, 192
 T, 451

I

iam files, 19, 323
ide files, 20, 630
Identical option
 Mirror Feature tool, 601
 patterns, 195–196
idv files, 20, 424
idw files, 20
iFeature (.ide) file type, 20, 630

iFeature Author Table tool, 637–639
iFeatures
 Application Options dialog box, 22
 creating, 630–634
 editing, 637–639
 exercise, 640–644
 inserting, 104
 iParts, 612–630
 placing, 634–636
 PunchTool, 712–713
 table-driven, 639
IGES files, importing, 90
iMates, 615
importing files
 assembly design view representations, 421–424
 AutoCAD, 82–89, 92–96
 other file types, 89–90
In-Place Activation tool, 372–373
In-Place Features From/To Extent option, 375
Include Work Features option, 193, 196
included files, 6
Incremental option
 circular patterns, 197
 Drive constraint tool, 411
independent features, copying, 597
inferred points, 54
Inheritance, BOM, 466
insert (type) constraint, 343
Insert AutoCAD File tool, 52
Insert iFeature tool, 104, 634–636
Insert Image tool, 52
Insert Point option, splines, 506
inserting
 AutoCAD files, 89
 iFeatures, 104, 634–636, 639
Interactive Contact option, 325
Interference Analysis tool, 404–405
Intersect option
 coils, 580
 extrusions, 105
 revolving, 112
 sweep features, 564
intersections, 3D paths from, 569–570
Interval option, 454
iParts, 612–630
ipj files, 20
ipn files, 19, 448
iProperties, 215, 606–608, 610–611 *See also* BOM (Bill of Materials)
ipt files, 19, 323
ISO (International Organization for Standardization), 160, 209
Isometric View, 30, 229
Item Numbering dialog box, 465–466

J

JIS (Japan Industrial Standard), 209
JIS Pipe Threads standard, 160
joggles, 692
Join option
  coils, 580
  extrusions, 105
  revolving, 112
  sweep features, 564
justification, view, 208, 234

K

keyboard shortcuts, 26–29
Keys tab, iPart placement, 621–622
keys/key values, 613–614, 623 *See also* hot keys
kFactor, 659, 666
knitting surfaces together, 103

L

Label option, drawing views, 232
Launchpad, 1–2
Leader Text tool, 212
leaders, text, 301
Left Hand option, Thread tool, 160, 168
Legs option, grounded work points, 183
Length option
  hems, 687
  Thread tool, 167
libraries
  iPart, 622
  search order, 9
  search paths, 7, 13
  style, 217–223
lighting, style standards, 220
Limit parameter, iFeatures, 632
Line Style option, drawing views, 233
Line tool, 50, 54
Line Weight Display options, 208
linear diametric dimensions, 74, 113
linear general dimensions, 289
Linear option
  Hole tool, 158
  Tweak Components tool, 452
linear patterns, 197
lines
  assemblies, 337
  Bend Line option, 691
  coincident constraints, 574–575
  construction geometry, 501
  dimensioning, 73
  mate constraints, 340
  sketching exercise, 59–61
linked parameters, 527, 530–533
links to files, resolving, 10
lists of parts
  adding custom parts, 482
  creating, 478
  editing, 480–482
  exercise, 489–491
  nested, 483
  splitting, 478–479
  spreadsheet view, 482
local styles vs. style libraries, 219
Location edit box (project), 13
Location options, threads, 167–168
loft features, 582–591
Loft tool, 102, 582
Look At tool, 32
loop measurement tool, 58
Loop mode, constant fillets, 145

M

main menus, 24
Make Independent parameter, iFeatures, 634
Manual explosion option, 450
Map Point option, lofts, 586
mapping points, 583
Margin option, drawing views, 233
mass properties, 608, 610–611
master positional representation, 433
Match Shape option
  extrusions, 108
  revolving, 113
mate constraints, 339–341
mate flush solution constraint, 341
Mate Plane and option, 376
materials *See also* BOM (Bill of Materials)
  changing part, 609
  styles, 220, 665
Measure Gap option, 710
Measure tools, 56–59
Mechanical Calculators, 649
Merge Tangent Faces option, lofts, 585
Method options, splitting, 592
Metric options, new file, 209
Microsoft Excel *See* Excel, Microsoft
Microsoft Windows shortcut keys, 26
Minimum Remnant option, sheet metal, 667, 673
Minimum Solution option, extrusions, 108
minimum-energy splines, 506
minor grid lines, 41
Mirror Component tool, 326
Mirror Feature tool, 103, 600–601, 721
Mirror individual features option, 600
Mirror Patterns, 721

Mirror Pattern dialog box, 600–601
Mirror Plane option, 600
mirror sketches, 517–518
Mirror the entire solid option, 600
Mirror tool, 51, 517
model (parametric) dimensions, 207, 233, 284–288, 293
model parameters, 527, 722
Model Value parameter, 528
Model Welding Symbols option, drawing views, 233
Modify Hatch Pattern option, 247
Modify List option, Sheet Metal Styles tool, 666
motion constraints, 339, 344–345, 430–438, 446–448
motion controls, 410–411, 449, 453–455
Motion options, 454
Move Component tool, 326, 348–349
Move Coordinate System option, 635
Move Face tool, 103, 327
Move Feature tool, 121
Move tool, 52
moving
   components, 348–350
   dimension text, 292
   drawing views, 276
   features, 121
   objects, 68
multi-user environment, 4, 6
multiple document environment, 20, 331
multiview drawings, 236–241

N

N hot key, 329
Name edit box, 12
Name parameter, iFeatures, 632–633, 635
naming
   drawing sheets, 213
   projects, 7–8, 12
   styles, 218
nested parts lists, 483
network environment, 7, 217
networks, rib, 552
neutral plane, 659
New icon, 3, 209
New Sheet tool, 210, 213–214
New Single User Project, 12
New Vault Project, 11
No Overlap option, 708
Nominal Diameter option, 381
nominal tolerance, 528
Nominal Value parameter, 528
non-Expert mode, Panel Bar, 25
normal vectors, 337

Notebook options, 22
numbering
   Item Numbering dialog box, 465–466
   Renumber function for parts lists, 482

O

Object Defaults style list, 223
Object Visibility option, 190
Occurrence Angle option, 196
Occurrence Count option, 196
occurrences of components, 328, 330
Offset option
   balloons, 475–476
   sheet metal tools, 671, 674–675
   text, 302
   Thread tool, 167
offset sections, 247
Offset tool, sketches, 51, 502
offset values, modifying constraint, 353
Offset/Angle option, assembly constraints, 339
Old Versions To Keep On Save option, 8
On Point option, holes, 158
Open dialog box, 3
Open icon, 3
open profiles, extruding, 547–548
Operation options
   extrusions, 105
   lofts, 585
   revolving, 112
   sweep features, 564
Optimized option
   Mirror Feature tool, 601
   patterns, 195
Options dialog box, 21–22
Options tab, drawing views, 232–234
Ordinate Dimension Set tool, 211
Ordinate Dimension tool, 211
Orientation options
   drawing views, 231
   rectangular patterns, 195
origin (center point)
   default, 47
   grounded work points, 183
   visibility, 47
Origin menu, 47
orthographic camera viewpoint, 31
orthographic view, 229
Output options
   extrusions, 107
   lofts, 585
   revolving, 112
   sweep features, 564
overconstrained dimensions/sketches, 42, 77
Overlap option, 708

Overlay View tool, 210
overlay views, 230, 446–448
overriding
  balloon styles, 466
  balloon types and values, 472–474
  color settings, 421
  corner relief sizes, 709
  mass and volume properties, 610–611
  object styles, 224
  visibility settings, 421
Owner, project, 8

P

P hot key, 328
Pan View tool, 32
Panel Bar, 25–26
parallel constraints, 41, 65
Parallel View on Sketch Creation option, 43
parameters, 527–533, 665–669
Parameters option(s)
  adaptive features, 380
  animations, 454
  iFeatures, 638
  Paste Feature dialog box, 599
Parameters table, iFeatures, 632
Parameters tool, 52, 104, 327, 528, 722
parametric dimensions, 78
parametric modeling, xxvii
parent-child relationships, 98
part (.ipt) file type, 19
part environment, 101
Part Feature Adaptivity option, 375
Part Features Panel Bar, 102–104
part libraries (iParts), 612–630
Part options, Application Options dialog box, 22, 42–43
Parts List tool, 212
Parts Only View, BOM View, 465
parts *See also* components; constraints; drawing views; features; lists of parts; sheet metal; sketches
  creating, 46–59
  Design Accelerator, 649–650
  highlighting on screen, 99
  iParts, 612–630
  materials, 609
  projecting edges, 133–134
  publishing, 644–645, 646–647
  splitting, 591–596
Paste Features options, 598–599
Path option
  rectangular patterns, 193–194
  sweep features, 564
paths
  3D, 566–573

patterns along, 197, 200–202
Pattern Component tool, 326, 391
Pattern Entire Solid option, 193, 196
Pattern Individual Features option, 193, 196
Pattern tools
  Centered Pattern, 298
  Circular Pattern, 51, 103, 327
  Flat Pattern, 661, 719–721
  mirror features, 600–601, 721
  options, 515
  Project Flat Pattern tool, 52, 701
  Rectangular Pattern, 51, 103, 327
patterns
  along paths, 197, 202–206
  assembly, 394–395
  circular, 185, 195–197, 200
  component, 324, 391–394, 395–397
  creating, 190–192
  flat, 52, 136, 658
  hatch, 247
  rectangular, 193–195, 197–199
  sketching, 514–515
pause controls, 410
perpendicular constraints, 41, 65
Perpendicular option, corner seams, 710
perspective camera viewpoint, 31
perspective views, 268–270
physical properties, materials, 220, 608, 610, 665
Physical tab, iProperties dialog box, 608
Pick Profile Plane option, 599
Pitch option, coils, 580
pitch, threads, 168
Place Component tool, 326, 328
Place Constraint dialog box, 338
Place Custom iPart dialog box, 623
Place Standard iPart dialog box, 621–622
Placed Component option, 325
placed features, definition, 142
placement modes for balloons, 468–469
Placement option, auto ballooning, 466
placing
  circular patterns, 196
  components, 328
  dimensions, 73
  holes, 157–158
  iFeatures, 104, 634–636, 639
  iParts, 612, 621–630
  published parts and features, 646–649
Plane option, holes, 158
planes
  active sketch, 129–133
  and axes, 43
  assemblies, 337
  default, 47, 364
  Face Draft tool, 176

mate constraints, 340
neutral, 659
options, 376, 558, 600–601
visibility, 47
work, 103, 183, 185–190, 326, 380–381, 520
plates, metal *See* sheet metal
Point option
  fillets, 147
  From Point option in break-out view, 254–255
  holes, 158
point sets, 586–587
Point, Hole Center tool, 51
points *See also* work points
  assemblies, 338
  loft features, 583
  mate constraints, 341
Polygon tool, 51
Position Geometry list, iFeatures, 633–634
Position option
  iFeatures, 634
  lofts, 586
Positional Representation option, drawing views, 233
positional representations, 432–448, *See also* angles
positioning
  circular patterns, 197
  dimensions, 77
Precise Input dialog box, 55
Precise Position options, iFeatures, 636
Precise View Rotation tool, 449
Precision settings, Measurement tools, 59
predefined styles, 217–223
Predict Offset and Orientation option, 339
presentation (.ipn) file type, 19, 448
presentation files
  creating, 448–463
  drawing views from, 463
  styles, 220
  views, 231
Preserve All Features option
  chamfers, 152
  constant fillets, 147
press brake-formed parts, 658–659
Previous View tool, 33
private design view representations, 424–425
PRO/E files, importing, 90
Profile option
  coils, 579
  Emboss tool, 557
  extrusions, 105
  revolving, 111
  ribs, 550
  sheet metal tools, 671, 674, 703

sweep features, 564
project (.ipj) file type, 20
Project Cut Edges tool, 52, 135
Project File Editor, 10
Project Flat Pattern tool, 52, 701
Project Geometry tool, 52, 135–142
project library, 217
Project tab, iProperties dialog box, 607
Projected View tool, 210
projected views, 229, 235–236
projecting features, 133–136
Projection options, automated centerlines, 300
Projects dialog box, 5–8
Projects icon, 4
projects
  creating, 10–13, 15–19
  editing, 14
  file management, 4–10
  setting up, 4
  types of, 11–12
Promote option, assembly components, 335
Promote tool, 104, 722
Prompt parameter, iFeatures, 633–634
Prompts options, 22
properties *See also* BOM (Bill of Materials)
  adaptive features, 378–381
  balloons, 469
  drawing sheet, 215–216
  drawing views, 276–277
  iFeature Author tool, 638
  iPart Author tool, 615
  iProperties (file), 606–608
  mass and volume, 608, 610–611
  physical materials, 220, 665
Property Field tool, text, 217
public design view representations, 416
publishing parts and features, 644–649
Pull Direction option, face drafts, 176
PunchTool tool, sheet metal, 712–719

Q

Quarter Section View tool, 326

R

radius general dimensions, 290
Radius option
  arcs, 42
  bends, 667, 671, 675, 678, 687, 694
  fillets, 148
rails, loft features, 583, 585
range of values, custom iParts, 618
Ratio option, motion constraints, 339

Rectangular Pattern tool, 51, 103, 327
rectangular patterns, 192–195, 199, 514–515
Redo tool, 30
redundant constraints, 351
Reference # option, holes, 158
Reference Data options, drawing views, 233
reference dimensions, 77
reference edges, 630
reference geometry, 187, 134
reference parameters, 527
Refresh Standard Components tool, 326
Release ID, 8
relief options, sheet metal tools, 668–669, 672, 709
Remove Color Overrides control, design views, 421
Remove Faces option, shelling, 172
Remove option, splitting, 592
renaming features and sketches, 123
Renumber function, parts lists, 482
reordering components, 336
reordering features, 604–605
Repetitions option, 411, 454
Replace Component option, patterns, 396
Replace Component tool, 326
Replace Face tool, 103
Representations folder, 417–418
Resolve Link dialog box, 10
restructuring components, 335, 397
Retrieve all model dimensions on view placement option, Drawing options, 207
Retrieve Dimensions tool, 212
retrieving dimensions, 284–286
Reverse Overlap option, 708
Revision Table tool, 212
Revolution option, coils, 580
Revolve tool, 102, 111, 327
revolved features, 380
revolving sketches, 75, 110–116
Rib tool, 102, 549
ribs, 549–556
Right Hand option, Thread tool, 160, 168
Roll Along Sharp Edges option, 146
rolling back features, 605–606
Rolling Ball Where Possible option, 147
rollup command dialog boxes, 29
Rotate Component tool, 326, 349–350
Rotate tool, 52
Rotation Axis option, circular patterns, 196
rotation constraints, 344
Rotation option
  coils, 579
  Tweak Components tool, 452
rotation tools, View Tools, 32
rotation-translation constraints, 345

rotational degrees of freedom, 336
rounds vs. fillets, 142, 146
Row Count option, rectangular patterns, 194
Row Spacing option, rectangular patterns, 194

S

SAT files, importing, 89, 384
Save tab, iProperties dialog box, 608
saving files, 20–21, 229
Scale from Base option, drawing views, 232
Scale option, drawing views, 232
scrubbing, 55
search paths
  library, 13
  projects, 5–7
Section All Parts option, 325
Section Standard Parts option, 208
Section View tool, 210, 230
sections
  loft features, 583–584
  section depth function, 263–265
  views, 230, 242–247
Select Edge option, sheet metal tools, 677, 687
Select Mode, corner rounds, 711
Select option
  iFeatures, 634
  splits, 593
Select Other tool, 130, 346–347
Selected Features, iFeatures, 631
selecting objects, 56, 145, 294–295
Selection(s) option, 338–339, 466
Setback option(s)
  chamfers, 152
  fillets, 149
Shaded Display tool, 33
shadows, setting, 31
Shape options
  bends, 694
  corner tools, 708
  extrusions, 105
  faces, 671
  flanges, 674, 677–678
  folds, 691
  hems, 687
  revolving, 111
  ribs, 550
  sweep features, 564
shared sketches, 516
shared templates, 46
Sheet Formats list, 213
sheet metal (.ipt) file type, 19
sheet metal
  bending, 692–694

common tools, 721–722
Contour Flange tool, 673–676
corner tools, 707–712
creating parts, 662–663, 678–685
cutting, 701–707
design methods, 661–662
detailing, 722–729
fabrication, 658–661
Face tool, 669–673
Flange tool, 676–678
flat patterns, 719–729
folds, 690–691
Hem tool, 685–690
introduction, 657
modifying parts, 694–701
punching, 712–719
styles, 220, 664–669
Sheet tab parameters, 665–666
sheets, drawing, 212–217
Shell tool, 102, 170–175
shells/shelling, 170–175, 662
SHIFT+F hot key, 144
SHIFT+K hot key, 150
SHIFT+O hot key, 192
SHIFT+R hot key, 192
shortcut keys, 26–29
shortcut menus, 26–29
Shortcut, project, 8
Show Constraints tool, 52
Show Me animations, 187
Show Preview option, assembly constraints, 339
Show Trails option, drawing views, 234
simple holes, 159
simulating mechanical motion, 409–415
single-user projects, 12–13, 15–19
size of features, editing, 119
Size options
   Hole tool, 160–161
   iFeatures, 635
   Thread tool, 168
Size Parameters table, iFeatures, 632
Sketch Doctor, 23
sketch environment, 100–101
Sketch on New Part Creation option, 43
Sketch option(s)
   adaptive features, 379–380
   Application Options dialog box, 22, 40–43
sketch planes
   active, 128–133
   default, 47, 364
Sketch tools, 50–55
sketches *See also* assemblies; features
   3D, 566–575
   break-out drawing view, 254, 256–258
   complex, 508–513

constraining, 64–72
construction geometry, 500–501
consumed vs. unconsumed, 97
creating, 47–55, 59–64
dimensioning, 72–82
draft drawing views, 210, 266–267
editing, 55–56, 124–127
ellipses, 502
importing external files, 82–92
in-place component method, 364–365
Measure tools, 57–59
mirror, 517–518
on another part's face, 520–524
open profile, 547–548
options, 40–42
pattern, 514–515
section drawing, 242–247
shared, 516
sheet metal, 663, 673
Slice Graphics option, 519–520
splines, 503–508
title blocks, 216
Slice Graphics option, 519–520
Smooth Radius Transition option, 148
smooth splines, 505
snaps, 67
Snap to Grid option, 42
Solid option, 107, 112
Solution option, constraints, 339
Sort function, parts lists, 482
sorting Drawing Resources headings, 217
Specification tab, Thread tool, 168
Spline Tension option, 508
Spline tool, 503–513
splines, 503–513, 572–573
split balloons, 469–470
Split Face method, 592
Split Part method, 592
Split tool, 103, 591–593
splitting
   lists of parts, 478–479
   parts and faces, 591–596
spreadsheet view, parts lists, 482 *See also* Excel, Microsoft
standard iParts, 613, 622
standard toolbar, 25
standards, threads for holes, 160
Start of Bend option, 691
Start option
   coils, 581
   rectangular patterns, 195
Start/End option, 412
Start/End/Start option, 412
startup screen, 1–4
stationary components, 333
Status Bar, 25

Status tab, iProperties dialog box, 608
STEP files, 89, 384
Stitch Surface tool, 103
Structured option, BOM View, 465
Style from Base option, drawing views, 232
style libraries, 8
Style option, drawing views, 232
styles, 217–224, 305–310, 466, 664–669
sub-styles, 218
subassemblies, 328, 334–335, 382–383, 421
subfolders, 5, 7
Summary tab, iProperties dialog box, 607
Suppress Features option, 379, 602
Suppress option
  constraints, 351
  Pattern tools, 515
suppressed features, 601–603
Suppression tab, iParts, 615
Surface option, 107, 113
Surface Texture Symbol tool, 211
sweep features, 563–578
Sweep tool, 102, 327, 563–565
Sweep tool;Sweep tool, 102
sweet splines, 506
Symbols tool, 212
symmetry constraints, 66, 517–518

T

T hot key, 451
Table Layout function, parts lists, 482
table-driven iFeatures, 639
tabulated parts (iParts), 612–630
Tangent Arc tool, 51
Tangent Circle tool, 50
Tangent Edges option, drawing views, 234, 278
tangents
  constraints, 65, 342
  dimensioning to, 75
  sketching with, 64
Taper option
  coils, 580
  extrusions, 108
tapped holes, 160
Tapped Hole option, 159
templates
  locations of, 8
  new, 3
  saving files as, 229
  styles, 220–221
  using, 45–46
tension, spline, 508
terminations
  extrusions, 106
  holes, 159

Text Box option, tweaks, 453
Text tool, 212
text
  creating, 52
  dimension, 292
  drawing annotation tools, 212
  embossing/engraving, 556–563
  hole and thread notes, 304–306
  leaders, 301
  positioning, 301–303
  style exercise, 307–310
  title blocks, 214–217
texture symbols, 211
Thicken/Offset tool, 103
Thickness option
  ribs, 550
  sheet metal, 665
  shells, 172
thin-walled parts (shells), 170–175
threads, 160, 211
Thread Feature options
  creating threads, 167–168
  drawing views, 234
Thread Notes tool, 304–306
Thread tool, 102
Thread Type option
  Hole tool, 160
  Thread tool, 168
threads, 166–170
Threads tab
  iFeature Author tool, 639
  iPart Author tool, 616
Three Point Arc tool, 50
Three Point Rectangle tool, 51
Threshold options, automated centerlines, 300
Through Part option, break-out view, 254, 261–263
Title Block Insertion option, 208
title blocks, 88, 215–217, 224–229
To Hole option, break-out view, 254, 258–260
To Next option
  extrusions, 106
  ribs, 551
To option, extrusions, 106
To Sketch option, break-out view, 254, 256–258
Tol. (tolerance option, parameters), 528
tolerance, nominal, 528
toolbars, customizing, 25–26
Top Face Color option, embossing, 558
top-down assembly approach, 323, 329–332
Total # of Steps option, 411
trails, 450
Trail Origin option, 452

Transformations options, tweaks, 452–453
Transition Angle option, coils, 581
Transition options
  lofts, 586–587
  sheet metal, 668
transitional constraints, 339, 346
translational degrees of freedom, 336
translucency settings, 43
Triad Move option, 122
Triad Only option, 453
triads, grounded work point, 183
Trim tool, 51
turret punch tools, 712
tutorials, 23
Tweak Components tool, 449, 451–453
tweaking components, 450–453, 457
Two Distances option, chamfers, 151–152, 712
Two Point Rectangle tool, 51
txt files, 660
Type option
  assembly constraints, 338
  coils, 580
  hems, 687
  motion constraints, 339
  transitional constraints, 339

U

unconsumed vs. consumed sketches, 97, 157
underconstrained adaptive features, 376–378
Undo tool, 30
Unfold Options tab, sheet metal, 671
unfolding sheet metal, 658–659, 664, 666, 671
Unique Face Thickness option, Shell tool, 172
Units options
  general, 44
  parameters, 528
  presentation files, 456–457
Update Mass Properties option, 609
Use Styles Library option, 218
User Defined button, balloons, 471
user interface, customizing, 24–25
user parameters, 527, 529–530
Using Unique File Names option, 8

V

Value parameter, iFeatures, 632, 636
Variable fillets, 147–148
Vault projects, 11, 14
vertical constraints, 41, 66
Vertical Placement mode, 468

video drivers, 22
views, 427
View Catalog tool, 104, 630
View Justification option, drawing views, 208, 234
View Tools, 32–33
viewpoint options, 30–40, 43, 415–429
Visibility Control tool, 374
visibility
  axes, 41, 47
  center point (origin), 47
  components, 391
  component patterns, 397
  coordinates indicator, 42
  design view representations, 421
  dimensions, 287
  grid lines, 41
  planes, 48
  shared sketches, 516
  spline features, 507
  toolbars, 25
  work planes, 188
Visible options for drawing views, 232
Visual Syllabus, 23, 187
volume properties, 610–611

W

webs, 549–556
Weight option, lofts, 586
Weld Annotations option, 234
welding representation, 212
Welding Symbol tool, 211
Weldment option, 232
Weldments standards, 220
Width option
  ribs, 550
  sheet metal, 675
Windows shortcut keys, 26
Wireframe Display tool, 33
work axes, 181, 183–185
Work Axis tool, 104, 181, 326
work features, 181–190, 721
Work Features option, 234
Work Features tab, iParts, 616
Work Plane tool, 103, 186–187, 326
work planes, 183, 185–190, 326, 380–381, 520
Work Points tool, 104, 181–183, 326
work points
  3D paths with, 570–573
  creating, 181–182
  grounded, 104, 182–183
workgroup search paths, 7
Workspace folder, 12
workspace, definition, 6

Wrap to Face option, embossing, 558

X

X-Ray Ground Shadow option, 31
X-Y plane, 43
X-Y-Z option, tweaks, 452
X-Z plane, 43
XML files, 29, 217–218

Y

Y-Z plane, 43

Z

z-bend, 692
zone labels, 214
Zoom Target for Place Component with iMate option, 325
zooming, 32, 99, 325

LICENSE AGREEMENT FOR AUTODESK PRESS
A Thomson Learning Company

Educational Software/Data

You the customer, and Autodesk Press incur certain benefits, rights, and obligations to each other when you open this package and use the software/data it contains. BE SURE YOU READ THE LICENSE AGREEMENT CAREFULLY, SINCE BY USING THE SOFTWARE/DATA YOU INDICATE YOU HAVE READ, UNDERSTOOD, AND ACCEPTED THE TERMS OF THIS AGREEMENT.

Your rights:

1. You enjoy a non-exclusive license to use the enclosed software/data on a single microcomputer that is not part of a network or multi-machine system in consideration for payment of the required license fee, (which may be included in the purchase price of an accompanying print component), or receipt of this software/data, and your acceptance of the terms and conditions of this agreement.

2. You own the media on which the software/data is recorded, but you acknowledge that you do not own the software/data recorded on them. You also acknowledge that the software/data is furnished "as is," and contains copyrighted and/or proprietary and confidential information of Autodesk Press or its licensors.

3. If you do not accept the terms of this license agreement you may return the media within 30 days. However, you may not use the software during this period.

There are limitations on your rights:

1. You may not copy or print the software/data for any reason whatsoever, except to install it on a hard drive on a single microcomputer and to make one archival copy, unless copying or printing is expressly permitted in writing or statements recorded on the diskette(s).

2. You may not revise, translate, convert, disassemble or otherwise reverse engineer the software/data except that you may add to or rearrange any data recorded on the media as part of the normal use of the software/data.

3. You may not sell, license, lease, rent, loan, or otherwise distribute or network the software/data except that you may give the software/data to a student or and instructor for use at school or, temporarily at home.

Should you fail to abide by the Copyright Law of the United States as it applies to this software/data your license to use it will become invalid. You agree to erase or otherwise destroy the software/data immediately after receiving note of Autodesk Press' termination of this agreement for violation of its provisions.

Autodesk Press gives you a LIMITED WARRANTY covering the enclosed software/data. The LIMITED WARRANTY can be found in this product and/or the instructor's manual that accompanies it.

This license is the entire agreement between you and Autodesk Press interpreted and enforced under New York law.

Limited Warranty

Autodesk Press warrants to the original licensee/ purchaser of this copy of microcomputer software/ data and the media on which it is recorded that the media will be free from defects in material and workmanship for ninety (90) days from the date of original purchase. All implied warranties are limited in duration to this ninety (90) day period. THEREAFTER, ANY IMPLIED WARRANTIES, INCLUDING IMPLIED WARRANTIES OF MERCHANTABILITY AND FITNESS FOR A PARTICULAR PURPOSE ARE EXCLUDED. THIS WARRANTY IS IN LIEU OF ALL OTHER WARRANTIES, WHETHER ORAL OR WRITTEN, EXPRESSED OR IMPLIED.

If you believe the media is defective, please return it during the ninety day period to the address shown below. A defective diskette will be replaced without charge provided that it has not been subjected to misuse or damage.

This warranty does not extend to the software or information recorded on the media. The software and information are provided "AS IS." Any statements made about the utility of the software or information are not to be considered as express or implied warranties. Delmar will not be liable for incidental or consequential damages of any kind incurred by you, the consumer, or any other user.

Some states do not allow the exclusion or limitation of incidental or consequential damages, or limitations on the duration of implied warranties, so the above limitation or exclusion may not apply to you. This warranty gives you specific legal rights, and you may also have other rights which vary from state to state. Address all correspondence to:

AutodeskPress
Executive Woods
5 Maxwell Drive
Clifton Park, New York 12065-2919